Zoophysiology Volume 17

Coordinating Editor: D. S. Farner

Editors:
B. Heinrich W. S. Hoar K. Johansen
H. Langer G. Neuweiler D. J. Randall

R. J. F. Smith

597.0525
Sm64c

WITHDRAWN

The Control of Fish Migration

With 74 Figures

Springer-Verlag
Berlin Heidelberg New York Tokyo 1985

Prof. R. J. F. SMITH
University of Saskatchewan
Department of Biology
Saskatoon S7N OWO, Canada

WITHDRAWN

Cover motif: A group of Pacific salmon moving upstream at dawn symbolizes the many interacting factors that typify the control of fish migration. The flowing water provides chemical, mechanical, thermal, and visual stimuli that direct the migration. The rising sun allows sun compass orientation and serves as an important timing stimulus. The fish travel in a social group that allows synchronization of migratory activity and protects the fish from predators.

ISBN 3-540-13707-6 Springer-Verlag Berlin Heidelberg New York Tokyo
ISBN 0-387-13707-6 Springer-Verlag New York Heidelberg Berlin Tokyo

Library of Congress Cataloging in Publication Data. Smith, R. J. F. (Reginald Jan Frederick), 1940- The control of fish migration. (Zoophysiology; v. 17) Bibliography: p. 1. Fishes – Migration. I. Title. II. Series. QL698.9.S65 1985 597'.052'5 84-14121 ISBN 0-387-13707-6 (U.S.)

This work is subject to copyright. All rights are reserved, whether the whole or part of the material is concerned, specifically those of translation, reprinting, re-use of illustrations, broadcasting, reproduction by photocopying machine or similar means, and storage in data banks. Under § 54 of the German Copyright Law, where copies are made for other than private use, a fee is payable to "Verwertungsgesellschaft Wort", Munich.

© by Springer-Verlag Berlin Heidelberg 1985
Printed in Germany

The use of registered names, trademarks, etc. in this publication does not imply, even in the absence of a specific statement, that such names are exempt from the relevant protective laws and regulations and therefore free for general use.

Typesetting, printing and bookbinding: Brühlsche Universitätsdruckerei, Giessen
2131/3130-543210

This book is dedicated to my wife Jean
and our children David and Tina

CAT Oct23'85

10-10-85 ABS 42.08

85-2358

Preface

Fish migration is important and spectacular. Migratory fish gather energy in one portion of the environment and transport it to other areas, where it often becomes available to humans or to other elements in the ecosystem. Migration brings fish into situations that allow easy harvest as they concentrate along migration routes. Their journeys also make them vulnerable to human intereference at critical points along their route. Salmon, for example, may harvest plankton in the open ocean and transport that food energy to coastal and inland regions, where it is captured by fisheries or deposited in inland streams and utilized by the flora and fauna of the region. These salmon are able to complete journeys of thousands of kilometers from their natal streams to oceanic feeding grounds and back to the same home streams, an accomplishment that strains our credibility.

We now understand some of the timing and guiding stimuli used in these migrations, and mechanisms can be logically proposed, on the basis of the established abilities of fishes, to account for the unexplained portions of the migrations. There is no single factor guiding these fish. Instead, they are dependent on the presence in their environment of a great variety of appropriate orienting and timing stimuli. These stimuli are vulnerable to human interference. The more widespread and easily available the information on these requirements, the more readily fish can be protected from such interference. Where interference with natural migration is inevitable, then greater understanding of the controlling mechanisms at work in fish migration will allow us to compensate for the interference more successfully.

I have approached this task by summarizing and explaining representative experimental studies that illustrate the mechanisms used by fish in the guiding and timing of their migrations. This approach has meant that not every study is reported, since in some well-worked subject areas their number and variety would have obscured the main theme of the book, without changing the broad conclusions reached on the basis of a few critical works. My approach has also meant that two large areas of literature related to fish migration have been to a great extent omitted. First, there are many descriptive studies detailing the routes taken by fish in their migrations. Unless these are accompanied by experimental work to determine the factors controlling the timing or direction of migration or bear on a specific problem related to these controlling factors, these have not been presented. Second, there are papers dealing with the physiological adjustments that migratory fish must make as they move from one habitat to another and as they prepare for the energy expenditure of migration. Again, unless these studies bear on some particular aspect of orientation or timing of migration, they have not been covered.

Fish sensory systems are sufficiently different from familiar mammalian systems for me to feel it desirable to present a brief review of the relevant sensory capabilities of fish at the beginning of each chapter. This provides the reader with some understanding of the abilities and limitations of fish with regard to sensing the particular type of stimuli under discussion.

The study of fish migration is entering a new era in which exciting new mechanisms such as magnetic orientation are being widely studied and the variety of adaptations, even among different regional stocks of the same species, is coming to be appreciated. I hope that this book will introduce readers to this field and increase their appreciation of the abilities and requirements of migratory fish.

Saskatoon, September 1984 R. J. F. SMITH

Acknowledgements

This book had a long preparation period. I am particularly indebted to Dr. W. S. Hoar for this encouragement, patience, and very useful advice throughout that period.

The book was started at the Pacific Biological Station of the Fisheries and Marine Service of Canada at Nanaimo, British Columbia. The bulk of the writing was completed at the Zoology Department of Louisiana State University in Baton Rouge, Louisiana. I am grateful to both institutions for providing me with space, library facilities, and uninterrupted time to work. My home institution, the Biology Department of the University of Saskatchewan, also provided support and sabbatical leave to complete the book.

Librarians at the Pacific Biological Station, Louisiana State University, and the University of Saskatchewan have been consistently helpful and cooperative. My wife Jean Smith has performed many hours of proof-reading and typing and has provided much valuable advice. Bev Garnett assisted in typing the manuscript and Dennis Dyck redrew some of the illustrations. Miles Keenleyside offered valuable advice on portions of the manuscript, as did Jean-Guy Godin, Jon C. Cooper, Jack Gee, Charles C. Coutant, Thomas Quinn, and Robin Liley. I gratefully acknowledge their assistance.

Contents

Chapter 1

Introduction

The term migration is often applied to adaptive, long-distance movements that occur predictably in the life cycle of a species. These movements take advantage of spatial and temporal differences in the distribution of resources, and thus increase the fitness of the migrants. For example, a species that breeds most successfully in the gravel of fast-flowing streams can take advantage of greater food availability in quiet, more productive waters by migrating from the breeding area to a feeding area after hatching and then migrating back to the gravel beds to breed. For such a system to evolve, the benefits of using two or more different areas during one life cycle must outweigh the costs of the migration. These costs include the energy cost of moving from one place to another and any increased risk of injury or death attributable to the migration. Migrants that survive the trip and have sufficient energy reserves for breeding will tend to leave more offspring. If the migration is not beneficial, then non-migratory individuals will be favoured.

This type of selection will lead to many different changes in the migratory population. Some changes will improve orientational or navigational accuracy, others will improve locomotor efficiency, while still other changes will reduce the risk of injury or death from predators, environmental extremes or other natural hazards. The variety of these adaptations reflects the complex nature of major migrations. These are elaborate activities with many constraints and requirements. There must be advance preparation. The timing must be appropriate. The movements must be accurate and sustained over long periods and yet must be terminated at the appropriate time and place.

This complexity does not mean that every component of a migration will be carried out with great precision or that every orientation mechanism will be elaborate or highly refined. In some cases simple mechanisms will achieve the best results. Simply moving up into midwater in a flowing stream will predictably carry fish toward the sea.

In addition to the better-known migrations, such as the migrations of salmon or eels between widely separated feeding and breeding areas, there are other situations in which movement from one region to another will confer a selective advantage. For example, foraging trips within a home range may be more efficient if the forager is capable of accurate orientation. Animals that derive significant advantages from knowledge of a home area or from previous "investment" in territorial defence may leave more offspring if they home successfully after accidental displacement than if they try to establish themselves in a new location.

Directed movements such as foraging trips or homing may not be considered "migration" by many, but they will be subjected to some similar selection pressures; selection for accurate orientation and safe and efficient travel, for example.

Thus they can be studied profitably as models of migration indicating the types of adaptation of which a group of animals is capable. Their relatively short duration and occurrence over a wide seasonal period often make these lesser directed movements more amenable to experimental study than major migrations. The homing of pigeons, for example, is not migration in the sense that I have used the term in the first paragraph, but it is an example of animal navigation that can be elicited over and over again at all seasons of the year and that can be completed in hours or days, allowing experiments which would be difficult with a truly migratory species. These directed movements deserve to be examined in any treatment of migration for the light they shed on the adaptations "available" to migratory species.

The aim of this book will be to understand the mechanisms that control fish migration, that is, the ways in which the timing, direction and distance of migrations are regulated by natural stimuli. In examining the controls or causes of animal behaviour it is often useful to distinguish between two levels of causation (Orians 1962). The adaptive or evolutionary advantage of a behaviour pattern can be considered an "ultimate" cause of migration. An example of ultimate causation might be the increased reproductive success of migrants that moved to more productive feeding areas. The controlling mechanisms which time, direct or terminate the migration can be termed "proximate" causes. These proximate causes include the specific stimuli that fish use in determining the direction of migration or in timing their migratory activity. Stimuli such as the position of the sun or the length of the photoperiod are examples of stimuli involved in proximate causation. Because the primary concern of this book is with control mechanisms, I shall deal mainly with proximate causation. Ultimate causation will also be discussed where appropriate, for example in dealing with the reasons why one stimulus might be favoured over another as a timing cue for migration.

The levels of ultimate and proximate causation are partially independent of each other. The ultimate cause of one migration might be the advantage conferred by feeding in a location remote from the breeding habitat, while another migration might have evolved in response to selection favouring movement to a safe winter habitat, yet both may share the need for accurate orientation toward the goal, and thus evolve similar proximate orientation mechanisms. Or species migrating for similar ultimate reasons may have different sensory abilities and thus evolve different orientation or timing mechanisms.

My approach to these control mechanisms will be to examine the various sources of information that fish might use in timing, directing or terminating their migrations, and the evidence for or against their use. The information available to the fish will be limited by their sensory capability, and hence discussion of the sense organs and their capability will be included. Because sense organs tend to be specialized for the reception of a particular type of energy, such as radiant, thermal or chemical energy, the book is organized into chapters dealing with each of the energy types that are thought to be involved in the control of fish migration. In each chapter there is a summary of the information available to migratory fish from a particular source, a description of the sense organs which could detect the information and a discussion of evidence of its use in timing, directing or terminating fish migration.

One chapter, apart from the introductory and concluding chapters, does not fit this mold. An important source of information available to migrants is the information carried forward from the past either in the memory of individual fish or encoded in the genes of a population. Chapter 7 will deal with "the past" as a source of information available for the control of migration.

Of course, fish often use more than one source of information in the control of an aspect of their migration. In the cases where such interactions have been examined in the literature this will be pointed out in each of the relevant chapters and cross-referenced. It is a weakness of the current literature that many of the studies deal with only one or two major sources of information. This is changing as the importance of multiple information sources and their interaction becomes better understood. Perhaps this book will encourage the trend by emphasizing the number and variety of potential sources of information available to migratory fish.

Terminology. The term migration is frequently used in biology but is, in fact, relatively difficult to define satisfactorily because it has different meanings in different fields (Baker 1978). Definitions based on the distance covered, the degree of accuracy, or the types of habitat occupied all fail to include movements that would be termed migration by some researchers. The definition that opens this chapter, "adaptive, long-distance movements that occur predictably in the life cycle of the species" is not ideal (e.g., how far is long?; how can you be sure it is adaptive?) but it does have the virtue of corresponding fairly well to common usage of the term by fish researchers. Thus it does not require as much adjustment on the part of the reader as more wide-ranging definitions such as Baker's (1978) "the act of moving from one spatial unit to another."

The inclusion of the term adaptive is intended to restrict our use of "migration" to cases that have evolved in response to natural selection and to exclude accidental displacement.

Long-distance movements occur widely in fish and are of considerable economic and biological importance. Often such movements are characterized by several different stages with different requirements for orientation or timing. The whole sequence is usually required for successful completion of the life cycle. It is probably best to think of the "migration" as the complete long-distance movement. This will emphasize the integrated nature of the several components of a complete migratory sequence.

The definition specifically avoids mention of the accuracy of orientation or of active directed movement, since some fish migrations depend on drifting with water currents. Despite the apparently passive nature of such migrations they may be highly developed and effective ways of reaching a goal, and are often characterized by sophisticated mechanisms for the selection of the appropriate water mass for drifting.

Finally the inclusion of "predictable occurence in the life cycle" emphasizes the importance of timing in most fish migration. The migration will only be adaptive if carried out at the appropriate stage in the life cycle and there is often a best season and time of day for successful migration as well. This means that mechanisms for the timing of migration confer an adaptive advantage.

In summary, the characteristics of a "typical" fish migration include a preparatory phase in which the animal develops characteristics suitable for its migration, such as adequate fat reserves or physiological capability for surviving changes in environment. Such changes are often controlled by "priming" stimuli which control preparation but do not actually initiate migration. Once prepared, the fish must initiate migration often within a fairly narrow time window if the migration is to succeed. The actual start of migration will be controlled by "releasing" stimuli which release or initiate the migratory behaviour. The migration must then be sustained in the appropriate direction, which often requires orientation to environmental stimuli. This phase of migration also requires economical locomotion, so adaptations to reduce energy use have evolved. Travel also entails risks. Hence predation and other natural hazards may constitute a major selective pressure on migrants. So we can expect adaptations that foil predators or reduce injury to be important components of the migratory syndrome. Finally, the migration must be terminated at the appropriate location. This can be thought of as migrating the right distance, although there are several different mechanisms that can terminate migration. These include goal recognition, limits on the distance travelled or on the duration of migratory behaviour. Termination may resemble initiation of migration in being characterized by changes that prepare the fish for the post-migratory phase of its life cycle, and are controlled by priming stimuli. The actual termination, however, will usually be controlled by releasing stimuli.

There are several other terms that have acquired specialized technical meaning in the field of migration research.

Pilotage occurs when an animal finds its way about by using its knowledge of a familiar area. As, for example, when a channel catfish moves between its resting site and its automatic feeding machine by following swimways worn into the bottom of its pond.

Orientation is used to refer to mechanisms by which an animal moves in a given plane or compass direction. When an animal using an orientation mechanisms is displaced laterally it continues to move in the same direction (Baker 1978). The situation is similar to the case of a human using only a compass to find direction. If one is displaced an unknown distance off course there is no information available from the compass that would allow one to select the correct course to compensate for displacement. Animals using a sun compass or magnetic compass mechanism in their orientation are in much the same position. A salmon smolt using sun-compass orientation to migrate along a migration direction, that is appropriate in its native lake, will select the same compass direction after an experimental displacement of many kilometers, that has rendered the direction inappropriate (Groot 1965).

Navigation, in contrast, refers to a situation in which an animal determines the position of a given point in space. When a navigating animal is displaced it can adjust to a new heading, corrected for the displacement. The homing pigeon is a good example. Homing pigeons, no matter in which direction they are displaced away from their home loft, can select the appropriate homeward direction even in totally unfamiliar regions.

Chapter 2

Light

2.1 Light in Water

Light, a narrow band of the electromagnetic spectrum between wavelengths of about 390 nm (nanometers) and 700 nm, is one of the primary sources of information about the external environment available to fishes. Underwater light is influenced by two main factors: first, the optical properties of the water itself; second, the optical effects of the materials suspended or dissolved in the water. The water surface reflects a portion of the incident light, while the light that does penetrate the surface is refracted downward so that the light from the surface seems to come through a circular window limited by a 97.6° cone with the apex at the observor's eye. After penetrating the surface, the different wavelengths are absorbed at different rates. Longer wave red light, 610 nm and above, and shorter wave violet and ultraviolet below 465 nm are absorbed more rapidly than blue light around 465–475 nm. Hence blue light penetrates farthest into the depths (Lythgoe 1966, 1979 a, b; Munz and McFarland 1977). The spectral quality of light under water is, however, quite variable, changing with time of day as well as with the chemistry of natural water bodies (Munz and McFarland 1977). Fish living near the surface will have almost the whole spectrum available to them, while those living in the depths will be faced with an increasingly monochromatic blue-green world. In many freshwaters and coastal seawaters there is a significant amount of dissolved yellow material from vegetable decay. This shifts the wavelength of maximum penetration from 475–480 nm toward longer wavelengths (Lythgoe 1966, 1979 a). In general, light tends to lose wavelengths and intensity as it travels through natural waters. Even the direction in which the fish looks will affect the amount and colour of the available light. More wavelengths are available in downwelling light from above than in the dimmer upwelling light reflected from objects below the fish, because of the longer underwater path taken by the upwelling light (Munz and McFarland 1977).

Much of the world's ocean water is very clear, but in productive coastal and surface waters and in many freshwaters the optical properties of the water are dominated by the presence of suspended particles that scatter light. This light scattering will reduce the penetration of light into the water and it will mask objects in the environment by scattering light into the visual pathway between the object and the observer (veiling brightness). Both effects will change the visual contrast of objects viewed underwater and may also interfere with the reception of information about the position of celestial light sources such as the sun, moon, and stars. Duntley (1962) and Hemmings (1966) have discussed the effects of light-scattering particles on underwater visibility. Turbid water with many light-

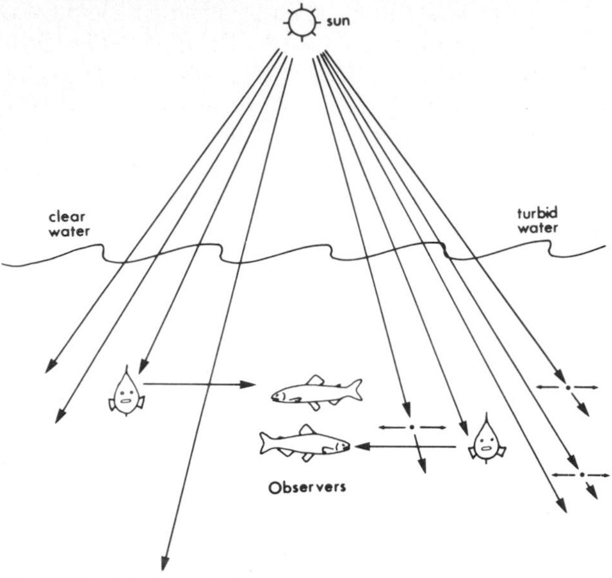

Fig. 2.1. Particles in turbid water tend to scatter light. This means that light scattered by particles behind an object will tend to make the background bright, while light scattered by particles between the object and the observer will tend to obscure the object with veiling brightness. Dark objects will tend to contrast with the light background of turbid water, while light objects will be most visible against the dark background provided by clear water

scattering particles will provide a light background against which dark objects will be easily visible (Fig. 2.1). Light objects will be hard to see against this light background. Veiling brightness is familiar to anyone who has been in fog. The light reflected from water droplets between the observer and the observed object veils the observor's view (Lythgoe 1979 b). In clear water, less light will be scattered toward the observer from the water behind an object, providing the appearance of a dark background. Light objects will contrast with the background and dark objects will merge with it, the reverse of the situation in turbid water (Fig. 2.1).

The scattering effects that disperse underwater light also polarize the light by differential scattering of some planes of polarization. These patterns of polarization are related to the direction of the light source and can potentially provide information about the position of the sun or moon in sky and hence about time and direction (see Sect. 2.2.1).

2.2 Fish Photoreceptors

2.2.1 Vision

The eyes of fish share a common basic plan with the eyes of other vertebrates. However, the eyes of each vertebrate species show specializations for the particular requirements of the biology of that species (Walls 1942). Each of the 20,000 to 30,000 species of fish will have somewhat different visual capacities. The many anatomical and biochemical specializations recorded by Walls (1942), Brett (1957), Munz (1971), Ali and Anctil (1976), Lockett (1977), Munz and McFarland (1977), and Lythgoe (1979 a) may be relatively minor compared with the differences between species or even between developmental stages of a single species in retinal or central nervous system analyzing mechanisms. The eyes of eels, *Anguilla* sp., for example, undergo dramatic changes between the sedentary and migratory stages of the life cycle (Tesch 1977; Pankhurst 1982).

2.2.1.1 Focusing

The lens is usually the only light-focusing structure in a fish eye. In order to refract light, a lens must be harder and have a higher refractive index than the surrounding medium. Fish are usually surrounded by water with a refractive index of about 1.33. The cornea, which does much of the focusing in a terrestrial vertebrate eye, has almost the same refractive index as water and so does not focus light underwater. The lens, therefore, must perform virtually all of the light-bending, and for this reason is very hard in comparison with the lenses of terrestrial vertebrates. Most birds and mammals focus light on the retina by changing the shape of the lens through a variety of techniques (Walls 1942). The fish lens is too hard to change its shape easily, so instead the focus is adjusted by moving the lens in or out relative to the retina, much in the same way that the hard, glass lens of camera is moved toward the film or away from it in order to focus an image on the film (Walls 1942).

2.2.1.2 Adaptation to Brightness

The fish cornea cannot serve as a light-gathering organ. Instead, the lens itself protrudes through the pupil into the space occupied by the aqueous humor, achieving a wide visual field. This means that the pupil cannot be easily opened and closed to change the amount of light entering the eye, so other mechanisms are used to adapt to changes in brightness. In teleosts much of this adaptation occurs through retinomotor movements (Brett 1957; Munz 1971; Lythgoe 1979 b). The pigments in the retina migrate in toward the center of the eye during the day to shield the sensitive rods from bright light and migrate to the rear of the retina at night to expose the rods. These pigment movements are accompanied by movement of the rods and cones themselves to facilitate the shielding effects (Fig. 2.2).

This retinomotor response may be important in fish migration. First, it is a relatively slow process. This led to the suggestions by Ali and Hoar (1959) that juvenile pink salmon *(Oncorhynchus gorbuscha)* lose visual contact with the river bottom when the light intensity in the evening declines more rapidly than the fish's retinal adaptation. This would allow them to be swept downstream. Second, the retinomotor responses of several fish seem to be governed by endogenous circadian rhythms (Schwassmann 1971). This may affect their ability to respond to light or darkness at inappropriate times in the diel cycle, an important factor to consider if we subject fish to artificial lighting during their migratory movements, at illuminated fishways for example.

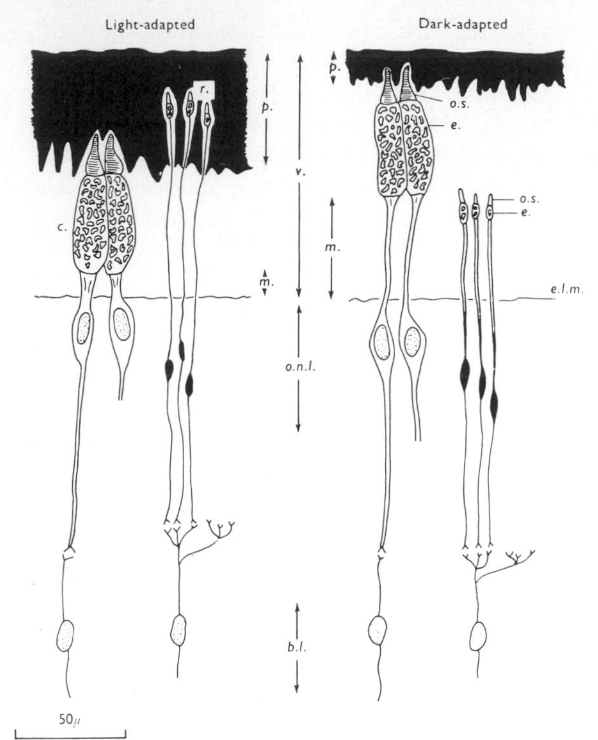

Fig. 2.2. In bright light the cones contract, drawing the photosensitive part of the cell closer to the front of the eye, while the rods extend to the back of the eye away from the light. At the same time pigments migrate forward to shield the sensitive rods. The process is reversed during dark adaptation (Blaxter and Jones 1967, also used by Munz 1971)

2.2.1.3 Vision Through the Water Surface

Vision through the surface must be considered if migrating fish use celestial bodies for orientation. The ability to look up through the water surface is also important to fish that are victims of aerial predators or that prey on aerial or terrestrial organisms. This subject has been clearly and concisely covered by Walls (1942). When a fish looks up at a calm surface, it can see through the surface into the air through a circular "window" bounded, in freshwater, by an angle of 97.6° (Fig. 2.3). Through this window ("Snell's window") the fish can see everything "from horizon to zenith" (Walls 1942), although the proportions of objects may vary in different regions of the window. Beyond the window the fish will see a reflection of an area below the surface, including a reflection of the bottom if the water is shallow and clear.

If the wind or some other factor disturbs the surface, the situation becomes considerably more complex. Each wavelet of a rippled surface may act as a new surface with its own little window, or the view available may change with each passing roller. Natural water bodies often have several sizes of waves on them at once, so vision through the surface may be complex.

It is not always necessary for a fish to see clearly through the surface in order to respond to the position of celestial bodies or even to terrestrial landmarks. The sun, and presumably the moon and stars, cast visible beams through a rippled sur-

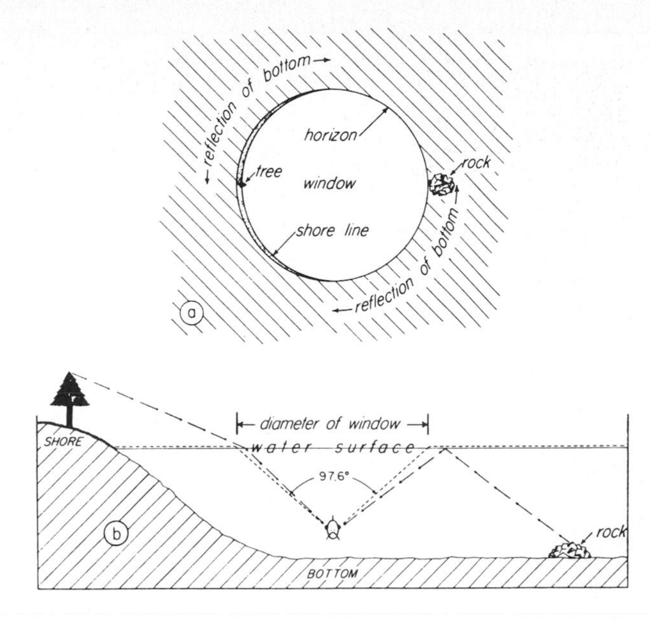

Fig. 2.3. The visual field of a fish looking up toward the surface. Within an angle of 97.6° the fish sees into the air with refraction compressing the aerial hemisphere into the surface window. Outside the window the fish will see a reflected view of objects beneath the surface (Walls 1942)

face. The angle of these beams is a reliable indicator of the position of the light source. Terrestrial objects that cast shadows into the water also indicate the location of the light source.

Migrating fish often break the surface, jumping or rolling, so that for a fraction of a second their eyes are above the surface. One of the most common explanations offered for this behaviour by casual observers is: "They're coming up to take a look around." This is unlikely. With the exception of a few species with special adaptations to aerial vision, fish would not get a clear image of their surroundings. The cornea, which does not focus light underwater, is effective in air and would render the eye extremely short-sighted. Furthermore, because the shape of the corneal surface is determined by streamlining rather than optics it will be astigmatic when it functions as a lens.

2.2.1.4 Sensitivity to Polarized Light

Characteristics of Polarized Light. Light becomes polarized when it is reflected, transmitted through material with a linear internal arrangement, or scattered by molecules or small particles. For example, the light reaching earth from the blue portion of the sky has been scattered by molecules in the air and polarized. Most of the remaining light is vibrating parallel to a plane defined by three points, the sun, the observer and the point in the sky being observed (Fig. 2.4). The degree of polarization of the sky light is greatest when the point of observation is at right angles to a line from the observer to the sun. When the point of observation is

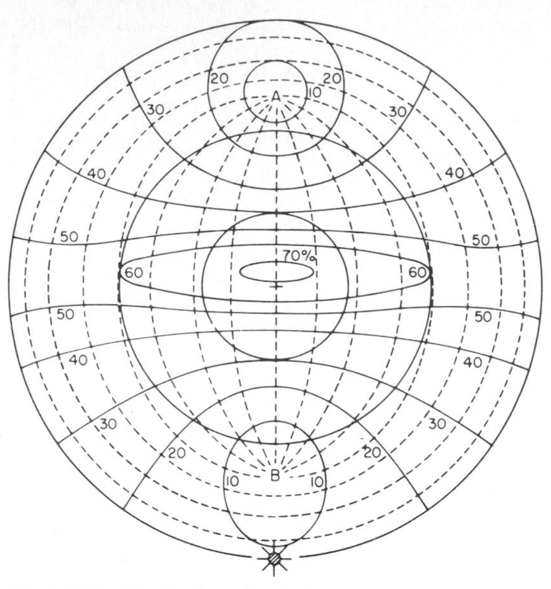

Fig. 2.4. The distribution of polarized light in the sky when the sun is on the horizon. *Solid lines* indicate the points of equal percentage of polarization and *broken lines* indicate points where the angle of polarization is the same. *A* and *B* are the two points where the light is unpolarized. The diagram represents the hemisphere of the sky, + being the highest point and the two rings concentric with + being the lines subtending angles of 30° and 60° with the horizontal (Hoar 1975, originally from Carthy 1958)

close to the sun or directly away from the sun, the light observed is virtually unpolarized (Waterman 1981).

The pattern of polarization in the sky, since it is based on the position of the sun relative to the observer, can be used by terrestrial animals to determine the sun's position, even when the sun itself is obscured by cloud. Bees, for example, can use their sun-compass orientation mechanism as long as there is a patch of blue sky equal to 2.5% of the sky visible to them (von Frisch 1949; Zolotov and Frantsevich 1973).

Polarized light perception could be useful to aquatic organisms for the same reason. Fish use sun-compass orientation (see Sect. 2.4.2) and polarized light is available underwater. First, the polarization pattern of the blue sky is visible through the surface window, if water depth and clarity permit the fish to see through the surface. Second, light penetrating water is polarized in a distinctive pattern related to the sun's direction. Reflection at the water surface does not play a significant part in this process. Instead, most of the polarization is accomplished by scattering in the water (Waterman 1954, 1958, 1981; Waterman and Westell 1956). The pattern that emerges is similar to the sky pattern with the plane of polarization (e-vector) typically at right angles to a plane defined by the sun, the observer and the observation point (Fig. 2.5) and the maximum degree of polarization perpendicular to the sun's apparent direction.

A fish that could detect polarization could determine the sun's direction from either the orientation of the e-vector in the surrounding water or the pattern of the percentage of polarized light. It could obtain information on the elevation of

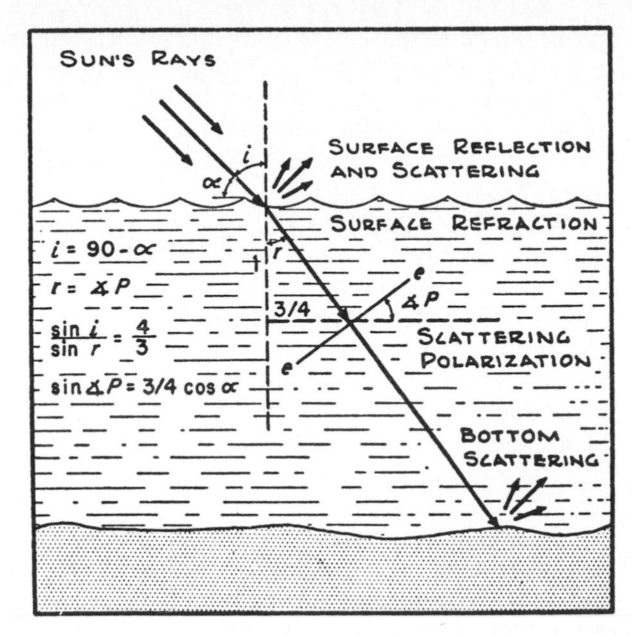

Fig. 2.5. Underwater polarized light originates from the scattering of directional light in the water. The overall e-vector pattern will tend to lie at right angles to the apparent line of sight to the sun (Waterman 1972)

the sun from the tilt of the e-vector or the tilt of the belt of maximum polarization. This information may be available at considerable depths. The imposition of such a variable medium as natural water on the clean geometry of optics leads to some unpredictable effects. Elliptical polarization has been observed underwater on some lines of sight (Waterman 1954, 1958; Ivanoff and Waterman 1957); turbidity and cloud cover can alter the pattern of polarization (Waterman 1954, 1958, 1981; Waterman and Westell 1956); anomolous polarization patterns at twilight have also been reported by Waterman and his colleagues. The importance of these effects to fish is not known.

Perception of Polarized Light. Fish must be able to detect polarization of light before they can use it to assist their orientation.

Indirect evidence that migratory salmon might be responding to polarized light came from two sources. Braemer (1959) observed that young coho salmon, *Oncorhynchus kisutch,* trained to orient by a sun compass, could maintain their trained directions for a few minutes after sunset when sky polarization is strongest. Groot (1965) found that the highest levels of sockeye, *O. nerka,* smolt migration occurrred during periods of strong sky polarization when the sun was just below the horizon. Changing the apparent position of patches of sky with mirrors caused the smolts to change their preferred direction to a direction that would be predicted if they were using some pattern in the sky for their orientation. Groot (1965) then went on to test the effects of polarizing filters on the preferred orientation of migrating sockeye smolts. Polarized light induced changes in preferred direction at dusk but not at other times of day (noon and afternoon). However, at dusk the greatest differences between fish under a polarizing filter and control fish without a filter occurred when the filter was reinforcing the existing polarization of the sky rather than when the filter was arranged to change the sky pattern; a difficult thing to explain, but still an indication that the fish were detecting polarized light.

Dill (1971) followed up with conditioning experiments. Sockeye smolts were successfully trained to select a feeding location associated with vertically polarized light in preference to three feeding locations equipped with horizontally polarized light sources. Dill also looked for a relationship between the preferred orientation of sockeye smolts and the plane of polarization of a vertical column of light. He found a significant preference for directions at 22.5° to 45.0° to the right of the e-vector. There is no ready explanation for this directional preference. The cichlid *Pseudotropheus macropthalmus* has also been trained to respond to the plane of polarization of light using a food reward (Davitz and McKaye 1978). Tilapia, *Sarotherodon mossambicus,* rainbow trout, *Salmo gairdneri*, and yellowtail, *Seriola quinqueradiata,* could be conditioned to show reduced heart rate in response to polarized light stimuli, but common carp, *Cyprinus carpio,* and sea bream, *Euynnis japonica,* could not be conditioned to respond (Kawamura et al. 1981).

Marine halfbreaks, *Zenarchopterus* spp., show a time-compensated ability to select a preferred direction relative to the e-vector of linearly polarized light (Waterman and Forward 1970, 1972; Forward et al. 1972). Response to polarized light was evident both in fish tested underwater without access to the surface and in fish tested with access to the surface. Maximum change in orientation occurred when the polarizing filter was adjusted to change the pattern of the sky rather than reinforce it. The relationship between the results of these studies and the biology of the animal is not clear, although Waterman and Forward (1970) suggest that the preferred directions may correspond to the orientation of the channel in which the fish were captured. Similar responses to polarized light occur in the related freshwater halfbeak, *Dermogenys pusillus* (Waterman 1972, 1975).

One of the unexpected results of the work on halfbeaks was the response of the fish to clouds covering the sun. They were disoriented even though 60%–65% of the sky remained clear (Waterman and Forward 1972; Forward et al. 1972). If the polarization pattern of the sky is a supplement to the normal sun-compass mechanism, as it apparently is in bees (Zolotov and Frantisevich 1973), these fish should have been able to orient.

Goldfish (*Carassius auratus*), tested in a radially compartmentalized tank under polarized illumination preferred the compartments that were parallel to the e-vector of the polarized light (Kleerekoper et al. 1973). Neurophysiological recording from cells of the optic tectum in the goldfish brain has shown that the tectal cells respond differentially to different planes of polarized light striking the eye (Waterman and Aoki 1974; Waterman and Hashimoto 1974). These results support the earlier reports of behavioural responses by fishes to polarized light.

There is, then, evidence that several fishes can perceive and respond to the plane of polarized light, although none of the studies unequivocably ruled out the possibility that the fish were responding to intensity differences or differences in patterns of reflection. One common thread does unite the results of all the work on responses of fish to polarized light: the absence of any clear-cut relationship between the behaviour of the animals in the experiments and their behaviour in nature. This is particularly striking in the sockeye smolt work, where there was considerable information available on normal patterns of migratory movement (Groot 1965). Waterman and his co-workers have revealed very little about the natural requirements for orientational information in their animals; it is not even specified whether or not animals were migratory at the time the studies took place.

Since negative results are seldom published, we will probably never know how many fish have been tested for response to polarized light and found wanting. Waterman (1959) did test "several" species without success and a study on Pacific herring, *Clupea pallasii,* in which an analyzer for polarized light was found associated with the eye, did not lead to any published report of a response to polarized light (Stewart 1962). Waterman (1981) briefly reviews possible mechanisms for detection of polarized light by vertebrate eyes.

Orientation is not the only possible function for polarized light sensitivity in fish. The ability to detect polarization may make it easier to distinguish light or dark objects from the background (Lythgoe and Hemmings 1967). For example, when a polarizing filter was oriented so as to cut out most of the (polarized) background light, bright objects like silver fishes were seen more easily, the reflected light from their sides apparently being less strongly polarized than the background. Turning the filter to allow the polarized background light to pass, and hence to lighten the background, made dark objects more visible. Perhaps many fish can see polarization patterns and respond to them, but use the information primarily for prey or predator recognition rather than for orientation. When they are placed in the deprived environment of a test apparatus they might respond to changing polarization patterns, but their responses would have little to do with migration.

2.2.1.5 Extraoptic Photoreceptors

Fish have been known to possess extraoptic photoreceptors (EOP's) since the studies of von Frisch (1911) on the pigment responses of the European minnow, *Phoxinus phoxinus*. Fish pineal organs show histological features that resemble the cones of a vertebrate retina. Indeed, a number of studies have demonstrated a photosensory capability for the pineal (Falcon and Meissl 1981; Tabata 1982) and in some cases for other extraoptic structures as well (Fenwick 1970). The question that has not been answered is: can fish orient themselves in space on the basis of information from EOP's? This does happen in amphibians and sometimes includes polarized light orientation (Adler and Taylor 1973). There are studies that indicate that the pineal may modulate fish phototactic responses. Breder and Rasquin (1947, 1950) found that the phototaxis of different forms of blind characins, *Anoptichthyes* sp., depends on the degree of shielding between the pineal and the external environment. Fish with a lot of melanin or guanine covering the pineal tend to avoid light in a choice box, fish with an exposed pineal are attracted to light; the responses can be reversed by covering exposed pineals or exposing shielded pineals.

Sockeye smolts are able to maintain their normal negative phototactic response after blinding, but become indifferent to light when the pineal is destroyed (Hoar 1955). Goldfish apparently require both pineal and eyes for their phototactic responses. Deprived of either or both, they become indifferent in a light gradient (Fenwick 1970). Larval herring, *Clupea harengus,* and larval plaice, *Pleuronectes platessa,* perform vertical migrations toward the surface as light intensity drops in the evening. Eyeless larvae of both species perform these oriented movements in the laboratory at the same time as intact controls. This indicates sufficient extraoptic photoreception to govern this vertical migration (Wales 1975). European eels, *Anguilla anguilla,* retain their nocturnal activity rhythm and photonegative behaviour even after blinding, pinealectomy, and both operations together. However, blinding combined with shading the brain with aluminium foil does interfere with the activity rhythm and photonegative behaviour (van Veen et al. 1976).

As well as these behavioural responses to light there is other evidence for extraoptic photoreception in fish. Several fish possess clear windows in the skull, which will facilitate entry of light into brain area (Breder and Rasquin 1950; Rivas 1953). Furthermore, the morphology of the pineal supports the idea that it is a photoreceptor. However, the presence of EOP's does not mean that they function in orientation. They may play other roles in behaviour or physiology, the matching of bioluminescence to external light, for example (Young et al. 1979), or the timing of seasonal reproductive cycles (Matty 1978).

2.3 Timing

2.3.1 Diel Timing

Earth's rotation provides animals with a changing photic environment, night and day. Light intensity varies in a predictable manner throughout a 24-h cycle and the pattern and degree of change varies with season and latitude. The diel rhythm of light and dark will affect the success of predators, the survival of prey and the availability of orientation cues such as the sun or the stars. Few animals are equally well adapted for activity at all levels of light intensity. Therefore, most restrict their activities to a portion of the diel cycle. These rhythms of activity must match the external illumination if they are to be adaptive. Although it may be modified by cloud cover, water clarity or eclipse, the daily cycle of light intensity is one of the most precise and regular cycles accessible to vertebrates. It is an appropriate timing stimulus for behavioural and physiological rhythms related to diel activity.

2.3.1.1 Detection of Diel Cycles of Illumination

The diel light cycle is detected by fish primarily through the eyes and EOP's, but there are secondary effects of the light cycle that could stimulate other senses.

These include daily cycles in temperature, pH, dissolved gases and other compounds produced by photosynthesis, and the other biochemical activities of autotrophs (Müller 1978 a). These cycles of chemical stimuli may serve as timing stimuli for the movement of adult and juvenile sockeye salmon in tributaries of the Paratunka River, USSR (Krogius 1954).

Many field studies of fish diel activity have not been controlled for responses to thermal or chemical stimuli. Laboratory studies in which temperature was controlled and in which chemical changes similar to those found in nature were highly unlikely have often indicated that behavioural rhythms can be kept on time by changing light intensity alone. The American eel, *Anguilla rostrata,* for example, is clearly nocturnal in nature and when tested in laboratory aquaria is more active in the dark than in the light (Bohun and Winn 1966). On the other hand, there are cases in which controlled light regimes have not been able to elicit the same rhythms as found in nature, or where conflicting results have been reported for the same fish by different authors (Eriksson 1978 b). Some of these cases may turn out to be examples of diel rhythms based on thermal or chemical stimuli. In many cases, however, the conflicting results probably reflect differences in recording methods or real duality in the response of the fish to light stimuli. Some species that are nocturnal or crepuscular in summer become diurnal during the short days of the arctic winter (Müller 1978 a) (Fig. 2.6). Rhythmicity may also change with light intensity. Bullheads, *Ictalurus nebulosus,* are normally nocturnal but become more active during the "day" if the light intensity is reduced (Eriksson 1978 b).

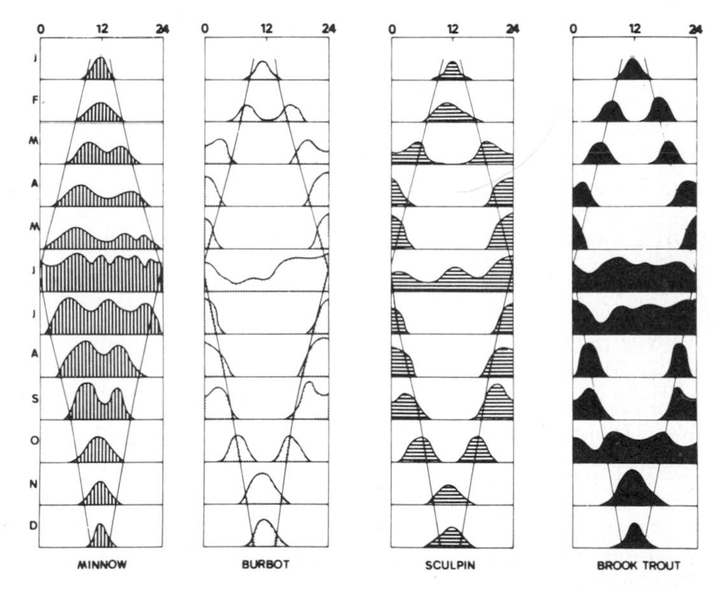

Fig. 2.6. Diel activity patterns during the year of four fish species in the River Kaltisjokk, near the arctic circle. Nocturnal fish such as the burbot and crepuscular species such as brook trout become diurnal during the short midwinter days. Legend: minnow, *Phoxinus phoxinus;* burbot, *Lota lota;* sculpin, *Cottus poecilopus* and brown trout, *Salmo trutta.* With permission from Müller K (1978 a) Locomotor activity of fish and environmental oscillations. In: Thorpe JE (ed) Rhythmic activity of fishes. Academic, London, pp 1–19. Copyright: Academic Press Inc. (London) Ltd.

2.3.1.2 The Priority of Migration Relative to Other Behaviour

Many fish have become specialized for activity in a particular portion of the daily photocycle. The timing of migratory behaviour will depend to some extent on the relative priority of migration compared with other behaviour such as feeding or predator avoidance in any particular situation. In situations in which migration supplants feeding in the animal's priorities, such as the freshwater portion of the breeding migrations of Pacific salmon, the animal may migrate during the period of peak visual ability.

Migratory behaviour may have low priority relative to feeding or predator defence for early fish in a particular run, fish with an undemanding migration route or without restrictive limitations on the time used to complete the migration. In other situations, such as fish that are "late" in a run, or that face severe seasonal or other limitations on the time for migration, migration may assume a higher priority than feeding or predator defense and replace these activities in their normal diel periods.

I have assumed that the best time for migration is also the best time for feeding, while the time normally given over to inactivity and/or predator defense behaviour (resting and hiding) is less suitable for migration and less likely to be usurped by migratory activity. This may not always be the case. In many small birds, migration replaces the nocturnal period of inactivity of the non-migratory phase rather than the diurnal feeding period (Griffen 1964). This situation may reflect a requirement for feeding to supply energy for migration or the necessity for avoiding high diurnal predation, or both.

There are a few cases of fish that undertake migrations during the time of day that is otherwise taken up by inactivity or defensive behaviour rather than the period normally occupied by feeding. Pink salmon fry, *Oncorhynchus gorbuscha,* for example, migrate downstream to the sea at night, then feed diurnally after arriving in the marine environment (Godin 1980 b). The downstream migration continues into daylight if the migration route is long and, once in the sea, feeding and migration alternate during the day (Healey 1967). So the downstream migratory phase occurs during the period that is not used for feeding a few days later, but migration supercedes feeding and rest for the downstream phase.

The tautog, *Tautoga onitis,* migrates offshore in the fall, apparently in response to shortening photoperiod and declining temperatures. Fish held in laboratory tanks are active during the day and inactive at night, like most labrids, except for two periods during the year. During the spawning period they remain active at night and during the fall migration period nocturnal activity actually exceeds diurnal activity for a few days at the same time and under the same photoperiod and temperature conditions as the natural offshore migration (Olla and Studholme 1978; Olla et al. 1980). This suggests that the natural migration is more nocturnal than the non-migratory behaviour shown at other times of year.

2.3.1.3 Physiological Control of Diel Rhythms

Selection pressures on diel rhythms provide the ultimate level of causation; but between the environment and the behaviour of the animals lie the physiological mechanisms, the mediators of proximate cause, that control behaviour. The study of these physiological rhythms is a complex and controversial subject. Orthodox thinking at the moment is that there are many endogenous rhythms in the functioning of the body. Endogenous diel rhythms tend to have a natural periodicity of nearly, but not exactly 24 h (circadian rhythms). They are constantly reset to coincide with local conditions by means of their response to external stimuli, the "Zeitgebers". The rapid increase in illumination at dawn and the rapid decrease in illumination at dusk are common Zeitgebers for diel rhythms. The steps between the internal physiological rhythms that may act as "biological clocks" and the rhythmic changes that occur in the migratory behaviour of fish are not understood. There have been studies that indicate that the activity rhythms of some fish are exogenous: the activity rhythm of immature American eels, *Anguilla rostrata,* is an example (Bohun and Winn 1966). Decreased light leads to increased activity regardless of time of day, and constant conditions lead to arrhythmic activity. Other studies indicate that the activity rhythms (sometimes of the same species) are endogenous and responsive to Zeitgebers. This subject has been reviewed by Woodhead (1966), Blaxter (1970), Schwassmann (1971), Thorpe (1978), and Godin (1982).

A physiological requirement for sleep does not seem to be a major factor in determining the daily rhythm of migratory activity in fish. There is no doubt that several species of fish normally show periods of inactivity during specific portions of the day, or that in the case of some of these species the inactive period is characterized by a state resembling the sleep of higher vertebrates (Karmanova et al. 1981). Parrotfishes, Scaridae, lie inert on the bottom at night. Juveniles or members of the smaller species often secrete a mucus envelope around themselves, which may protect them from the olfactory hunting of the moray eel, *Gymnothorax* sp. (Winn 1955; Winn and Bardach 1959, 1960; Winn et al. 1964b). Even in this strongly diurnal family, however, experimentally displaced fish will continue to move after dark until they reach a relatively safe location. When Winn et al. (1964b) released parrotfish, *Scarus guacamaia* and *S. coelestinus,* at dusk or after dark, with lighted balloons attached to the fish by a long nylon line, the parrotfish tended to keep moving until they reached a shoreline, where they stopped.

2.3.1.4 Physiological Rhythms and Natural Migrations

Major problems occur when one tries to relate fish migration to laboratory studies on the physiological control of activity rhythms. The main problem lies in comparing the various indices of "activity" used by experimenters to the actual migratory behaviour of a species. One could assume that when a migratory species was studied during a migratory phase of its life cycle the period of maximum "activity" would reflect the period of maximum migratory effort – migratory restlessness. However, unless this point has been specifically considered by the experimenter, such an assumption could be quite misleading.

Students of rhythmic behaviour seem to have a penchant for automatic recording devices, many of which require a relatively barren tank. Yet studies by Jones (1956) on minnows, *Phoxinus phoxinus,* Bregnballe (1961) on flounders,

Pleuronectes flesus, and Edel (1975b) on silver eels, *Anguilla anguilla,* indicate that, at least in these species, high activity levels at some times of day will disappear if hiding places are provided – shelters for the minnows and eels and sand for the flounders to bury themselves in. Much of the activity recorded in the barren tanks was, in fact, a fright or escape response in the exposed conditions. Similarly, the plaice, *Pleuronectes platessa,* showed nocturnal activity in experimental tests that did not correspond to the diurnal activity patterns indicated by information on catch and stomach content from the commercial fishery (de Groot 1964; Verheijen and de Groot 1967). Replacing bright "day" illumination with lower levels of illumination, closer to the levels actually found on the sea floor during the day, and use of more sensitive recording apparatus did confirm daylight feeding activity by the plaice. This type of change in rhythm with changing experimental conditions could probably be demonstrated fairly often. However, in many cases the prime aim of a study of activity rhythms has been to learn about the physiological control of rhythms in general. Functional relationships between the rhythm studied and the natural behaviour of the animals may be given little consideration.

2.3.1.5 Diurnal Timing of Salmon and Trout Migration

Fortunately, some situations occur that are conducive to the production of studies of the rhythms of activity shown by migratory fish in their natural habitat. The salmonid fishes, particularly the genera *Salmo* and *Oncorhynchus,* provide one such situation. In this group we have a combination of high economic value, complex patterns of migration and accessibility for study, at least during the freshwater and coastal portions of their migrations. The obvious vulnerability of the salmon to human interference and their strong emotional appeal to the public have also been factors encouraging the study of their natural biology. The timing of salmon migration has recently been reviewed by Godin (1982). Other economically important diadromous fish, such as the Atlantic eels, genus *Anguilla,* and American shad, *Alosa sapidisissima,* provide similar opportunities but have generated smaller amounts of information, although both are the subject of active research by several investigators.

2.3.1.5.1 Emergence from the Gravel (The Swim-Up Stage)

The eggs of salmonid fishes are usually buried in gravel by the female. When the young hatch and have absorbed most of their yolk, they emerge from the gravel and swim up to the water surface to gulp a mouthful of air, filling their swim bladder so that they may assume natural buoyancy and a normal swimming posture. This procedure has, in miniature, all the characteristics of true migration. It is a precisely timed and oriented movement from one major habitat, the gravel, to another, open water. It is related to a definite physiological requirement, air for filling the swim bladder.

Bams (1969) describes the mechanisms that determine the timing of this "swim-up" in fry of the sockeye, *Oncorhynchus nerka.* As the fry in the gravel develop, they tend to move up toward the gravel surface. This upward movement increases as water temperatures rise during the late afternoon. (They will also move upward as an emergency response to high carbon dioxide concentrations). But the fry are photonegative, so their upward movement in the gravel is halted just below the gravel surface during the light portion of the day. As darkness falls in the evening, the inhibitory effect of light is removed, while the stimulatory effect of high water temperature is still present. The fry that have been accumulating just below the gravel surface all day can emerge and swim up to the water surface, giving an evening peak in swim-up activity. Bams (1969) links this timing mechanism, which restricts swim-up to the hours of darkness, to the vulnerability of the fry during this stage. At the time of emergence, they have negative buoyancy (they sink) and must swim vigorously to reach the surface. Their lack of buoyancy forces them to assume orientations that destroy the effectiveness of their natural counter-shading. Since fry are seldom successful on their first attempt, they often make a number of trips to the surface. Since these factors combine to make the fry conspicuous, and since they do not show a fright response to predatory fish (Bams 1969), there seems good reason to suspect that predation pressure may be an important factor in the evolution of nocturnal timing for swim-up behaviour (Ginetz and Larkin 1976; Godin 1982). There may be a further advantage if the fry are synchronized in their emergence, potentially swamping resident predators (Godin 1982).

Fry of pink salmon, *O. gorbuscha,* also emerge from the gravel primarily at night (Godin 1980a), and the general assumption in much salmon literature has been that most fry emergence occurs at night. On the other hand, emergence also occurs in daylight under some conditions (Godin 1982). For example, Heard (1964), using traps set in the gravel to catch emerging fish, found that about the same number of sockeye fry emerged under continuous artificial illumination as under a normal diel cycle. Artificial illumination did tend to reduce night emergence but this was balanced by increased daytime emergence in the lighted half of the traps. This seems to be in direct conflict with Bams' (1969) observation that 50 fry in a continuously illuminated tank died of starvation in the gravel without emerging.

2.3.1.5.2 Fry Migrations

After the young salmonids have emerged from the gravel and filled their swim bladders, they move from the adult spawning habitat to a habitat where they feed and grow. The conditions that favour egg survival, a rapid flow of oxygenated water through unproductive gravel beds, are not always conducive to easy feeding by juvenile fish. A common response of emergent fish is to move to a quieter and more productive habitat. If they remain in the main stream they tend to occupy riffles and similar regions where reduced flow rates and turbulence are available. There are, however, differences among species in how they achieve this result (Table 2.1). This gravel-to-nursery migration varies from a few meters between the gravel and a quiet portion of the natal stream to a migration of many kilometers to the sea.

There are many variations on the general patterns summarized in Table 2.1 (Northcote 1978; Godin 1982). Some coho fry, for example, migrate upstream

Table 2.1. Nursery habitats of eight species of trout and salmon

Species	Nursery area
Salmo gairdneri	Streams, lakes
S. clarki	Streams, lakes
S. salar	Streams
Oncorhynchus kisutch	Streams, lakes
O. nerka	Lakes
O. tshawytscha	Streams, lakes, estuaries
O. keta	Sea, estuaries
O. gorbuscha	Sea

shortly after emergence (Gribanov 1948, personal observation), others disperse downstream (Neave 1949). Sockeye fry may hatch above their nursery lake in an "inlet" stream, below the lake in an "outlet" stream or in the lake itself as a result of shore spawning (Foerster 1968). The orientation required to reach the nursery area will be different in each case. A similar situation occurs in rainbow trout as studied by Northcote (1958, 1962, 1969 b).The most common response, however, is downstream movement and a mechanism, which depends on the response of young salmonids to light, has been proposed for this downstream migration.

The passive Drift Hypothesis. It is proposed that downstream migrants lose visual contact with their surroundings as light levels fall in the evening. Without visual reference points they are unable to maintain their position in the stream and are swept downstream by the current (Hoar 1953). In support of this idea, studies on the rate of dark adaptation in the eye of juvenile salmon indicate that dark adaptation may occur more slowly than the rate of decline in illumination that occurs in the evening. This would lead to a period of night blindness until the dark adaptation processes catch up with the falling illumination (Brett and Ali 1958; Ali and Hoar 1959; Ali 1959) (Fig. 2.7). It is presumably during this period of night blindness that downstream displacement would occur.

This relatively simple mechanism for achieving downstream migration probably operates under some conditions. There is, however, evidence that simple loss of visual contact with the substrate does not account for all downstream migration by juvenile salmonids (Godin 1982). For example, MacKinnon and Brett (1955) found that chum and pink fry moved through a small impounded water basin more rapidly than did drifting floats, indicating active migration through this area of quiet water rather than passive drift with the current. Wickett (1959) also observed pink fry moving downstream more rapidly than the current speed. Pink fry have been directly observed swimming vigorously downstream in the same direction as the current (Neave 1955). In some longer coastal streams downstream migration of pink fry continues until dawn (Fig. 2.8). If the migration depended on the period of temporary night blindness at dusk, then movement of fish should cease as soon as their eyes are fully dark-adapted. This situation may be approached in very short coastal streams where the fry clear the stream and enter the sea within a few hours after dusk, but even in these situations significant numbers of fish continue to move downstream throughout the night (Fig. 2.8).

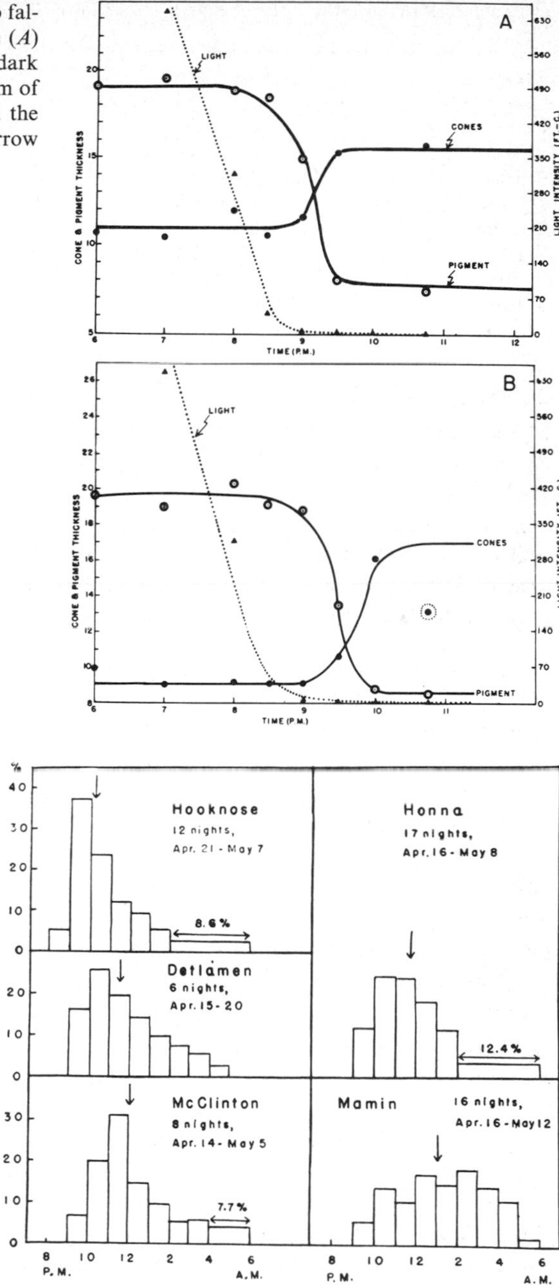

Fig. 2.7. Dark adaptation in relation to falling light intensity at dusk in sockeye (*A*) and coho (*B*) fingerlings. During dark adaptation cones move to the outer rim of the retina and pigments, which shield the rods during the day contract into a narrow band (Brett and Ali 1958)

Fig. 2.8. Percentage of pink salmon caught by hours, during nightly migrations, summed over several nights (Neave 1955). *Arrows* indicate mean time at which 50% of the migration was completed. In Hooknose Creek pinks spawn for about 1.5 km above the counting station. In Detlamen Creek, McClinton Creek, and Honna River they spawn up to 6 km, above the counting site. The Mamin River has 13 to 16 km of productive pink spawning area above the counting station. Note that even in the short creeks a significant portion of the migration occurs after midnight when dark adaptation should be complete

Another form of counter-evidence comes from the work of McDonald (1960), who found that fewer sockeye fry migrated out of his experimental troughs on cloudy nights (56–60%) than on a partly cloudy night (78%) or on clear nights (94%–86%). If migration is a function of loss of visual contact, then one would expect more fish to migrate on the darker cloudy nights. McDonald's (1960) data on the migration of sockeye fry in William's Creek, B.C. also show that between 40% and 72% of the migration occurred after midnight, depending on the season. Pink fry were found to migrate throughout the full 24-h cycle once they had traveled about 16 km from their hatching site. This effect is, of course, more probably one of time since hatching than of distance traveled. Some daylight migration occurred in clear shallow streams, although the highest levels of daylight migration were recorded in turbid water (McDonald 1960). The activity of pink fry in laboratory tanks shifts from an irregular daily swimming rhythm with a tendency to swim near the surface at night to a diurnal rhythm with an increasing tendancy to swim near the surface in daylight 7 to 13 days after emergence (Godin 1980 b).

Most evidence indicates that, once their eyes are fully dark-adapted, juvenile Pacific salmon can see fairly well at the levels of illumination normally encountered at night. Ali (1959) places the rod threshold of juvenile salmon at 10^{-5} lx candles, while Brett and Groot (1963) report cessation of feeding at 10^{-6} lx in coho fry. Above these thresholds the fish can see well enough to school, feed, and avoid obstacles. For example, young salmon moving downstream at midnight are able to avoid chains hung in glass tubes but not empty glass tubes, indicating a visual response to obstacles (Brett and Groot 1963).

In spite of the evidence against the "passive drift with loss of visual contact theory," it has had an active life as the orthodox explanation of downstream travel at dusk. The initial movement downstream may start with loss of visual contact during a temporary period of night blindness. Once initiated, however, the movement continues in spite of full dark adaptation, which occurs after about 40 to 50 min (Brett and Ali 1958; Ali 1959; Ali and Hoar 1959). It seems reasonable to assume that much of the downstream migration of juvenile salmonids is a directed response and occurs as a specific behavioural response rather than as a passive accident resulting from the different rates of change of dark adaptation and falling evening light intensities. One of the important stimuli releasing the downstream movement is low light intensity. Other factors such as temperature and maturation are no doubt involved (Bams 1969).

Predation pressure may account for the restriction of peak downstream migration to early darkness. First, visual predators such as birds and some fish will be less effective at night. Rainbow trout, for example, catch fewer sockeye fry under low illumination levels than under high illumination (Ginetz and Larkin 1976; Godin 1982). Second, it will be advantageous to the fry if they all move at the same time so that the pulse of migrating fish will swamp the response of resident predators as the fry move through each stretch of river along the migration route, although many fry predators may not reach saturation levels in natural systems (Peterman and Gatto 1978). Initiating migration at dusk will coordinate movement of the group and insure a maximum period of darkness for travel. In cases such as the pink fry, that continue to migrate during the day, if the trip exceeds about 10 miles of travel (McDonald 1960) reaching the sea presumably overrides any advantage of ceasing migration until the next night.

Upstream Fry Migration. Some young salmonids hold position or migrate upstream after emergence (Godin 1982). Coho fry, for example, move to quiet portions of their natal stream or sometimes move upstream into productive headwaters. They have a lower cone threshold than chum, pink or sockeye fry, while showing the same rate of dark adaptation (Ali 1959). Coupled with this is a difference in their behavioural response to low light intensities. Instead of swimming up toward the surface at night like the other species, they tend to settle to the bottom with approaching darkness (Hoar et al. 1957). Both the improved vision at low light intensities and maintaining contact with the bottom will help prevent displacement by the current. The diel timing of upstream migration by coho fry has not been extensively studied (Gribanov 1948; Mundie 1969), but on at least some occasions it takes place in bright sunlight and clear water (personal observation, Robertson Creek, B.C.). This sort of upstream migration by daylight is also characteristic of the more intensively studied upstream migration of rainbow trout fry and sockeye fry (Northcote 1962, 1969 b; McCart 1967; Clark and Smith 1972).

Migration of Inlet and Outlet Fry Populations. Loon lake, British Columbia, contains a population of rainbow trout that spawn in the inlet stream at the head of the lake and other populations that spawn in the outlet river and in Hihium Creek, a stream that runs into the outlet river (Fig. 2.9) (Northcote 1962, 1969 b). Young trout from each of these populations make their way back to Loon Lake to feed and grow. This means downstream migration by the inlet fish, upstream migration by the outlet fish and a combination of first downstream, then upstream movement by the fish from Hihium Creek. These differences in behaviour are probably based on some combination of genetic differences and responses to differences in water temperature and water chemistry (Sects. 5.4.3.1 and 7.7.4).

The diel cycle of migratory activity in these three groups is interesting. Inlet fish moved downstream at night, while outlet fish moved upstream during daylight (Fig. 2.10) (Northcote 1962, 1969 b). The Hihium Creek fry moved downstream to the outlet at night, then they presumably moved upstream in the outlet during the day since there were no appreciable levels of upstream night migration of fry recorded from traps on the outlet. Outlet fish also tended to emerge sooner, due to higher water temperatures, but to wait about a month before starting their upstream migration, while inlet and Hihium Creek fish emerge later, but start their downstream migrations immediately. The same pattern of diurnal upstream migration and nocturnal downstream migration is repeated in yearling and juvenile trout, that sometimes stay in the rivers before moving in the lake.

Rainbow trout show similar patterns of migratory activity in other lakes. In the Sixteen Mile Lake system, for example, most trout spawning occurs in the outlet and the upstream migration to the lake occurs largely during the day. Downstream migration from inlet streams in the Finger Lake system in New York State was primarily nocturnal (Northcote 1969 b). Similarly in Pothole Lake, B.C. with both inlet and outlet spawning populations there was again a greater tendency for downstream migration at night, while upstream migration

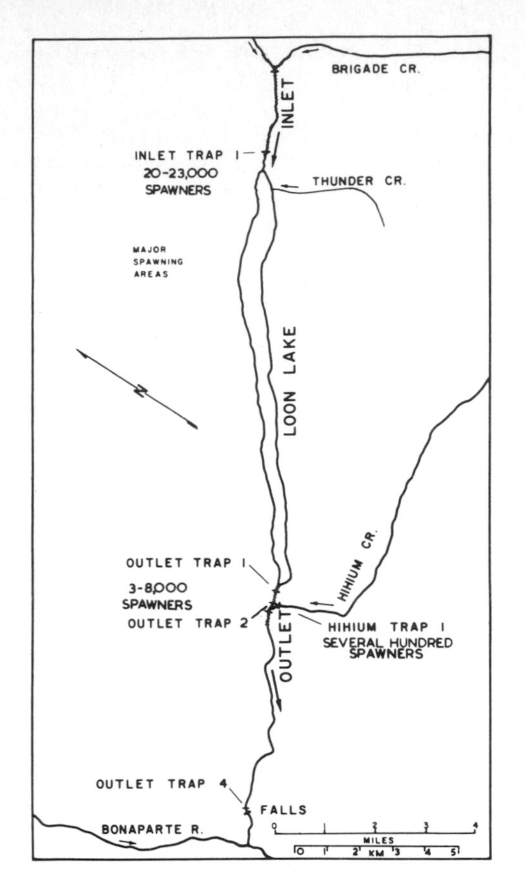

Fig. 2.9. The Loon Lake system showing major rainbow trout spawning areas and the location of fish traps used in assessing the number of migrants (Northcote 1969 b)

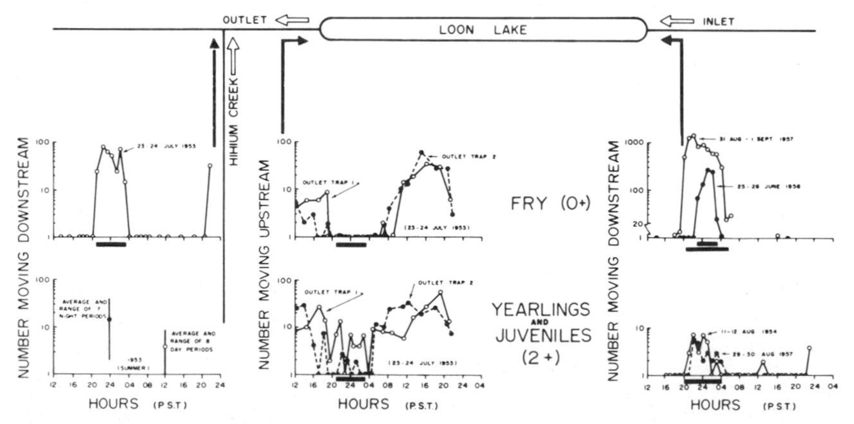

Fig. 2.10. Diel pattern of lakeward migration of young rainbow trout in the Loon Lake stream system. *Horizontal black bars* indicate periods of darkness (Northcote 1969 b)

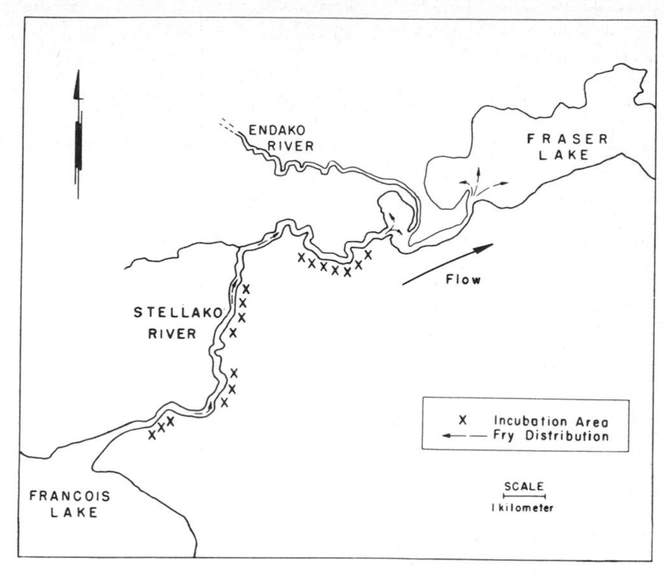

Fig. 2.11. Hatching sites and migration routes of sockeye fry from the Stellako River stock (Brannon 1972)

occurred during the day, although the differences between day and night movement were less pronounced than in the Loon Lake system (Northcote 1969 b).

Sockeye fry usually move downstream into nursery lakes from "inlet" rivers. There are, however, several locations where sockeye spawn in the outlets of lakes and the fry move upstream into the nursery lakes. In these cases, including the Babine and Chilko Rivers, British Columbia, fry of outlet spawners initially move downstream a short distance at night immediately after emergence. Then they hold position for a period of days to weeks before moving upstream during the day into Babine Lake (McCart 1967; Clark and Smith 1972; Brannon 1972).

At least one other case occurs in which downstream migration is nocturnal while upstream migration occurs during daylight. Fry of the redside shiner, *Richardsonius balteatus,* move downstream at night while adults on their spawning migration move upstream during the afternoon and early evening (Lindsey and Northcote 1963).

Why should upstream and downstream migration occur at different times of day? Hartman et al. (1967) make the point that night migration may serve to reduce predation, and McCart (1967) describes the upstream migrant sockeye fry in the Babine River as being particularly vulnerable to predation. Even American robins, *Turdus migratorius,* were found preying on sockeye fry as they migrated in broad daylight, close to shore. There must be some powerful reason for upstream migration to occur during the day in the face of increased predation. One suggestion is that upstream migration requires greater energy expenditure and is thus restricted to the warmer portion of the day (see Chap. 5). Another possibility is that daylight allows visually mediated positive rheotaxis. Godin (1982) argues that diurnal upstream migration may occur in spite of high predation because visual rheotaxis allows faster upstream movement than rheotaxis based on mechanical cues.

Oceanic Fry Migration. Once they leave the streams, the diel activity rhythm of young salmonids may change again. Pink fry in the early portion of their ocean migration migrate during the day near the water surface (Healey 1967) although they were nocturnal in their downstream freshwater migration. These fish continue to migrate although they also feed and grow rapidly during this portion of their migration. They migrate in a saltatory manner, periods of migration alternate with periods during which the fish hold in bays and feed (Healy 1967). In orientation tests, they show their strongest preference for the direction of migration at midday and are most strongly oriented on clear days. Since these fry were poorly oriented in early morning tests, they may use a sun-orientation technique that functions best when the sun is high in the sky, unlike sockeye smolts, that orient most accurately at dawn and dusk (see Sect. 2.4.1) (Groot 1965).

2.3.1.5.3 Smolt Migrations

After a period of 1 to 3 years in freshwater, some young anadromous salmonids undergo a physiological and morphological transformation into smolts. The smolt is a migratory stage with more silver pigment, a leaner body shape and a greater affinity for seawater than the non-migratory phase (Hoar 1976; Schreck 1982). The smolts migrate downstream to the sea. In many cases this means migrating out of nursery lakes, and occasionally, through lakes or artificial impoundments that lie downstream from their nursery areas. This migration through the relatively still water of lakes is most pronounced in sockeye with their obligatory period of lacustrine existence, and it is only in sockeye that the mechanisms of seaward migration through lakes have been studied to any extent. The other species will often face the same problems as sockeye.

An intensive study of the guidance mechanisms involved in the migration of sockeye out of Babine Lake, British Columbia (Groot 1965) provides some data on the diel pattern of migratory movement. The most active migration through the lake occurs at dawn and dusk, primarily dusk (Johnson and Groot 1963). A similar pattern occurs in other lakes. Furthermore, the frequency of migratory behaviour patterns shown in an experimental orientation apparatus also showed dawn and dusk peaks (Groot 1965). When the migrating smolts were observed in the lake at all times of day and night by means of SONAR, it was found that the speed and directness of migration were greatest at twilight (Groot and Wiley 1965; Groot 1972). Sockeye smolts may use a technique of orientation that functions best in twilight, perhaps orientation to the polarization pattern of the sky (Groot 1965; Dill 1971).

Once the sockeye smolts are out of the lakes and into the rivers they migrate largely at night (Foerster 1937a; Kerns 1961; Hartman et al. 1967). Hartman et al. (1967) state that the night-migrating Alaskan sockeye smolts they studied were negatively rheotactic and schooled, as they moved downstream. There are exceptions. Meehan and Siniff (1962), for example, found no significant diel cycle in the downstream migration of sockeye smolts in the Taku River, Alaska.

As in the case with fry, once the smolts reach the sea we know very little about their behaviour until they return as adults. Groot (1965), however, did make some observations on sockeye smolts during the period in their development when they would be moving down an inlet of the sea on their way to the open Pacific. In his apparatus the smolts maintained directional preferences that were appropriate to seaward migration but one of the migratory patterns, fluttering, became less frequent and no longer showed dawn and dusk peaks when these fish were transferred to seawater. The time of the peaks of feeding activity also shifted, which may be appropriate since the fish would utilize different prey in the sea.

2.3.1.5.4 Adult Salmon Migration

When the salmon are in the sea there may be a tendency for an upward movement at night and movement to deeper water during the day, as Manzer (1964) found for salmon in the Gulf of Alaska. This is a very common pattern among marine and lacustrine organisms (Woodhead 1966). Madison et al. (1972), tracking returning adult sockeye by means of ultrasonic transmitters, found that migration is more direct during the day.

As the fish return to the river mouth they become more accessible to biologists and the amount of data on their diel activity is large. River entry has largely been studied in terms of response to such factors as tide, freshets, and onshore winds rather than diel timing (Banks 1969).

In the rivers, the pattern of diel movements by adult salmon is quite variable. The safest generalization is that upstream migrants can migrate at any time of day or night, but at a given point in the river they often show a characteristic diel rhythm. Much of the data on the upstream migrants is collected at barriers to migration such as wiers, traps or fishways (Banks 1969). Fish may accumulate below a barrier at night then attempt to surmount it when the light intensity increases.

Stuart (1962) noted that Atlantic salmon stopped jumping at a waterfall during the night, and I have personally observed (by SCUBA diving with flashlights) sockeye accumulating during the night below an area of fast water, then moving upstream with the first light of dawn. Similarly, less than 10% of the adult salmon passed through the fishway at Dalles Dam on the Columbia River between 20:00 and 04:00 h, regardless of water flow or illumination (Fields et al. 1964) and at McNary Dam salmonids did not enter the fishway at night, regardless of whether it was illuminated or not. However, adult sockeye, chinooks and coho all preferred to enter a fish ladder at the University of Washington at night rather than during the day and on dark nights rather than bright nights (Fields 1954; Fields et al. 1955). When this short fishway was illuminated at night the number of fish using it was reduced. So salmon certainly can surmount obstacles at night, a fact further attested by the 10% that did pass through the Dalles Dam at night.

There are relatively few published observations on the diel migration patterns of fish moving through unobstructed parts of a river. Ellis (1962, 1966) gives detailed observations on the daylight migration of adult sockeye and coho in a clear river, although nocturnal migration also occurs in the same system. With the possible exception of waterfalls where fish are required to jump, it appears that upstream migrant salmon are physiologically capable of migration at any time of day or night.

2.3.1.5.5 Summary of Salmonid Diel Migration Patterns

Downstream migration of young salmonids usually occurs at low light intensities, although there are some exceptions. Several lines of evidence indicate that this is a directed downstream movement, timed by the daily illumination cycle rather than a passive drift of fish that have lost visual contact with their surroundings. Passive drift may occur under some conditions, perhaps during the initial few minutes of migration by newly emerged fry. Upstream migration by young fish usually occurs during daylight, even under conditions of heavy predator pressure. Upstream migration may require better visibility than downstream migration. Daily temperature maxima in natural waters often occur during daylight. The effort required for upstream migration by young fish may restrict these poikilotherms to movement during the warmer period of the day (see Chap. 5). There also seems to be a link between small size and upstream migration during the day. In Northcote's (1962) studies, more fish migrated upstream at night as the fish became larger. An interesting parallel occurs in the upstream migration of European elvers, *Anguilla anguilla*. Small elvers migrate upstream during daylight, closer to the daily temperature maximum, while larger elvers are more nocturnal (Sorensen 1951).

Where studied, the migration of young salmonids through open water has been crepuscular (e.g., sockeye, Groot 1965) or diurnal (e.g., pink salmon, Healey 1967) rather than nocturnal. These authors suggested that the activity rhythms reflect the use of celestial orientation cues related to the sun's position. There is little information on migration of adults in the open ocean.

Upstream migration by adult salmonids occurs at any time of the day or night. Some move through difficult areas during daylight, but this is not universal. There is no obvious simple mechanism determining the diel pattern of upstream migration. When we consider the variety of genetic stocks and environmental conditions that are lumped together under this heading the confusion is not surprising. Detailed observation on single populations is probably required to work out the timing mechanisms, which may well differ between species and between populations within a species.

2.3.2 Seasonal Timing

There are many changes in the daily pattern of illumination that could serve as indicators of seasonal change. The relative length of day and night (photoperiod), the rate of change in the sun's azimuth, the sun's zenith, the intensity and spectral quality of light all change regularly with season. Each of these variables could serve as stimuli to time seasonal cycles of migration and reproduction. Photoperiod, however, seems to be the stimulus most often used by animals.

I will consider it an established fact that photoperiod influences the endocrine cycles of many temperate zone fish via the hypothalamic-hypophyseal axis

(Schwassmann 1971; Thorpe 1978). Many fish migrate in association with their breeding cycles and, since the pioneering work of Rowan (1926) on the effects of photoperiod on bird migration and gonadal development, it has often been assumed that gonadal development and/or associated endocrine events might be one of the proximate causes of fish migration.

Despite the obvious functional relationship between migration and spawning in some fish, there is very little direct evidence that gonadal changes time migration. Gonadectomized three-spine sticklebacks, *Gasterosteus aculeatus,* show the same migration-associated changes in salinity preference as do intact animals and migrating juvenile salmonids often show a range of gonadal activity from complete gondal quiescence to precocial maturity (Baggerman 1962, 1963). Similarly migratory behaviour has been described in hybrid fish (*Colisa labiosa* × *C. lalia* and *Salmo trutta* × *S. alpinus*) that lack gonads (Forselius 1957). On the other hand, gonadal steroids have induced increases in activity, which may be associated with migrations. Van Iersel (1953) reported an increase in fluttering against the sides of the aquaria in male three-spined stickleback treated with testosterone propionate. He interpreted the fluttering as migratory behaviour. Similarly, Hoar et al. (1952, 1955) demonstrated increased activity in fish treated with gonadal steroids.

Endocrine changes can influence migratory behaviour through several possible mechanisms. For example, the sensory response of fish to olfactory stimuli can be altered by hormone treatment (e.g., Oshima and Gorbman 1966a, b, 1968, 1969; Hara 1967) and both the learning of homestream odour and the subsequent response to it by adults are under endocrine control (Scholz 1980; see Chap. 3). Endocrine aspects of fish migration have been reviewed by Woodhead (1975).

Photoperiod and hormonal control of migration is more difficult to study than control of reproduction, which can be assessed by a variety of morphological changes. Migration has fewer morphological correlates and the behavioural changes involved are often difficult to observe or even recognize in captive animals. Rowan's admirably simple procedure of tagging and releasing his birds after subjecting them to photoperiod changes has only been used successfully on fish by one investigator. Wagner (1970, 1974) released juvenile steelhead into streams after exposing them to a variety of photoperiods and temperatures, then trapped the marked fish if they moved downstream. Other investigators, working on hormonal or photoperiod control of fish migration, have relied on observing captive fish and recording behaviour that they assumed to be associated with migration, such as fluttering (van Iersel 1953), movement in an artificial stream (Northcote 1962) or changes in salinity preference (Baggerman 1957, 1960a, b, 1962, 1963). It is often difficult to determine the relationship between these behaviour patterns and migration in the field.

2.3.2.1 Photoperiodic Control of Salinity Preference

One of the earliest and most comprehensive studies on photoperiod and fish migration was carried out by Baggerman (1957) on the three-spined stickleback (*Gasterosteus aculeatus*). She found that long photoperiod (16L : 8D) changed the preference of adult sticklebacks from seawater to freshwater. Many populations of three-spined stickleback migrate to sea as juveniles and return to freshwater and spawn the following spring. Untreated wild sticklebacks changed preference twice a year to match the spawning migrations from saline to freshwater in the spring and fresh to saline in the fall (Baggerman 1957).

Gonadectomy did not influence preference. Thyroxine, however, increased freshwater preference (and fluttering), while thyroid inhibitors (thiourea, thiouracil, KCNS) increased saltwater preference (Baggerman 1957, 1962, 1963). Thyroid-stimulating hormone (TSH), secreted by the pituitary, also increased freshwater preference, presumably by stimulating thyroxine release. Long photoperiod (and rising temperature) induce the pituitary to release TSH, which in turn increases release of thyroid hormones. The high levels of circulating thyroxine change the fish's preference from saltwater to freshwater preparatory to the spring migration. Then, in the fall, shortening photoperiod leads to decreased thyroid activity and a reversion to seawater preference. Appropriate changes also take place in the salinity tolerance of wild stickleback populations, although prolactin and other hormones are also involved along with thyroxine (Lam and Hoar 1967).

Baggerman (1960 a, 1962, 1963) subsequently examined the salinity preference and thyroid activity of juvenile Pacific salmon. The young salmon move from freshwater to seawater in the spring. Appropriately, they change preference in the spring from freshwater to seawater, even sockeye fry that would not normally migrate to the sea in the first spring (see Chap. 3). If held in freshwater beyond the migration period, they revert to freshwater preference. In the case of yearling coho, short photoperiod delays the spring onset of seawater preference and long photoperiod advances it. However, unlike the stickleback, seawater preference in juvenile salmon is associated with high thyroid activity and freshwater preference with low thyroid activity (Hoar and Bell 1950; Eales 1963). Salmon migration is initiated in freshwater and maintained through many kilometers of freshwater, so the change in salinity preference is only one facet of the control of their migration.

2.3.2.2 Photoperiodic Timing of Smolt Migrations

Wagner (1974) applied a variation of Rowan's (1926) technique by subjecting juvenile steelhead, sea-run *Salmo gairdneri,* to various annual cycles of temperature and photoperiod (Fig. 2.12).

He then marked the various groups of fish and released them in streams. Traps located at 4.8 to 6.5 km downstream from the release site were used to sample the number of downstream migrants, and one stream was checked by electrofishing at the end of the experiment. Increasing daylength, and, to a lesser extent, long days as such and accumulated hours of light exposure stimulated the parr–smolt transformation. Wagner (1974) used two indices of smoltification: downstream migration and the change in condition factor (K).

Downstream migration occurred earlier in fish subjected to advanced or accelerated annual photocycles than in fish subjected to normal cycles. Delayed photoperiodic cycles and constant darkness tended to delay migration and other parr–smolt changes although they did eventually occur.

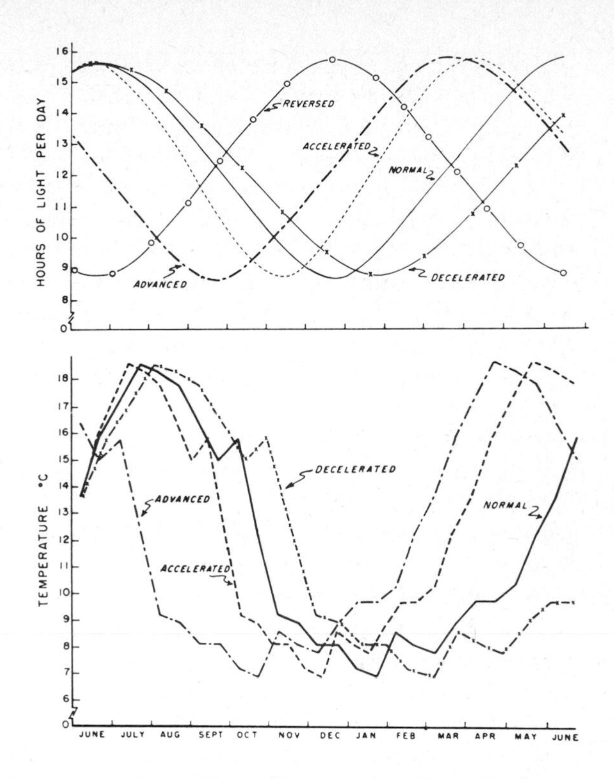

Fig. 2.12. Annual photoperiodic and temperature cycles used in experiments on juvenile steelhead trout (Wagner 1974). *Normal* the normal temperature and photoperiod cycle at the Alsea River in central Oregon. *Advanced* a cycle of normal annual length with a given photoperiod or temperature occurring on an earlier date than normal. *Accelerated* a cycle of less than annual length so that a given temperature or photoperiod is reached earlier in the calendar year. *Decelerated* a cycle of greater than 1 year in duration so that a given temperature is reached later in the calendar year. *Reversed* a cycle of annual duration with conditions reversed so that summer photoperiods occur in December

The change in K is one of the many physiological and morphological changes that occur during smoltification (Hoar 1976). "K" is an index of the body shape of a fish. Non-migratory freshwater parr are relatively fat. As they change to migratory smolts they become thinner and develop more silver pigment in the scales, which hides the parr marks on the side, and they undergo many changes in their osmoregulatory physiology. The transition to a thinner more streamlined body shape is reflected in lower K values. As with several other parr–smolt changes, K reverts to freshwater levels if the fish are prevented from entering saltwater.

The change in K served as a more reliable index of smoltification than downstream migration, which was influenced by seasonal differences in conditions in the wild streams into which the fish were released (Wagner 1974). The change in K could be advanced or delayed by changing photoperiod in essentially the same way as downstream migration.

Wagner's (1974) data indicate that photoperiod influences the timing of smolt transformation and downstream migration in juvenile steelhead. They also indicate that photoperiodic effects can be modified to some extent by temperature: the period of downstream migration was prolonged in fish subjected to temperature cycles that were delayed relative to their photoperiodic cycle and the migration period was shortened in fish subjected to temperature cycles that were advanced relative to their photoperiodic cycle. Wagner suggests that the earlier cooling in the advanced cycle may terminate migratory readiness. Smoltification and migratory behaviour also occurred in fish held in constant darkness but at a later date and at larger body size than in other groups treated, suggesting a circannual cycle that is longer than the calendar year. Wagner did not establish the mechanism by which photoperiod alters the levels of migratory activity in steelhead, but it seems likely that it acts through the changes in endocrine activity that occur at this time.

There are differences between the various species of salmon and trout and even between populations within one species. In *Salmo gairdneri,* for example, Northcote's (1958, 1962) juvenile rainbow trout tended to move downstream at times of short photoperiod and upstream at times of long photoperiod, while Wagner's (1974) steelhead delayed downstream migration with short photoperiods and advanced downstream migration with longer photoperiods. Similarly, while photoperiodic treatment induces normal smolt transformation in steelhead, it does not induce silvering in *Salmo salar,* the Atlantic salmon (Saunders and Henderson 1970). In the salmon, silvering is controlled more by temperature and body size, with photoperiod having no appreciable effect (Johnston and Eales 1968, 1970).

2.3.2.3 Photoperiod and Sun-Compass Time Compensation

Sunfish, *Lepomis cyanellus* and *L. macrochirus,* change their angle of orientation to the sun's azimuth more rapidly when subjected to long photoperiods and more slowly when subjected to short photoperiods (Schwassman and Braemer 1961) (Sect. 2.4.2.2.1). This adjustment is appropriate, since the rate of change in the sun's azimuth with time of day is more rapid near midday during summer than during winter; it is easier to advance the fish's seasonal response than to delay it.

2.3.2.4 Photoperiodic Timing of Rheotaxis

Common shiners, *Notropis cornutus,* move upstream prior to spawning in the spring. Positive rheotaxis in an optomotor apparatus could be increased by spring photoperiods and warm water temperatures (Dodson and Young 1977), indicating that photoperiod times the upstream migration.

Juvenile rainbow trout from the outlet spawning population at Loon Lake, B.C. migrate upstream later in the spring than the downstream migration of the inlet fish. In an experimental stream tank more downstream migration occurred under short photoperiod and more upstream migration under long photoperiod (temperature also has an effect, see Chap. 5) (Northcote 1958, 1962).

2.3.2.5 Summary of Photoperiodic Timing

Seasonal changes in photoperiod probably serve as major stimuli for synchronization of annual physiological changes with the seasonal environmental changes found in temperate zones. These photoperiod effects are probably detected by either the eyes or EOP's, and alter the animal's physiology through the CNS. Where migration is seasonal and is coincident with other physiological changes such as preparation for reproduction or alterations in osmoregulation, some of the hormonal changes required for these functions may also affect behavioural aspects of migration. Several authors, including Baggerman (1957) and Wagner (1974) emphasize that, while photoperiod-controlled changes may bring the animal into a state of preparedness, "prime" it for migration, other, "releasing" stimuli may initiate and maintain migration.

2.3.3 Lunar Timing

The 29.5-day cycle of moonlight does have a timing function in some fish behaviour. Since the position of the moon is the major factor determining the tidal cycle of the oceans, one could expect lunar rhythms in intertidal animals. There will also be abundant mechanical and chemical cues associated with tidal cycles available for these animals to use in timing their activity. The phase of the moon and the time of moonrise and moonset also determine, at least partially, the amount and duration of light available on a given night. Since both tide and night illumination follow regular monthly cycles, they can be used as timing cues to allow anticipation of lunar-linked conditions affecting an animal.

2.3.3.1 Lunar Timing of Activity Cycles Related to Tides

Many fish are affected by the rhythm of the tides. Many species, such as *Bathygobius soporator* (Aronson 1951, 1971), wooly sculpins, *Clinocottus analis,* and opaleye, *Girella nigricans* (Williams 1957) live in tide pools and forage over the surrounding area at high tide. Others, including juvenile plaice, *Pleuronectes platessa* (Gibson 1973) move inshore with each flood tide to forage over the freshly covered area. But, there are few data on the persistence of tidal rhythms under constant conditions or on what effects moonlight might have on these cycles if they exist. *Blennius pholis,* shows a persistent tidal rhythm under constant conditions (Gibson 1967, 1971). However, Gibson (1971) feels these rhythms are entrained by pressure changes, on the basis of comparing fish in tanks in the intertidal zone with fish in tanks on a wharf (see Chap. 4). Persistent tidal rhythms have also been demonstrated in juvenile plaice (Gibson 1973) and adult hogchokers, *Trinectes maculatus,* (O'Conner 1972), but no direct correlation with moonlight has been demonstrated.

The grunion, *Leuresthes tenuis,* is a well-known example of lunar timing of a tidal rhythm in fish. It spawns at a high tide near the beginning of a descending series of high tides (Fig. 2.13) and the eggs develop in the sand and the larvae are washed out about 2 weeks later on the next series of peak high tides. Hatching

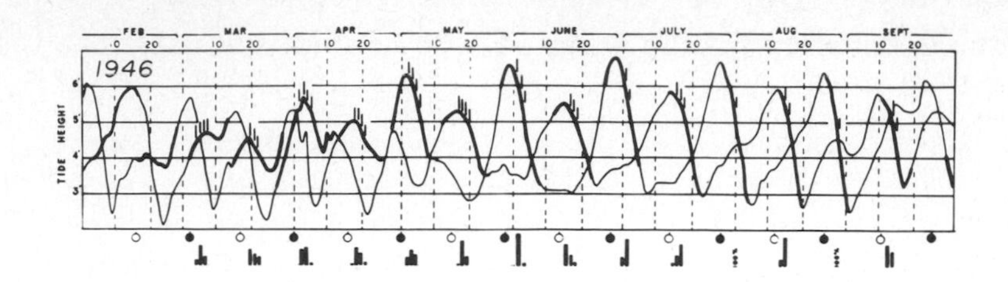

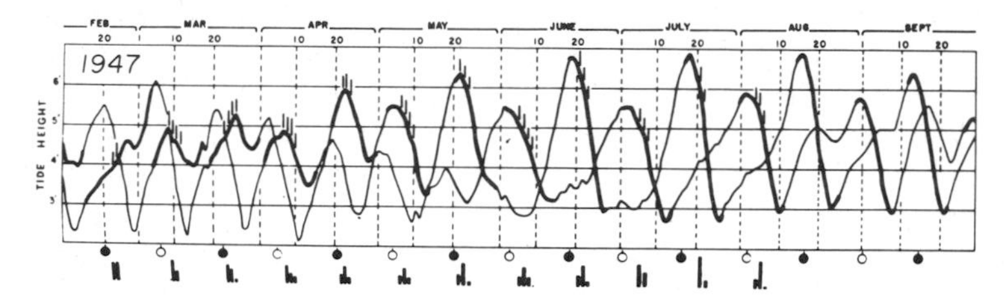

Fig.2.13. Grunion spawning periods observed at La Jolla, California, in 1946 and 1947, plotted in relation to variations in observed high tide heights at La Jolla. The heights of high tides only have been plotted. The high tides about 24 h apart have been connected by *smooth lines*. The two tides each day yield the *two series of curves*. Tides occurring during darkness are indicated by the *heavier lines*. The occurrence of grunion spawning is indicated by the *short vertical lines* above the tide curves. The moon phases are indicated at the bottom of each graph, by a *solid circle* to indicate new moon and an *open circle* to indicate full moon. *Histograms* at the bottom portray the percentage intensity of spawning runs in each series. Seasonal variation in strength of spawning is not indicated. All data are based on observations made at Scripps Beach, La Jolla. Data for time and height of tides are from records of the tide recording machine maintained for the Coast and Geodetic Survey on Scripps Pier (Walker 1952)

of the eggs is triggered by the action of the surf as it uncovers and agitates the eggs. If they are not uncovered on the first series of peak high tides following spawning, then they remain unhatched but viable for another 2 weeks until the next series of peak tides (Walker 1952). The grunion spawn only at night but another "closely related" Californian species, *Hubbsiella sardina,* has spawning habits almost identical with those of the grunion, with the exception that it will often spawn during daylight (Walker 1952).

The grunion's short inshore spawning migration may be timed by the moon. To quote Schwassmann's (1971) account of Walker's (1949) argument: "The time of spawning seems to be determined by the interaction of a physiological rhythm of gonadal maturation, showing a period of about 2 weeks, with some factor related to the second preceding full or new moon." What the factor or factors are remain unknown. Diel photoperiodic changes time reproductive cycles, acting through the photoreceptors, CNS, and endocrine system. Tidal reproductive cycles might be governed by the cyclical changes in moonlight acting through much the same route.

Lunar spawning rhythms also occur in yellowtail damselfish, *Microspathodon chrysurus,* which spawn at sunrise during the period of full to new moon (Pressley 1980). Since the spawning correlates poorly with the small tidal changes off the Caribbean coast of Panama, the actual adaptive significance of this and other lunar cycles in damselfish spawning is not clear. Perhaps the timing is adapted to increase survival by synchronizing the emergence of the young with tidal currents or with dark nights (Pressley 1980).

2.3.3.2 Moonlight

The moon may extend the migration time available to diurnal species or limit the migration of the nocturnal species to the nights with little moonlight. The migratory South American characin, *Prochilodus platensis,* for example, migrates upstream more actively on days of full moon than at other times and these peaks of migratory activity apparently do not correlate with factors such as air temperature, flow volume, turbidity or weather (Bayley 1973). Water temperature did affect the catch in fish traps along the river, Bayley's criterion for migratory activity; he speculated that moonlight might allow migration at night or that darkness might inhibit migration. Since his trapping technique did not distinguish between day and night migration, the possibilities are difficult to evaluate.

2.3.3.2.1 Moonlight and Eel Migration

The European eel, *Anguilla anguilla,* illustrates the effect of moonlight on a nocturnal species. The eels become readily accessible to investigators as newly metamorphosed elvers when they move inshore and migrate up the streams and rivers of Europe and North Africa. After a period of growth in freshwater as yellow eels, the adults go through a second metamorphosis to become migratory silver eels. As with smolting salmonids, there are a number of physiological and morphological changes associated with this transformation, including colour change, enlarging eyes and fat deposition. The general effect is to change a resident, feeding, freshwater fish into a migratory, non-feeding, deep-sea fish.

The inshore migration of the elvers is strongly correlated with tides, but apparently this is controlled by chemical stimuli, the smell of ebb tide water, rather than light (see Chap. 3). The migration of the elvers upstream also seems unaffected by moonlight, even through the migration is largely nocturnal (Sorensen 1951; Meyer and Kuhl 1953). Yellow eels, *A. rostrata,* are apparently attracted to dim lights but repelled by bright lights (Cox 1916) but there are no records of their response to moonlight. The downstream migrations of the silver eels, however, have long been thought by fishermen to be strongly affected by moonlight (Frost 1950). Studies on the River Bann in Northern Ireland (Frost 1950) and on English streams (Lowe 1952) indicate increased silver eel runs on nights when there was no moon in the sky during the period around midnight when the eels prefer to migrate. This situation tended to occur during the last lunar quarter (Fig. 2.14).

Both Lowe (1952) and Frost (1950) relate this effect to light intensity, since eel migration also increased in turbid water and on cloudy nights and could be inhibited with artificial lights of relatively low intensity (Lowe 1952). There was also a relationship between river discharge and migration, with strong eels runs occurring during periods of rising and high river levels in both studies. The river discharge changed in response to rain or the opening of dams and was indepen-

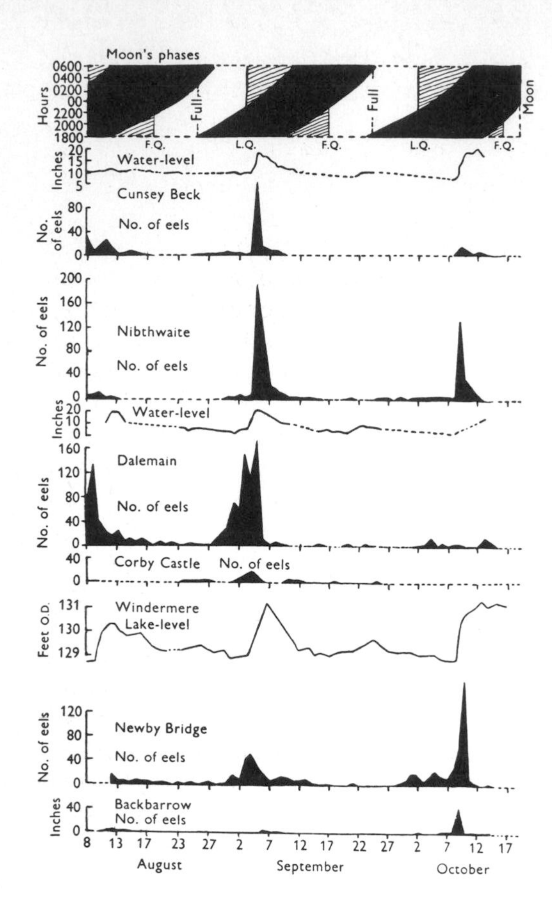

Fig. 2.14. Eel runs at several English streams in relation to phases of the moon and rain-induced floods during 1942 (Lowe 1952, p. 280)

dent of the moon. Silver eels held in tanks in a laboratory also tended to escape more frequently on nights when strong eel runs occurred in local rivers (Lowe 1952).

Records from the Rhine River show better catches of silver eels (and hence more downstream migration?) during periods of rising or high water level and during the third quarter of the moon (Jens 1953). The effect was independent of moonlight per se, cloud cover also had little effect, and occurred in regions free of tidal influence.

Boetius (1967) kept male silver eels in permanent darkness in a container which allowed escape through outflowing tubes into a sack. More eels escaped "downstream" through these tubes during the third quarter of the moon than at other times, even under constant darkness. Boetius returned the eels to the central chamber of his apparatus after each escape and kept them in captivity long enough to record a spring peak in escape behaviour as well as an autumn peak corresponding to the usual migratory season for silver eels.

When comparing the work of Frost and Lowe with that of Jens and Boetius, there does not seem to be any renal difference in the timing of the eel runs in relation to the moon phases, although there is some difference in the interpreted mechanism. Frost and Lowe favoured direct illumination effects as an explanation. Jens and Boetius preferred endogenous rhythms. Both mechanisms may be at work. The similarity between Lowe's observation of laboratory eels escaping on nights of peak migration and the results of Boetius' escape experiments is remarkable and would seem to indicate that at least some effects other than direct illumination are involved.

These data (with the exception of Cox 1916) deal with the European eel, *Anguilla anguilla,* rather than the less-studied American eel, *A. rostrata.* The life cycles are basically similar, in fact some authors argue that they are the same species (Tucker 1959). The colour change associated with seaward migration is less pronounced in downstream migrant American eels. Vladykov (1955) suggested that they should be termed bronze eels rather than silver eels. On the basis of the Canadian eel fishery, Eales (1968) states that bronze eels migrate downstream on dark moonless nights or when the moon is in the first quarter. This agrees with the observation by Hain (1975) that bright moonlight inhibited downstream migration by American eels and that black plastic covers over the tanks could induce early migration. Two New Zealand anadromous eels, *Anguilla australis* and *A. Diefenbachii,* migrate downstream with a lunar rhythm (Todd 1981). The maximum migratory activity of the seaward-migrating adults occurs just before the last quarter.

Several authors report that yellow or partially yellow eels are found migrating downstream on the same nights as silver eels (e.g., Lowe 1952), but it is not known whether these eels are migrating in the early stages of transformation or whether they will later return upstream. If one can use smolting for comparison, then the change in external body colour in the eels may to some extent be independent from other physiological and behavioural changes related to migration.

2.3.3.2.2 Lunar Timing of Juvenile Salmonid Migrations

A link between lunar periodicity and social displacement in coho fry has been suggested (Mason 1975). Non-migratory fry show greater downstream movement on dark moonless nights. This effect could be caused by individuals being forced into portions of the stream where they can hold position in moonlight but not in the dark. In contrast, migratory coho smolts tend to move downstream more on bright moonlight nights than on dark nights.

The timing of the thyroxine surge, which coincides with smoltification in salmon and trout, may also be timed by the moon (Grau et al. 1981). When tested through the year for circulating levels of thyroxine, a number of different stocks were found to show peaks in thyroxine levels at the same time. In 1979, for example, five out of six Washington and Oregon coho stocks had peaks within the same week and in 1978 eight of nine stocks peaked within a few days of each other. The thyroxine peaks also coincided with the new moon (Fig. 2.15). Grau et al. (1981) speculate that lunar timing of the thyroxine surge could benefit the fish by coordinating their movements so as to take advantage of the benefits of group behaviour, and that timing the initiation of migration to coincide with the dark nights of the new moon could also benefit these nocturnal downstream migrants by reducing their vulnerability to predators. They suggest that the coho ob-

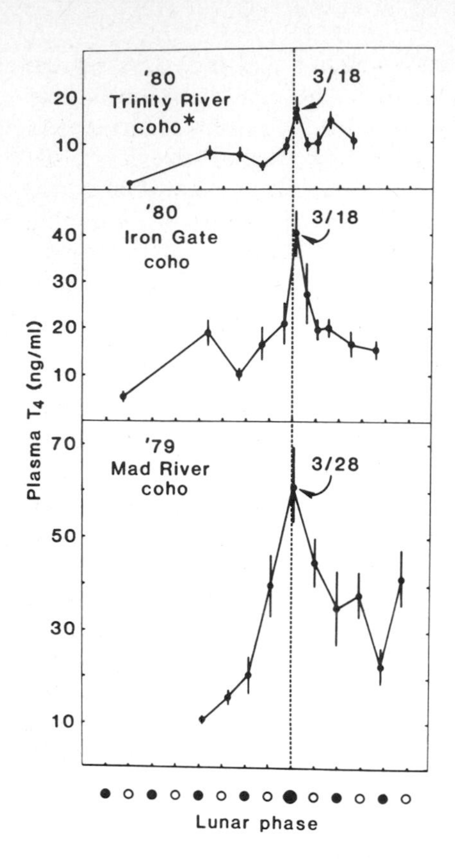

Fig. 2.15. Plasma thyroxine levels in smolts of three Californian coho stocks plotted against the lunar calendar. *Closed circles* new moon, *open circles* full moon (Grau et al. 1981)

served migrating downstream on the nights of the full moon by Mason (1975) would be at peak seawater readiness, having completed their thyroxine surge about 2 weeks after the peak levels, which coincided with the new moon.

2.3.3.3 Summary of Lunar Timing

There are apparently some lunar activity rhythms in fish. Some of these may be linked with tidal changes rather than changes in illumination, others are apparently independent of tide and perhaps of moonlight as well. Lunar rhythms seem to have received relatively little study perhaps because they tend to be important in nocturnal behaviour of fish, which is less easy to study, or because of the difficulty in separating lunar and tidal effects in marine species.

2.3.4 Social Interaction as a Visual Timing Mechanism

It will often be to an animal's advantage to travel with a group of conspecifics, or even with members of similar or closely related species. Groups of fish traveling together probably achieve metabolic savings by using the turbulence, and schooling probably increases the chances of finding food (Pitcher et al. 1982) and supplies some degree of protection against predators (Breder 1967). Schools may also have an advantage in orientation accuracy over solitary fish (Larkin and Walton 1969). If individuals jointly orient toward a group mean, then increased group size should decrease the error in orientation direction. Individuals can also benefit by moving through an area at the same time as many other fish, swamping the resident predators in a region (Peterman and Gatto 1978).

2.3.4.1 Social Facilitation as a Releasing Stimulus

The advantages of being in a group may account for the observation that migrations are often characterized by mass movements within a short period of time. Several timing mechanisms conspire to achieve the restriction of this movement to a short period, including the diel, lunar, and seasonal effects mentioned above. However, the peak of migration often appears to be restricted to a narrow period within the broader range set by these timing mechanisms. As Groot (1965) says, sockeye smolts may start to migrate on the same day, all over the lake.

There are several ways in which this sort of short-term coordination of activity could be achieved. One mechanism would be a common response to some stimulus such as loss of ice cover, a threshold photoperiod or lunar phase (e.g., Grau et al. 1981). Social facilitation is another type of mechanism found in animals that synchronize their activity; animals performing a behaviour pattern serve as a stimulus for other animals to do the same thing, leading to a synchronization of activity as more and more animals start to perform the behaviour.

2.3.4.2 Social Facilitation as a Priming Stimulus

Social facilitation can act as a priming stimulus on the endocrinological or developmental state of an animal as well as on its behaviour. In birds, for example, the behaviour of one bird often influences the reproductive endocrinology of its mate serving to synchronize their reproductive cycles (e.g., Hinde 1965; Lehrman 1965) and the behaviour of nest mates may influence the rate of development in young birds prior to hatching (Vince 1969). In male three-spined stickleback the presence of other individuals facilitates nest-building and reproductive behaviour (Tinbergen 1942; Reisman 1968), possibly by stimulating increased androgen secretions.

As well as these long-term "primer" effects of social facilitation, that operate over days or weeks, there is also a strong tendency for grouping animals such as schooling fish to synchronize their activity in the short term. Individual schooling fish, by their very nature, tend to swim in the same direction and at the same speed and generally to do things at the same time. Individuals, for example, match their velocity with one another. When individual juvenile jack mackerel, *Trachurus symmetricus,* are stimulated electrically to increase speed, other fish in the same school speed up too, especially if the stimulated fish was beside them and in full view (Hunter 1969). Fish also show social facilitation in such behaviour as maze swimming performance, e.g., green sunfish (Hale 1956) and perhaps in reproductive behaviour (Tinbergen 1942; Reisman 1968).

These physiological and behavioural synchronizing methods are probably essential if the group is to remain together and retain the advantages of group life. They will also tend to restrict activities like migration to a short time span by delaying the onset of migration in precocious individuals and advancing it in individuals that might otherwise start later due to incomplete physiological readiness or other reasons. The action of such synchronizing mechanisms is suggested by reports of migrating populations that contain individuals in a variety of states of physiological preparedness, e.g., Frost (1950) and Lowe (1952) report yellow and silver eels migrating together.

On a larger scale, once a critical number of fish begin migration in a system, the social facilitation may draw others into the migration. Chum salmon fry, for example, may be drawn out of coastal streams in part by their schooling response (Hoar 1976). Similarly, hatchery rainbow trout, 9–12 months old, migrating out of an experimental stream, left more quickly and completely when released in large than in small groups (Jenkins 1971). Jenkins postulated that, in some instances, fish in large groups undertook movements that they would not have made in small groups or as individuals. This may be an example of social facilitation of dispersal movements.

I do not know of any studies that definitely establish that social facilitation plays a role in the timing of fish migration. This may just be a matter of lack of study. The large masses of individuals that characterize many fish migrations indicate that suitable conditions for social synchronization are present. The sharp temporal peaking observed in many studies of fish migration may be partially a result of such a mechanism.

2.3.4.3 Social Interaction and Dispersal

Just as there are situations in which grouping together confers advantages, there are others in which dispersal of conspecifics over a wide area may be advantageous. Such dispersal is usually linked with periods of intensive feeding or reproduction rather that with migration. By its very nature, however, dispersal involves the movement of individuals. This is especially true of cases where fish disperse through a large feeding area from a smaller nursery area or from a restricted "arrival area" after a migration.

Dispersal occurs in the life cycle of salmonids, for example, when fry emerge from the gravel (Godin 1982). Fry such as coho, rainbow trout or Atlantic salmon that feed in the streams often disperse from the adult spawning areas into regions that provide food but lack the bottom characteristics necessary for salmon breeding and incubation. This dispersal may be accomplished in part by the displacement, through agonistic behaviour, of behaviourally subordinate fish from suitable feeding areas near the point of emergence. Chapman (1962) and Mason (1969) both found that coho fry that "emigrated" downstream from a populated stream (Chapman) or from a stream tank (Mason) tended to be smaller than fish that remained resident. In most cases small fish are less successful in agonistic encounters than larger fish (Chapman 1962; Frey and Miller 1972). In both studies the "emigrants" become resident when they are placed in stream aquaria or artificial streams with reduced competition. These workers suggest that some coho fry at least move downstream in response to the agonistic behaviour of conspecifics, which forces them out or excludes them from suitable habitat near their point of emergence.

Chapman (1962) indicates that aggressive interactions are one factor causing downstream movement of coho fry in spring, perhaps in combination with other density-dependent factors. Initial post-emergence movements may be due to "migration" or displacement. Chapman and Bjornn (1969) and Mason (1975) point out that exclusion from night resting locations may be important in downstream dispersal.

Mason and Chapman (1965) found a significant positive correlation between the amount of agonistic activity as measured by the number of nips per fish per 10 min and the rate of egress from artificial stream channels. There was also a positive relationship between temperature, aggression, and emigration, but it was not determined what, if any, causal relationships applied between these variables. Agonistic interactions between coho fry began about 7–10 days after emergence in early March, increased until early May, then declined.

Periods of social dispersion probably also occur when massed migrants arrive in a feeding area, for example sockeye fry entering a lake from spawning streams, salmon reaching the ocean feeding grounds after their offshore migration or elvers spreading through a freshwater drainage system after arriving in masses at a river mouth.

2.4 Direction and Distance

2.4.1 Landmarks

Humans use landmarks for orientation in three different ways: first, we use familiar objects, learned in our movements through an area, as indicators of position within a familiar range; obviously this pilotage is restricted in its usefulness to places where one spends some time in residence getting to know the spatial relationships between objects. Second, in longer-range orientation or navigation we use a compass or other instrument to determine the direction of travel, then we locate a landmark that lies on the appropriate heading and move toward the landmark, as a convenient method of staying on course without continual reference to the compass. Finally, we learn to respond to classes of objects as indicating particular types of environmental conditions, e.g., trees in a desert indicate water, even if the particular trees have never been seen before. There is some evidence for the use of each of these types of landmark in fishes.

2.4.1.1 Pilotage by Familiar Landmarks

We are not surprised that a fish should know its surroundings well enough to find its way about. This type of landmark orientation (using non-visual landmarks as well) must be common in fish that remain resident in one area for a period of time.

It is well illustrated by the work of Aronson (1951, 1971) on the small gobiid fish, *Bathygobius soporator*. These little fish often live in tide pools and, when chased by a collector, they will jump from their home tide pool and land, with impressive accuracy, in adjacent pools.

Aronson postulated that these fish learn the topography of the surrounding region when they are able to swim over it at high tide, then use the information to make accurate escape jumps at low tide. He used a set of experimental pools to find that fish with no experience in an area at high tide would usually refuse to jump and, when they did jump, were often inaccurate. However, after being allowed one night of artificial high tide to explore the flooded experimental pool area, their tendency to jump and their accuracy were greatly improved. The improved accuracy persisted for at least 40 days after only one night of experience (Aronson 1971).

The exact landmarks that the gobies remembered as orienting cues were not clear, but overcast skies did not interfere with the process so they are presumably not using the sun-compass direction of adjacent pools. The gobies did sometimes jump correctly even with no opportunity to explore the area at high tide and Aronson (1971) concluded, on the basis of experiments, that they were jumping over the lowest parts of the pool rim.

Intimate knowledge of home surroundings is probably widespread, but the oriented jumping of the gobies brings home the accuracy of that knowledge, the rapidity with which it can be learned and the tenacity of its retention.

2.4.1.2 Use of Landmarks to Stay on Course

Sockeye salmon smolts are reported to transfer orientational information from other mechanisms to landmarks in the manner of a human woodsman using a compass (Groot 1965). In Groot's experimental apparatus they sometimes remained oriented to marks on the tank when the tank was rotated after celestial viewing conditions were changed, even though they had initially taken up the "correct" orientation. Quinn and Brannon (1982) found that sockeye smolts were able to maintain their direction under overcast skies in spite of disruption of the magnetic field, and suggested that the fish were perceiving celestial orientation cues as the weather permitted and maintaning an established direction of movement. Their results could also have been due to fish only selecting outlets of the apparatus during intervals when celestal cues were momentarily present.

2.4.1.3 Use of General Landmarks as Indicators of Conditions

As well as learning specific landmarks, fish might respond to classes of landmark as general indicators of suitable migratory routes. This type of response has already been seen in Aronson's gobies. In unfamiliar pools, they jumped toward the lowest part of the rim. Ellis (1962, 1966) found a similar response in the upstream migration of adult sockeye salmon. They followed the contours of the river channel on their first ascent through a region and only after circling in a dead end did they switch to some other cue, such as strength of water current.

Fish Diversion at Dams and Similar Obstacles. The idea that fish might prefer certain visual configurations of the channel has led to numerous attempts to direct migrants by means of lights, louvers, bubble curtains, and similar manipulations. Brett and Alderice (1958) found that sockeye smolts, migrating downstream in schools, were best diverted by a slanted barrier of moving chains, which may have entrained the fish to follow them, through an optomotor response. Coho smolts, and pink, chum and coho fry, which tended to migrate as individuals in this study, were, however, not effectively diverted by the chains or by various combinations of light, bubbles or chemicals. The general problem of the diversion of migrant fish around hazards such as hydroelectric installations has generated a large literature, much of it in government reports (Bibliography: Sharma 1973). In many cases it is not clear to what extent the response to barriers is visual and to what extent it is mediated through other senses such as the lateral line system (Arnold 1974). Of the wide variety of diversion techniques attempted, the most frequent choice has been a slanting array of louvers across the stream flow (Arnold 1974). The spacing between the louvers may be much larger than the length of the diverted fish, e.g., 17.7- to 19.3-cm Atlantic salmon smolts were effectively diverted by louvers spaced 30.5 cm apart (Ducharme 1972). The effectiveness of louvers on night migrants, 80–89% efficiency in Ducharme's (1972) study, may indicate that the response is not visual (see Chap. 4 and Sect. 2.1.3.1). Similarly, diversion of downstream migrant chinook juveniles by perforated screens into the gatewell openings in the top of lowhead Kaplan turbine intakes may be a mechanical or a visual response (Marquette and Long 1971).

Most of the attempts to divert fish are made near the hazard, often under conditions of strong flow and high turbulence. Under these conditions mechanical stimuli may predominate in governing the fishes' response, requiring screens, chains or louvers that interfere with the flow of water, in order to divert the fish. Fish may be more susceptible to visual diversion, with an attendant reduction in interference with flow, when they are in relatively quiet portions of the migratory route; the diversion of adult migrant salmon by bottom contours (Ellis 1962) is an example. There would be the attendant disadvantage of leading the fish some distance from the quiet water diversion area to the region of the hazard.

It often appears that the objective of diversion studies is to repel fish from a portion of the river. Attracting them to the safe area could be an alternative. Sockeye smolts, for example, seek shade along the river shore, and provision of such shade may enhance the effectiveness of diversion channels (Brett and Alderdice 1958). Visual attraction methods have not been intensively studied in stream migrants.

Attraction of Fish to Objects in the Environment. Floating objects often attract fish, and this phenomenon has been used by fishermen, who either fish around naturally drifting objects or release such things as grass mats, bamboo and palm frond rafts or slabs of cork (Gooding and Magnuson 1967). Experimental studies, assessing the fish in the region of artificial structures anchored or drifting in open water, have consistently shown that fish are attracted almost immediately and that congregations of fish remain around the objects as long as they are being studied. Over 10,000 baitfish, round shad, *Decapterus punctatus;* Spanish sardine,

Sardinella anchovia; and scaled sardine, *Harengula pensacolae,* congregated around tent-shaped structures on the first day after their placement in the northern Gulf of Mexico (Klima and Wickham 1971). In a similar study in the same region, Wickham and Russel (1974) found that daily seine hauls around artificial structures in midwater yielded about 400 kg of fish per set, primarily round shad and Spanish sardine, when the structures were fished daily.

The attraction of fish to physical objects is also illustrated by the success of artificial reefs made from a variety of materials, including rock rubble, old automobiles, rubber tires and scrapped liberty ships (Unger 1966; Steimle and Stone 1973; Parker et al. 1974; Ditton et al. 1979; Parker et al. 1979; Stone et al. 1979). For example, Stone et al. (1979) found that a reef made of old car tires near Miami, Florida, matched the population levels of nearby natural reefs within a few months. The artificial reef did not seem to attract fish away from the natural reefs, but rather to increase the number of fish in the region. Similarly, Parker et al. (1979) found that when one artificial reef was experimentally depopulated it was quickly recolonized, but did not draw tagged fish from another artificial reef 46 m away.

These artificial structures may be successful because they draw what terrestrial ecologists refer to as the "floating population," those individuals that have been unable to establish themselves in the available natural habitat. These animals would be equivalent to the socially displaced salmon fry described earlier (see Sect. 2.3.5). The competition for living spacc on reefs may be very intense (Sale 1975, 1977).

Artificial structures of a variety of configurations are obviously very effective in modifying the distribution of fish, both in the short term and in the long term. Do such structures modify the migratory patterns of fish moving through an area? This is not an idle question. The surveys of fish species attracted to artificial structures often include migratory pelagic species as temporary visitors. Artificial structures with established fish attraction capability, such as offshore oil rigs and storage facilities, are being constructed in many of the productive continental shelf areas of the world. The presence of these structures might delay or deflect migratory fish through their attractive properties. They might also concentrate resident predators along a route that had formerly been safe for migratory fish. These suggestions are speculative, however, and most objective evidence indicates that the addition of structures to water improves fish habitat by providing hiding places, spawning surfaces, feeding areas and perhaps visual reference points.

2.4.2 Celestial Orientation

The celestial bodies (the sun, moon, and stars) provide predictable reference points that can be used by animals to orient themselves relative to the surface of the earth. Fish possess the basic requirements for this type of orientation: sensory receptors capable of detecting celestial bodies and a central nervous system capable of using the celestial reference point to maintain course.

2.4.2.1 Time Compensation in Celestial Orientation

A key element in maintaining a heading or direction is compensation for the apparent movement of the celestial body across the sky which results from the earth's rotation (Fig. 2.16). This "time compensation" is a common feature of animal celestial orientation.

In sun-compass orientation, for example, an animal that does not compensate for the sun's movement will gradually change heading as the sun moves across the sky. This may be acceptable if the orientation only needs to be maintained for a short period of time during which the sun's movement would be insignificant, or if the animal's requirements for accuracy are very low. These conditions might

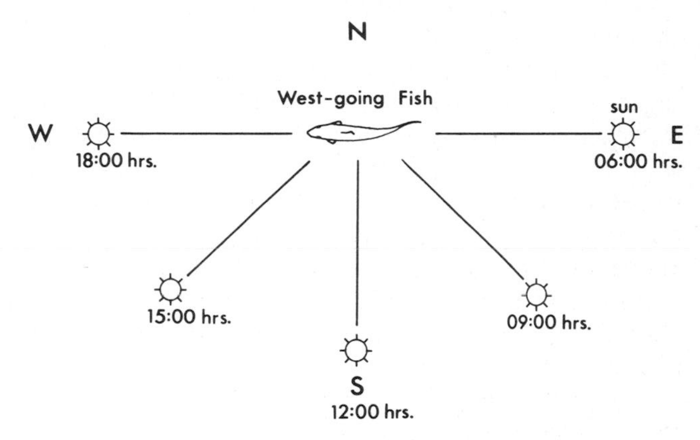

Fig. 2.16. Time-compensated sun-compass orientation. A westward-migrating fish, using a sun-compass mechanism for orientation at the equinox in the northern hemisphere. It can maintain a westward heading by changing the angle of orientation to the sun's azimuth throughout the day at an average rate of 15° per h (180° divided by 12 h)

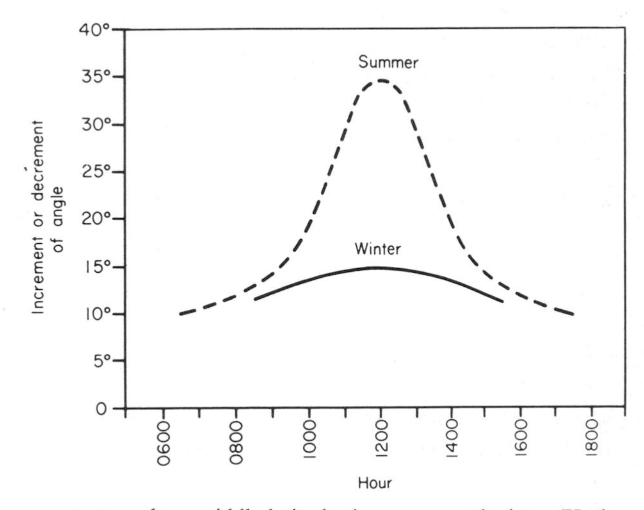

Fig. 2.17. Pattern of the sun's movement at northern middle latitudes in summer and winter (Hasler 1971)

occur in short feeding excursions. There are reports of fish that showed sun-compass orientation without time compensation. For example, some of the experimental subjects, mostly centrarchids and cichlids, used by Schwassmann (1962) showed orientation but not time compensation in his training experiments. There is no direct evidence, however, that such uncompensated orientation is actually used in natural behaviour.

There are several ways in which time compensation can be accomplished. In the case of the sun, an arbitrary compensation of 15° per hour would correct for the average movement of the sun's azimuth and would provide orientation accurate enough for many purposes. (The "azimuth" is the compass direction to the point at which a vertical line from the sun would intersect with the horizon.) However, the rate of movement of the sun's azimuth is not constant through the day (Fig. 2.17). The azimuth changes slowly while the sun is rising at a steep angle relative to the horizon, for example near dawn and dusk on a summer day. As the angle of movement approaches the horizontal, the rate of change of the azimuth increases, e.g., near noon in summer or throughout the day in the winter in temperate zones. This means that an animal that requires perfect orientation must adjust the rate of change of the angle of orientation during the day and must also change the diel pattern from season to season. In the case of north–south migrants, the rate of diel change would also have to be adjusted with latitude.

2.4.2.2 Sun-Compass Orientation

Two basic approaches have been used in the study of fish celestial orientation: (1) training fish to select a particular direction. (2) performing experiments on fish that are orienting as part of their natural behaviour. Once the directional preference is established, through either method, one can experiment by restricting or modifying the information received by the fish.

2.4.2.2.1 Training Experiments

The study of fish sun-compass orientation arose from the observation that the orientation of white bass, *Roccus chrysops,* returning to their spawning ground after experimental displacement, was more accurate under clear skies than under cloudy skies or when the vision of the fish was obscured by eye caps (Hasler et al. 1958, 1969). The implication that sight of the sun was important for orientation was then tested by a series of training experiments using a variety of species.

The training apparatus consisted of a release chamber surrounded by a circular arrangement of 16 hiding places immersed in a container with sloping opaque sides to hide the horizon and reduce shadows (Fig. 2.18). Fish were first acclimated to aquarium conditions, if they were wild-caught. They were then trained to move from their release point in the centre of the tank to one of the hiding compartments by chasing them with an electric prod. During this training period only one box was left open to the fish and the others were blocked by a metal band. The compass direction of the open box was kept constant while the whole experimental apparatus was rotated regularly to keep the fish from keying on some feature of the apparatus. The fish were thus forced to rely on information from

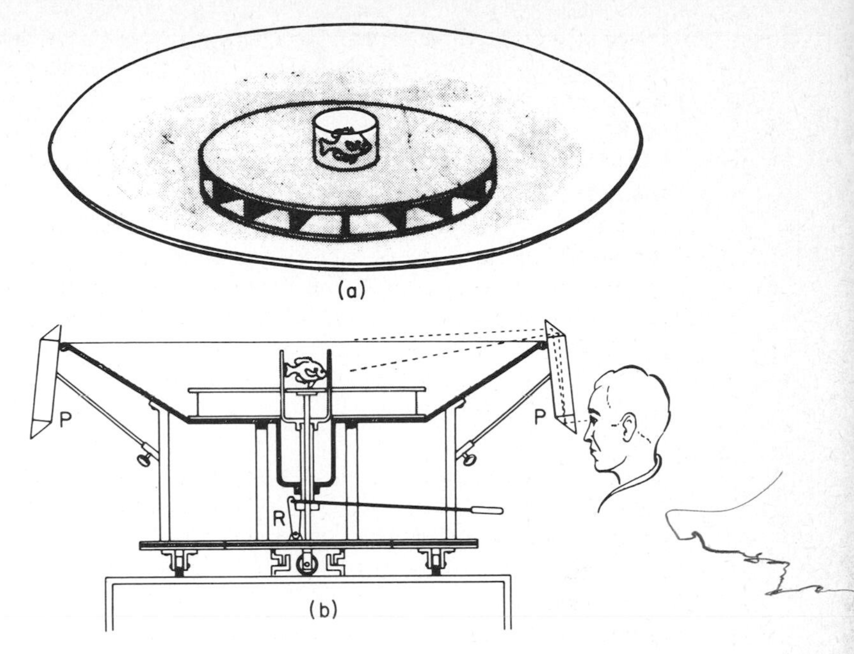

Fig. 2.18. Apparatus for training fish to show a compass reaction. The fish were released in the middle of the tank and chased to an open hiding place that was always in the same compass direction even though the apparatus was rotated. The fish was then tested by being released with all hiding places open and nobody chasing it (Hasler 1971)

the sky in order to learn the direction of the hiding place. Their accuracy in this task could be assessed by releasing them in the centre of the tank, with all the hiding places open and no electric prod, and then recording the compass direction of the chosen hiding place. With this basic apparatus and procedure a variety of fish species (primarily cichlids and centrarchids) were tested in several different experimental situations.

Some individuals could be trained to orient according to the sun (Fig. 2.19) and to orient toward an artificial light source as if it were the sun (Hasler 1971). They could then be tested for their responses under different photoperiods and sun altitudes and at different latitudes.

Since the rate of change in the sun's azimuth during the day varies with season and latitude, a fish could improve the accuracy of its orientation if it had some indicator of season or latitude that it used to adjust its rate of compensation. Photoperiod is widely used by fish and other animals as a seasonal timing mechanism for breeding and migration. Schwassmann and Braemer (1961) therefore tested the effects of photoperiod on orientation in their training apparatus.

They found that juvenile green sunfish, *Lepomis cyanellus,* and bluegill sunfish, *Lepomis macrochirus,* compensate with smaller changes in angle per unit time when they were subjected to short photoperiod and compensate at a faster rate after being kept under long photoperiod. These are the appropriate adjustments if the fish use photoperiod as an indicator of season. Summer (long photoperiod)

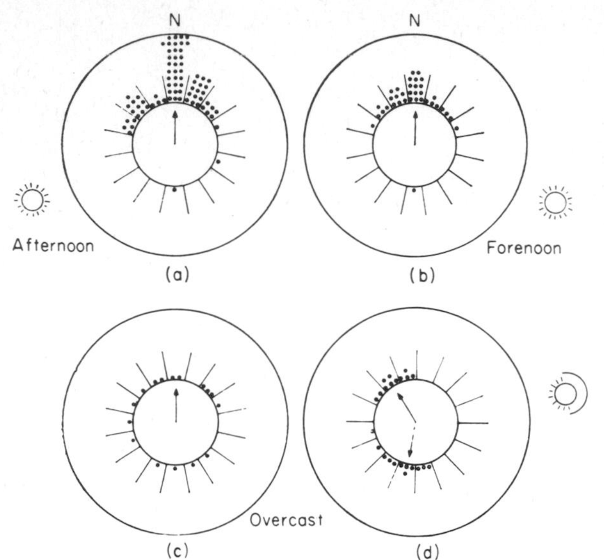

Fig. 2.19. Test scores of bluegill sunfish trained to seek cover in the north compartment of the apparatus illustrated in Fig. 2.18. The fish are tested with all 16 hiding places open: *a* tested in the afternoon; *b* tested in the morning; *c* under overcast sky; *d* using a fixed artificial light instead of the sun. *Solid arrows* fish tested in the morning; *dashed arrow* same fish tested in the afternoon (Hasler 1971)

requires a more rapid change in angular velocity near midday because of the higher angle of the sun's arc across the sky. Winter (short photoperiod) requires a slower change near midday.

These adjustments were affected by the photoperiod conditions under which the fish were held prior to the experimental photoperiods (Schwassmann and Braemer 1961). Fish that had been under a declining (autumn) photoperiod were more likely to show appropriate compensation under short photoperiod (winter) conditions than were fish that had been under increasing photoperiod. Conversely, fish held under increasing photoperiod prior to the experiment showed better compensation under long photoperiod. In other words, it seemed easier to advance the fish in their seasonal cycle than to delay or reverse the cycle.

Photoperiod changes with latitude as well as with season and these results could also be interpreted as adjustments to different latitudes (Schwassmann and Braemer 1961). Centrarchids do not undertake long migrations, so it seems more reasonable to consider the results as seasonal adjustments.

As an alternative to using photoperiod, a fish could use the sun's altitude or the rate of change in altitude as an indicator of season, or latitude. When they are first exposed to the natural sun after being reared indoors and trained under artificial light immature green sunfish adjust their rate of compensation as if the sun's altitude indicated season or latitude (Schwassmann and Hasler 1964). "Experienced" fish that had been reared outdoors under natural conditions use the sun's altitude as if it indicated time of day rather than season or latitude. It appears that the sun-compass mechanism is initially "set" to the appropriate season and latitude by a response to the sun's path (azimuth plus altitude) when it is first used. Once this initial setting has taken place, the sunfish then respond as if latitude were constant and the sun's height is therefore a reliable indicator of the time of day.

2.4.2.2.2 Role of Sun-Compass Orientation in Predator Avoidance

The methods of Hasler and Schwassmann and their co-workers illustrated that several species are able to learn an appropriate orientation direction to avoid unpleasant stimuli. This resembles the predator avoidance orientation of fish that dwell along shorelines, including some centrarchids. Juvenile bluegill sunfish, for example, orient toward vegetation along the shoreline or toward similar areas of cover when threatened by predators. This has been termed "Y-axis orientation." The fish orient at right angles to their home shoreline, which is designated as the X-axis. That they use a sun-compass mechanism is indicated by their disorientation under cloudy skies and their shift of orientation in response to a 6-h advanced diel cycle (Goodyear and Bennett 1979). A similar form of Y-axis orientation, in this case toward deep water, occurs in juvenile largemouth bass (*Micropterus salmoides*), and is also disrupted by cloud cover (Loyacano et al. 1977).

Time-compensated, sun-compass orientation along the Y-axis had been found earlier in mosquitofish, *Gambusia affinis,* under natural sunlight, and under fixed artificial lights but not under complete cloud cover or diffuse artificial lights (Goodyear and Ferguson 1969). These small fish move toward the shoreline in the presence of aquatic predators but seem to prefer to be away from shore when aquatic predators are absent. Naive fish, captured in areas that lacked predatory fish, move into deeper water, while those from areas with predatory fish move toward the home shoreline (Goodyear 1973). *Gambusia* that had been exposed to bass in the wild learn a new shoreward direction in 1–2 days and do not show subsequent improvement over the next 4 days. Naive fish take 10–30 days to learn shoreward orientation in the presence of a predator and up to 6 days to learn a new shoreward direction when transferred to a new location. The difference between naive and experienced fish seems to be based on early learning experience in this case, since young of both types of parents learn shoreward orientation if exposed to bass when they are 12–18 h old. Adults, but not young, revert to a deepwater preference 6–12 days after the removal of the predators (Goodyear 1973).

In a similar antipredator response, the starhead topminnow, *Fundulus notti,* jumps out onto the shore when pursued by predatory fish, then returns to the water after danger is past. It apparently uses sun-compass orientation to determine the direction of the shoreline during this procedure (Goodyear 1970). Goodyear tested the orientation of these cyprinodonts in wading pools and on plastic covered tables and found that they oriented when they had a clear view of the sun but not when the sky was overcast.

2.4.2.2.3 The Sun Compass and Homing

The experiments of Winn et al. (1964b) on the rainbow parrotfish, *Scarus guacamia,* and the purple parrotfish, *S. coelestimus,* probably deal with a situation intermediate between the Y-axis orientation of small inshore fishes and long-distance migration. Adult parrotfish were captured and displaced out to sea, then released with balloons attached to them by fishline so they could be followed. They oriented toward shore under clear skies but were disoriented under cloud.

Even a small cloud covering the sun was sufficient to stop the fish or disorient them; they would then revert to oriented movement when the sun was visible again. Winn et al. argued that the disruption of orientation by small clouds covering the sun indicates that parrotfish are not able to use polarized light as a substitute for direct observation of the sun. Opaque eye caps disrupt the oriented movement, as did dusk. Clock shifting (6 h) produces an appropriate change in orientation. Interestingly, parrotfish from a different population do not show sun compass orientation when tested under similar conditions, although they move toward shallower water.

2.4.2.2.4 Sun Compass and Migration

The study of migratory fish on their first trip along a route would provide some indication of whether the orientation had been learned through trial and error or had evolved as an genetically programmed response to conditions that had remained similar for generations.

Sockeye Fry. Brannon (1972), studying the lakeward migration of sockeye fry (see Sects. 2.3.1 and 3.4), tested the orientation of two stocks of fry in six-armed orientation tanks (Fig. 2.20). The tanks were regularly rotated and water changed to reduce landmark and chemical cues. Groups of 25 fry were released from the central chamber and the numbers trapped in each arm were recorded. Chilko fry normally migrate southwest along the shore of Chilko Lake after entering the lake. In Brannon's apparatus native Chilko fry captured in the field preferred a southerly direction when tested at Chilko Lake or after transport to the laboratory (Fig. 2.21). This southerly tendency was reduced in Chilko fry that had been in-

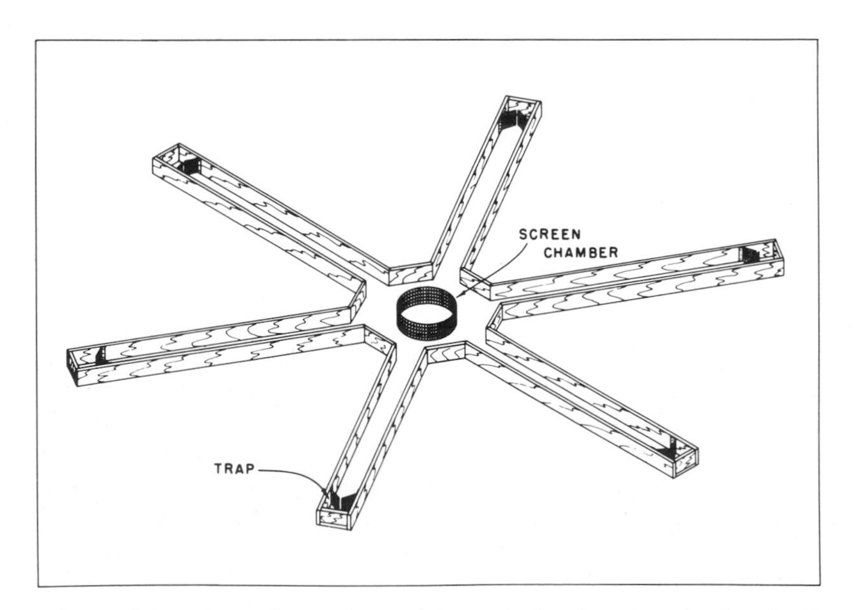

Fig. 2.20. Orientation-testing apparatus, used to study the orientation of sockeye fry (Brannon 1972)

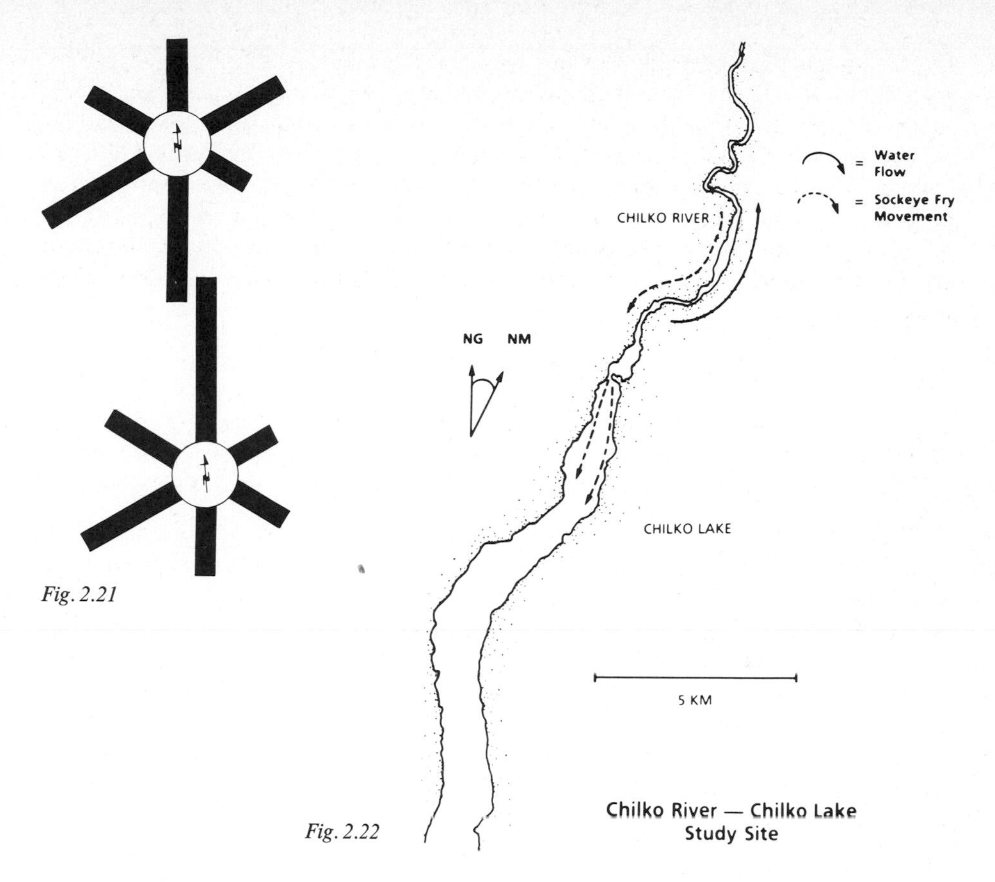

Fig. 2.21

Fig. 2.22

Fig. 2.21. Chilko experimental stocks: orientation of experimental stocks of Chilko fry tested in the orientation apparatus in Fig. 2.20. *Upper diagram* orientation on clear days; *lower diagram* on cloudy days. *Arm lengths* indicate mean response from all tests (Brannon 1972)

Fig. 2.22. Map of Chilko River and Chilko Lake showing waterflow and direction of sockeye fry migration. *NG* geographic North, *NM* magnetic North (Quinn 1980)

cubated in a hatchery. Under overcast skies the wild fish tended to show reversed orientation. At night they failed to enter the traps, and appeared to become disoriented at light levels near the limits of human vision.

Quinn (1980) used a four-armed orientation tank with traps on each arm to study migratory sockeye fry from Chilko Lake, B.C. and Lake Washington, Wash. In each of these lakes the fry emerge in a river, then move along the axis of the lake to feeding grounds. The Chilko Lake population spawns in the outlet river and the fry move upstream during the day, then migrate up-lake in a SSE (magnetic) direction (Fig. 2.22). The Lake Washington fish emerge at night and swim downstream in the same night, then migrate NNW in the lake. They are descendants of fish transplanted to the Cedar River in 1935 from Baker Lake, Wash., which is aligned in a similar compass direction to Lake Washington (Fig. 2.23).

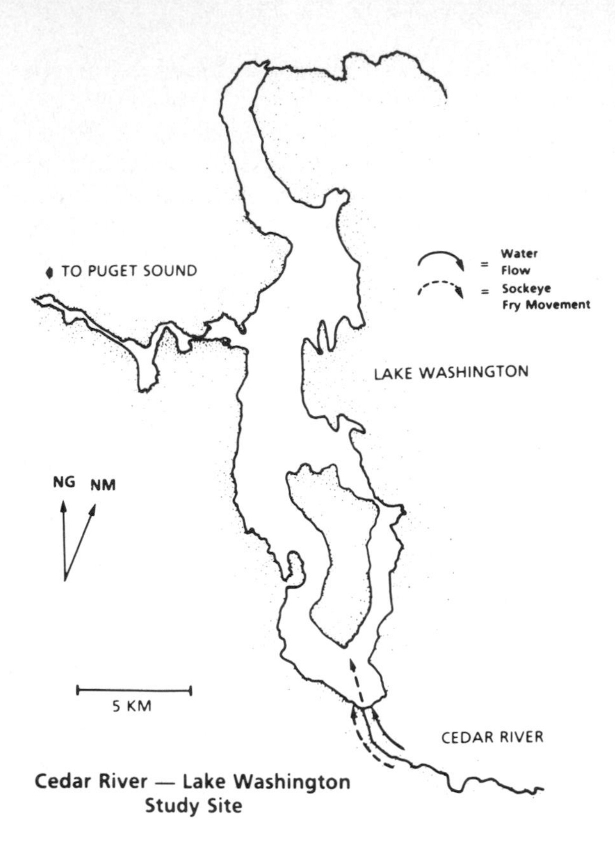

Fig. 2.23. Map of the Cedar River and Lake Washington showing water flow and direction of sockeye fry migration. *NG* geographic North; *NM* magnetic North (Quinn 1980)

Quinn (1980) captured migrating fry in the rivers and tested them the same day. He released groups of about 30 together by lifting a round Plexiglas chamber after a 5-min acclimation period in the centre of the tank. Forty-five min later the arms were blocked off and the number of fry in the terminal traps was counted, as well as the number still in the central chamber. Three tanks were used. They were rotated and exchanged between tests and drained and refilled to control for visual or chemical cues in the tanks. The tanks could be fitted with magnetic coils which, when on, left the intensity and inclination of the field intact but redirected the field by 90° counter-clockwise. They could be covered by black plastic at night or by translucent plastic during the day. The translucent plastic distorted the polarization pattern of the sky but allowed a human to see the sun's position.

Six different conditions were tested.

1. Clear view of the sky, normal magnetic field.
2. Clear view of the sky, magnetic coil present but off.
3. Clear view of the sky, coil on.
4. Tank covered, normal magnetic field.
5. Tank covered, coil present but off.
6. Tank covered, coil on.

In addition, tests could be run at night or during the day and during clear or overcast conditions.

The night-migrating fry from Cedar River, Lake Washington preferred an appropriate migration direction of 352° when tested at night with a view of the clear or partly clear sky and normal magnetic field. Under overcast skies the preferred direction, 347°, was not greatly altered. With the coil present but turned off the preferred direction was 287°. Turning the coil on resulted in a 79° counter-clockwise change in the bearing selected by the fish. Quinn (1980) felt that the effect of the coils when off was an artifact caused by the coils restricting the direction of approach to the observer. When the tanks were covered, the results were similar; the fish maintained the appropriate direction although with greater scatter in headings and there was a 103° counter-clockwise difference between "coil on" and "coil off" tests, as would be predicted if the fish were using a magnetic compass.

The day-migrating Chilko fish behaved differently. They maintained an appropriate direction when tested on overcast or clear days and when tested with the coils present but off. However, unlike the Lake Washington fish, turning the coil on did not materially affect their orientation under clear or cloudy skies. Under the plastic cover, they maintained their orientation but altered their preferred direction when the coils were turned on. Chilko Lake fish tested at night preferred a westsouthwest direction, perhaps a shoreward orientation, since these West-shore fish apparently rested near the lakeshore at night. When they were tested with the coils on at night their orientation was shifted 105° counter-clockwise.

These results indicate that the lakeward migration of sockeye fry is oriented by means of at least two mechanisms. A magnetic mechanism is used by night migrants and by fish with their view of the sky blocked. A second mechanism is used by day migrants when they have a clear view of the sky, even if the sky is cloudy and the magnetic field is altered. This shows an interesting parallel with the earlier work of Groot (1965) with sockeye smolts. Groot found a celestial orientation mechanism and a second, unknown mechanism that came into action when the celestial cues were not available.

Sockeye Smolts. Groot (1965) studied the migration of sockeye smolts as they left their nursery lakes on their way downstream to the ocean. On these migrations the 8-cm smolts must migrate through lakes, which may have dimensions measured in tens of kilometers, and they must find outlets, which are small and easily confused with blind bays and channels. Most of Groot's work was carried out at Babine Lake, B.C. (see Sect. 2.3.1.5.3). This large lake (Fig. 2.24) supports three demes, genetic populations, of sockeye (Johnson and Groot 1963). A "main lake deme" spawns in streams in the southeastern part of the lake and the fry spend their year of lacustrine life in that portion of the lake. Fish spawning in the outlet, the Babine River, on either side of Nilkitwa Lake produce fry that migrate upstream into the north arm of Babine Lake, forming a "north arm deme" while fish spawning in the streams running into Morrison Lake give rise to a "Morrison Lake deme" that spends its year in Morrison Lake. When these three demes of fry transform into smolts, they must migrate out of the lake system and into the Babine River to reach the sea.

Observation of the surface movements of smolt schools and trapping and experimental tagging all indicate that most of the smolts move out of the system in

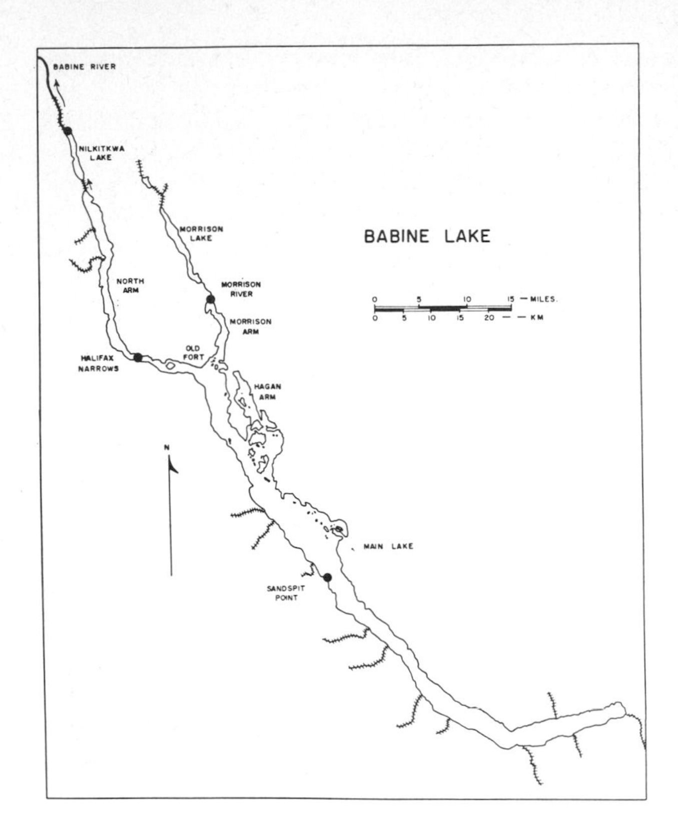

Fig. 2.24. The Babine Lake system in British Columbia. Sockeye spawning streams are marked by *cross-hatching* (Groot 1965)

a direct manner, despite the fact that the water currents in the lake are determined primarily by wind direction (Johnson and Groot 1963; Groot 1965). The apparent unsuitability of rheotactic information for this migration led Groot to consider celestial orientation as a directing mechanism. They tested sockeye smolts from two different lakes for their ability to maintain the appropriate orientation for their route through the lake when they were provided with only a view of the sky.

The apparatus used to test the smolt orientation response consisted of circular plastic containers (Fig. 2.25). Single smolts were released into the container and their preferred direction was recorded by an observer resting in the darkened chamber beneath the orientation tank. Since the observations generally continued for 15 min, the behaviour recorded was presumably migratory restlessness in confinement rather than simple migratory behaviour. I have largely followed Groot's interpretations of his data. There is some question about the appropriateness of his statistical analysis (Healey 1976) and the patterns of orientation are often obscure.

In Groot's apparatus the majority of the smolts showed a preferred direction that corresponded to the appropriate migratory direction for their deme, or they oriented directly opposite to the appropriate direction. This phenomenon, known as reversed orientation, is a commonly observed trait of migratory animals. Two main problems were investigated using the orientation response observed in this

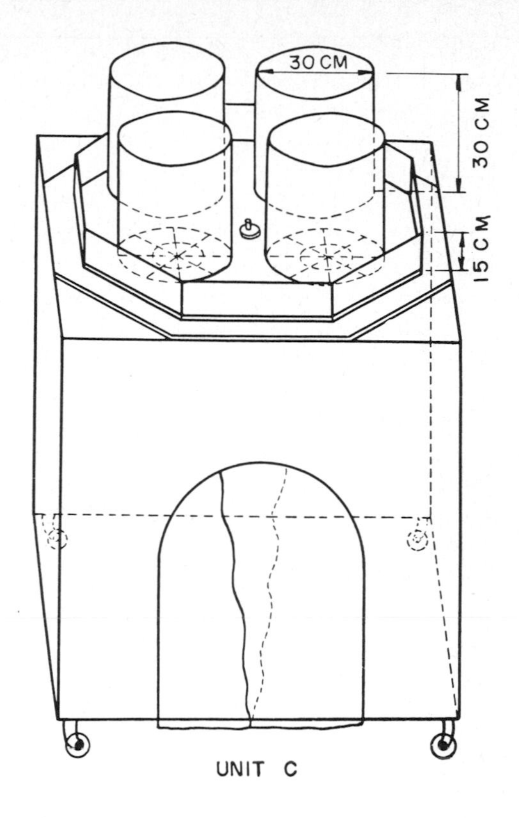

Fig. 2.25. Apparatus used for orientation tests on sockeye smolts. The observer lay in a darkened observation chamber beneath the four orientation tanks and looked up through the bottoms of the tanks (Groot 1965)

apparatus. First, how do migrants such as those in the Morrison Lake deme find their way to the Babine River when they must make substantial changes in orientation over the course of the journey? Second, what orientation mechanisms were being used by the fish to maintain the correct course?

The question of the selection of appropriate direction is important. Inappropriate orientation could lead to long detours or failure to get out of the lake. Each deme can face a different orientation requirement and in some cases, like the Morrison Lake deme, the direction may have to change over the course of the migration for the fish to reach the outlet (Fig. 2.24).

Nilkitwa Lake smolts and north arm smolts preferred a plane of orientation roughly parallel to the lakes, with little change during the migratory season. Morrison and main lake fish, however, showed changing orientation preferences during the migration period. The Morrison Lake fish changed from orientation roughly parallel to Morrison lake to orient parallel to the North Arm–Nilkitwa complex. Main lake fish initially preferred a compass direction somewhat to the west of the orientation of the main lake, but they shifted gradually around to a more northerly preference as the season progressed. This shift may lead the fish along the west shore and away from the blind alleys formed by Hagan Arm and Morrison Arm.

In both these cases the seasonal shift in orientation occurred whether or not the fish had actually moved the appropriate distance along their route. Groot

(1965) tested the orientation of three categories of fish from each deme, "run sample fish," "tagged-recovered fish," and "holding test fish." "Run sample fish" were captured along the migration route, the outlet for north arm fish, Sandspit Point for main lake fish and Morrison Arm for fish from Morrison Lake (Fig. 2.24). The fact that "run sample fish" from the main lake and Morrison Arm changed preferred direction during the season indicates that even fish that were still uplake late in the season had shifted orientation. In the case of Morrison fish they had even assumed a preference that was inappropriate for their geographic location. "Tagged-recovered fish" had been trapped, tagged, and released near the start of their route and then recovered farther along their migration and tested for preferred orientation. These fish that had made part of the journey also changed preferred migration with season. "Holding test fish" were captured early in the migration, held in plastic containers near the capture site, with a view of the sky, and tested for preferred orientation through the migration season. Again, both Morrison and main lake fish gave evidence of shifting orientation.

Smolts change their directional preference during the migration season without regard for their position on the migration route. Since the vast majority of the fish leave the lake in a few bursts of migration, most fish are probably correctly oriented most of the time. A similar phenomenon occurs in the migration out of Great Central Lake, B.C. (Groot 1965). In the Great Central study a further test was made by transporting smolts to a location 75 km away. Their orientation was still approximately correct for the original lake and season rather than for the new test location, indicating that they were not able to compensate for the displacement.

Groot suggested that the different seasonal patterns of preferred direction were genetically based and maintained by the home stream fidelity that allows sockeye demes to diverge genetically from one another. When fish from these three Babine Lake demes were later reared from the egg to the smolt stage in captivity and tested for their orientation preferences, the stocks differed in their preferred orientation (Simpson 1979) (see Sect. 7.4). However, the directions chosen did not clearly correspond with the migratory directions in the home lake, with the exception of the Morrison stock, which oriented appropriately and changed direction with season.

The other main question about sockeye smolt orientation, after that of preferred direction, was the mechanism(s) used by the migrants to maintain the appropriate heading. The apparatus eliminated rheotactic information, landmarks and probably most information about wind direction. The smolts did have a clear view of the sky. The traditional tests for sun-compass orientation, checking the response to cloud cover and to mirror images of the sun, were performed. The smolts showed greater concentration in their mean vectors of orientation when cloud cover was between 0 and 30% than when the sky was more overcast (40–60% cloud cover), consistent with use of celestial information when the sky was relatively clear. However, mean vector concentration improved again when cloud cover exceeded 70%, a result that Groot attributed to switching to some non-celestial orientation mechanism.

When the smolts were shown mirror images of the sky with the sun visible in the mirror, they changed orientation as if the image was the real sun. However, when only the blue sky was visible in the mirror, the smolts changed direction as if the polarization pattern of the sky had changed rather than as if the sun had changed position. With increased cloud cover the smolts tended to orient to features of the box rather than to the celestial information that might be available from the mirror. The difference between the response to the mirror image of the sun and mirror image of blue sky suggested that some patterned feature of the sky such as the pattern of polarization might be used by the smolts in their orientation (see Sect. 2.2.1.4).

Quinn and Brannon (1982) re-examined the orientation of Babine smolts using a circular tank with eight radially arranged outlets and a central water inlet. Orientation was assessed on the number of fish choosing the outlet oriented in each direction. In this apparatus, smolts captured at the outlet to Babine Lake oriented toward the lake outlet as long as they had a view of the sky, even if the magnetic field was rotated 90° or if the sky was obscured by overcast. In Quinn and Brannon's procedure the fish could leave the tank at any time over a 22.5-h period and were counted at 2-h intervals. In this situation brief availability of celestial orienting cues might allow the fish to take up orientation to local landmarks in the tank and maintain that orientation until they left the tank. When the tanks were covered, the fish took up a bimodal direction preference along the plane of migration and were deflected 56° by a 90° rotation of the magnetic field. This indicates the presence of a magnetic compass with lower priority than the celestial compass.

Pink Salmon Fry. Healey (1967) studied the orientation of pink salmon fry, *O. gorbuscha,* on their seaward migration, using the same type of apparatus as had been used by Groot (1965). Pinks migrate to sea as fry and, in British Columbia, must often negotiate convoluted channels through coastal inlets and archipelagos in order to reach the open sea. Healey sidestepped the troublesome reverse orientation, which had plagued Groot, by dividing the results into two categories, pointing along the axis of migration in either direction and pointing across the axis of migration. He took the axis of migration to be the axis of the channel the fish were migrating through. Using this method he found a significant preference for the migration axis during noon, afternoon, and evening tests. During the early morning, however, the fish tended to orient across the axis of migration. Orientation was compared under different degrees of cloud cover, indicating a reduction in the accuracy of orientation along the axis of migration when cloud cover exceeded 50%. Both Groot and Healey used percentage cloud cover in such comparisons. The observation of Winn et al. (1964b) that even a small cloud covering the sun was sufficient to disorient parrotfish would suggest that percentage cloud cover might be less important than whether or not the sun was actually visible.

Adult Pink Salmon. Pink salmon tend to follow the shoreline on their spawning migration through the Kurile Strait in the Sea of Okhotsk. When they were captured and tested for orientation in maze (Fig. 2.26) they took up an orientation appropriate for following the shore at the point of capture, if they were tested on a clear day (Churmasov and Stepanov 1977). The fish oriented randomly on cloudy days or when blinded, indicating a visual response to celestial stimuli. The shoreline changes direction in this region and when the fish were towed in open

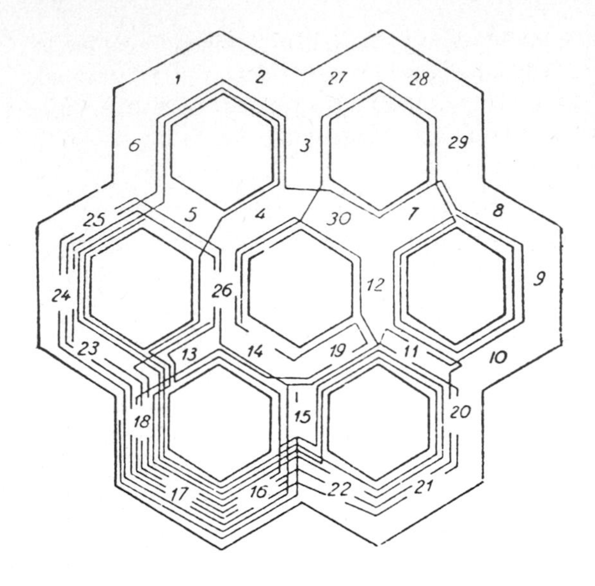

Fig. 2.26. Maze used to test the orientation of adult pink salmon in Kurile Strait. Each fish swam for 30 min in the array of numbered hexagonal channels. Orientation was assessed by the number of passages through each of the outer six channels. *Figures* indicate the code number of each channel. *Line* traces the trajectory of movements in one experiment (Churmasov and Stepanov 1977)

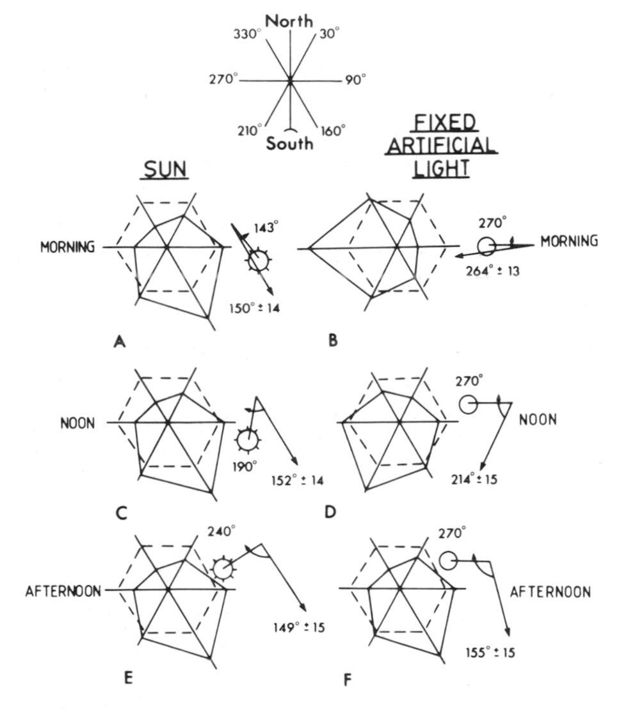

Fig. 2.27. Diagrams of orientation vectors by adult pink salmon in maze illustrated in Fig. 2.26. *A, C,* and *E* orientation to the sun under clear skies. *B, C,* and *F* orientation to a fixed light source in a closed maze with no view of the sky. *Dashed lines* equal orientation in all directions. *Solid lines* orientation of the experimental fish (Stepanov et al. 1979)

cages to a test location further along the shore, they changed orientation appropriately. If they were towed in closed cages they retained their original orientation. This may be a case of fish using a sun compass to maintain a heading that is initially based on other stimuli. Stepanov et al. (1979) compared the response of adult pinks to the sun on clear days with their response to a fixed projection lamp in a covered maze (Fig. 2.27). The fish showed appropriate time compensation in their orientation to the sun and to the lamp, indicating classic time-compensated sun-compass orientation. These studies indicate that adult salmon are capable of sun-compass orientation. This orientation mechanism is, therefore, available for the return migration across the open ocean.

Cutthroat Trout. Adult cutthroat trout, *Salmo clarki,* tracked with floats in open water, oriented toward shore during the first hour after release, rather than toward the stream from which they had been displaced (Jahn 1966). On cloudy days the orientation was less pronounced and migration speeds were slower. Blind, anosmic (unable to smell) and unoperated control cutthroats all homed successfully in the same situation, but from different release points, in Yellowstone Lake (McCleave 1967). Although the blind fish were slower in returning, all categories showed directed swimming when tracked by means of floats. Jahn (1969), using home stream recovery of tagged individuals and float tracking, also found that blind cutthroats homed as accurately as controls, while anosmic and blind-anosmic fish did not perform as well as controls. Some of Jahn's (1969) results seemed to indicate a bias toward swimming toward the sun's azimuth. An attempt to train hatchery-reared juvenile (150–213 mm) cutthroats to orient to a point source, using the same techniques and apparatus as Hasler et al. (1958), was partially successful. A few individuals did learn to orient to the light source (Jahn 1969).

McCleave and Horrall (1970) used ultrasonic transmitters to track displaced spawning adult cutthroats in Yellowstone Lake. Blinded and unblinded individuals moved to the East or South East toward the same shoreline as their homestream, then turned and moved along the shore to their stream. Blinded fish were as accurate as unblinded individuals. Further ultrasonic tracking again found that blinded and anosmic fish are capable of homing and that a 6-h time shift (over 4–8 days) did not appear to interfere with orientation (McCleave and LaBar 1972). If the cutthroat were using a time-compensated sun compass, and if the 4–8 days were sufficient to alter their time sense significantly, then this treatment would be expected to lead to a change in preferred direction of as much as 90°. The lack of change and the absence of disorientation on cloudy days was interpreted to be evidence against the use of sun-compass information by homing cutthroats. The retention of orientation by blinded fish in these studies does not, of course, eliminate the possibility of a sun compass based on extraoptic photoreceptors.

Artic Charr. The return of displaced mature charr, *Salvelinus alpinus,* to Floods Pond, Maine, was not affected by cloud cover, again indicating that sun-compass orientation was either not used or at least not essential (McCleave et al. 1977).

2.4.2.3 Lunar and Stellar Orientation

Some fish migration takes place at night when the sun is not visible (see Sect. 2.3.1). Birds have been shown to be able to orient according to the pattern of the stars (e.g., Emlen 1975) and amphipods to orient by the moon (Papi 1960). Surprisingly, there is virtually no evidence of lunar or stellar orientation in fish. I know of only one planetarium experiment on fish, although this technique has been very revealing when applied to birds. Johnson and Fields (1959) tested the orientation of eight sexually mature male chinook salmon under simulated star patterns that resembled sky patterns well to the North and South of the fish's expected route. They did not find significant orientation under these conditions. The fish were in the riverine phase of migration. It is conceivable that stellar orientation might be restricted to the oceanic phase of migration or be based on star patterns characteristic of the normal migration route.

The moon seems to be important as a timing signal in some fish migration (see Sect. 2.3.3), but reports of its use for orientation are rare. Mosquitofish were able to return to their home shore on moonlit nights, but this does not establish the orientation mechanism (Goodyear 1973).

The report of de Veen (1963) that "great numbers" of soles, *Solea solea* L., are occasionally seen swimming at the surface at night may indicate some form of nocturnal celestial orientation, although the report mentions that the phenomenon is associated with low light intensity. The initial reports were followed up with extensive polling by a questionnaire of commercial fishermen in Dutch North Sea ports (de Veen 1967). Analysis of the questionnaires indicated that the soles were observed swimming on the surface almost exclusively at night and that they were most likely to surface during the period of gonadal development prior to spawning. Two types of orientation seemed to be at work. In the first, the soles timed their excursions away from the bottom to correspond with periods when the tidal currents were setting toward the East and South where the inshore spawning areas of this stock lay. As well as selecting appropriately directed tidal currents, the soles also maintained an oriented body axis facing the spawning

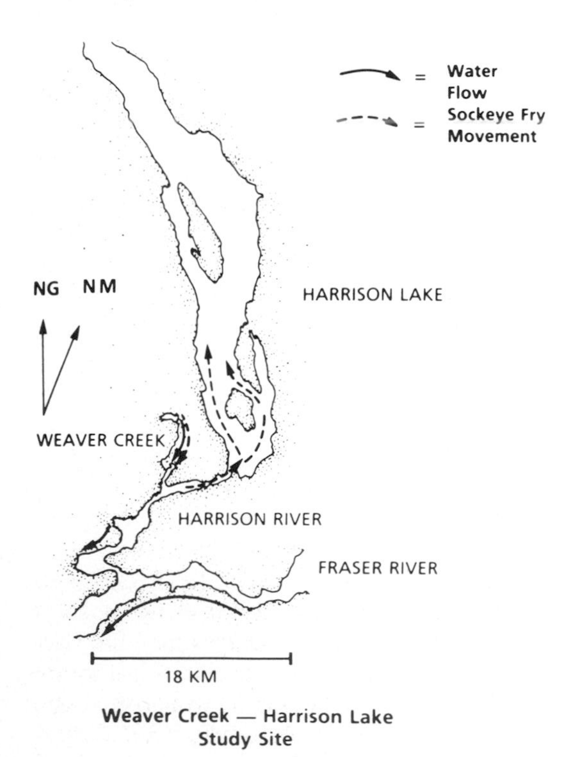

Fig. 2.28. Map of the Weaver Creek-Harrison Lake system, showing water flow and direction of sockeye fry movement (Brannon et al. 1981)

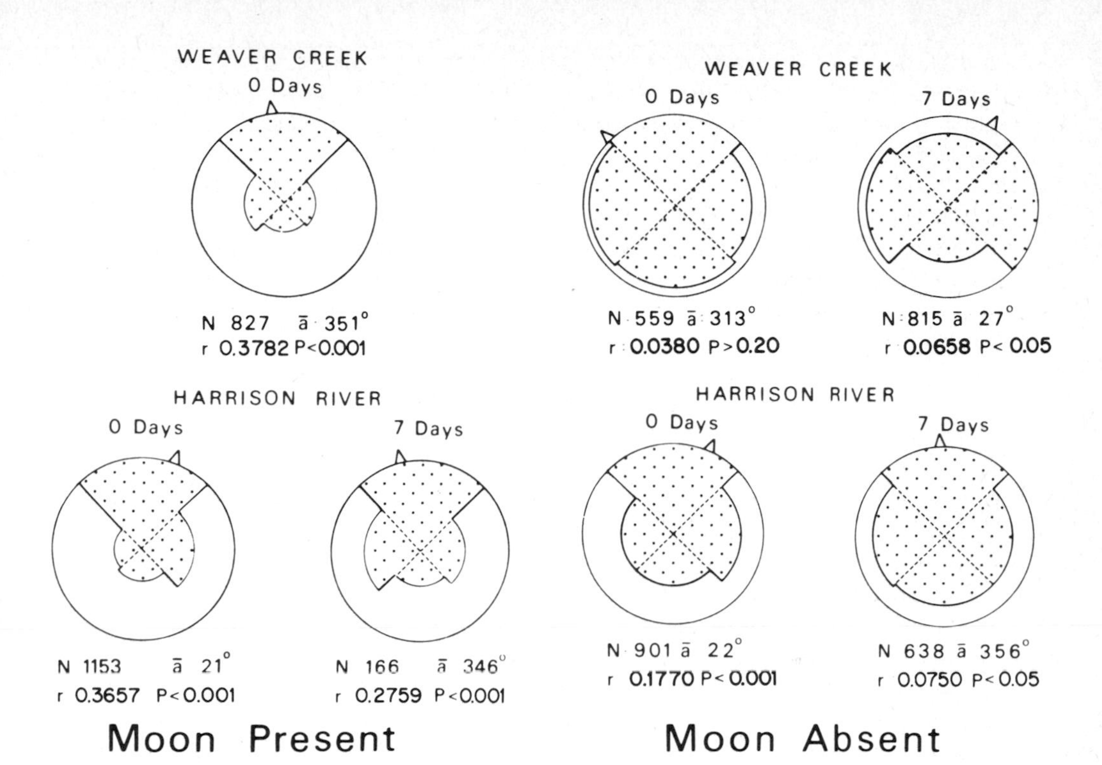

Fig. 2.29. Compass orientation of sockeye salmon fry from Weaver Creek. Fry were tested in 1980 at Weaver Creek and Harrison River on the night of their capture and 7 days later in the presence of the moon (Brannon et al. 1981)

areas, although they apparently drift quietly at the surface, rather than swim actively. De Veen (1967) found some evidence that these two processes were independent of one another. Fish that were in tidal currents setting the "wrong" way still had their body axis correctly oriented. Complete cloud cover seemed to interfere with the selection of the appropriate tidal current, but did not interfere with the orientation of the body axis. The presence of the moon did not have any noticeable effect on either orientation mechanism. De Veen (1967) felt that some form of celestial compass information might be used in the selection of appropriate tidal currents, because of the effect of cloud cover.

The moon may be used in the orientation of sockeye fry in their migration to the nursery lake or in their dispersion in the lake, although field observations indicate that fry migration is reduced on bright moonlit nights (Brannon et al. 1981). Fry from the population that spawns in Weaver Creek, B.C. (Fig. 2.28) must first swim down Weaver Creek then up Harrison River to reach Harrison Lake, where they disperse up the lake. During the last two stages of this process a roughly northward orientation would be appropriate and the fry oriented to the north when tested in a four-armed orientation tank outdoors at night. This orientation was strongest when the moon was visible and weaker when it was below the horizon or obscured by clouds (Fig. 2.29). This was not a simple effect of light and shadow because orientation was improved regardless of the moon's direction. Increasing age and exposure to Harrison River water also improved the northward orientation, the appropriate direction of movement after the fish left Weaver Creek.

2.4.2.4 Summary of Celestial Orientation

The use of celestial bodies as reference points for orientation occurs in a variety of fish. Celestial orientation is apparently used in predator-avoidance behaviour in which the direction of a safe location is learned within the animal's life span and the sun, for example, is used to select and maintain that direction (Goodyear 1973). It is also used in homing, as in the parrotfish tracked by Winn et al. (1964b), and in a number of migratory situations. In some of these cases the direction selected seems to be influenced by the genetic stock to which the fish belong (e.g., Simpson 1979; Brannon et al. 1981). Celestial orientation is often one in a hierarchy of mechanisms used in orientation (Quinn 1980; Quinn and Brannon 1982). This makes sense, as celestial orientation is capable of great precision and is therefore often the mechanism of first choice when it is available. On the other hand, an animal's view of the sky, particularly an aquatic animal, is often obscured by clouds, turbidity or by water depth. Furthermore, celestial cues are by their nature only available about half the time, the sun during the day and the moon and stars at night. Thus alternate orientation mechanisms, even if they are inherently less accurate than celestial orientation, will often be adaptive.

2.5 Summary

Light penetrates water best in the blue-green area of the spectrum. The red and ultraviolet wavelengths attenuate rapidly with depth and even blue-green light only penetrates the upper layers of deep water bodies. Natural waters contain dissolved and suspended materials that further alter underwater illumination. Turbidity can decrease the total light available and increase backlighting and veiling brightness. Dissolved materials can alter the wavelength of maximum transmission. Underwater polarization occurs and could provide reliable information on the sun's direction. In water, light is reduced and variable when compared to terrestrial conditions. This is a natural outcome of the presence of an additional medium, water, between the light source and the observer.

Fish eyes follow the basic vertebrate plan of cornea, lens and a duplex retina of rods and cones surrounded by a pigment shield. In fish the cornea does not focus light and the lens is relatively hard compared to the terrestrial vertebrate lens. Focusing is largely based on movement of the lens relative to the retina. Adaptation to brightness is accomplished by retinomotor movements of pigments and of the rods and cones within the retina. This is a relatively slow process and is often timed by circadian rhythms. As with other vertebrates, fish eyes and central analysis mechanisms are probably specialized in each species for the particular habitat and behaviour of that species.

Sensitivity to polarized light is present in fish, but the mechanism is not yet clear. Some terrestrial animals use polarized light in sun-compass orientation to locate the sun's position when it is hidden by clouds. Sun-compass orientation in fish is disrupted even by small clouds obscuring the sun, so it is possible that sensitivity to polarized light may not be used as an adjunct to sun-compass orientation, but rather may improve visual discrimination in turbid water.

Fish, like other vertebrates, possess extraoptic photoreceptors (EOP's), light receptors outside the eyes. These EOP's may be involved in phototactic responses and in the response to light cycles, including photoperiod. In amphibians, and perhaps in fish, EOP's can mediate compass orientation responses.

Natural light is rhythmic on both daily and seasonal time scales. The daily cycle of light and dark has profound consequences for fish in terms of feeding and predator avoidance, with some species specialized for diurnal activity and others for nocturnal or crepuscular activity. This specialization and the activities of specialized predators will determine the optimal diel timing for migration. Fish often use the diel cycle of light and dark as a Zeitgeber or timing stimulus to synchronize circadian activity rhythms with local illumination conditions. Such circadian rhythms allow physiological anticipation of periods of migratory activity.

Seasonal changes in the length of day, photoperiod, are reliable indicators of seasonal change in temperate regions. Many animals, including fish, use photoperiod to synchronize physiological cycles with seasonal cycles. Photoperiod, acting through the central nervous system, can control the timing of changes in hormone secretion. Hormones in turn affect the timing of physiological changes important in migration such as development of osmoregulatory capability in fish that will migrate between freshwater and saltwater. Behavioural changes are also timed by hormones.

Lunar cycles have a major influence on tides and also determine the timing of dark moonless nights that may be optimal for nocturnal migrants. Tidal cycles in activity occur but they are probably not timed by moonlight. Lunar spawning rhythms are well known in beach-spawning fish like the California grunion, and moonlight may play some role in their timing, although this has not been demonstrated experimentally. Lunar timing is important in the downstream migration of adult eels, which tends to occur on dark nights. It has also been suggested that the thyroxine surge controlling downstream migration in juvenile salmonids is timed in part by a lunar cycle. The adaptive advantage might be coordination of group activity and synchronization of migration with dark moonless nights.

Social interactions based on visual displays may play a role in the timing of migration, particularly in the synchronization of activity between individuals. Migrants may benefit, in terms of predator defence and orientation accuracy, by traveling in a group. Visual displays are appropriate for this sort of social facilitation, but there is no direct evidence of visual displays synchronizing migration.

Light can be used to direct migration through visual responses to the environment. Landmarks, for example, can be used in three different ways: to recognize familiar regions, to recognize general types of habitat, or to maintain headings initially set on the basis of other information.

Directional information is also available from the position of celestial bodies such as the sun and stars. Several species of fish use sun-compass orientation in homing or migration. Effective use of a sun compass requires sensory ability to detect the sun's position and a clock mechanism to compensate for the sun's apparent movement during the day. Precise long-term use of a sun compass may also require adjustment for seasonal or latitudinal changes in the sun's path.

Centrarchids can be trained to orient by a sun-compass mechanism and to show seasonal adjustments in their time-compensation response. This probably reflects their use of the sun compass in orienting predator avoidance responses near a home shoreline, a situation in which learning of local directions will be appropriate. Juvenile salmonids orient along migration routes using a sun compass. In these cases in which the fish is traveling the route for the first time, the direction of orientation may be genetically determined. In sockeye smolts, seasonal changes in preferred orientation occur that are appropriate to the changes in direction required for movement through complex lakes or coastal waterways. These seasonal adjustments in orientation are typical of particular genetic stocks.

Stellar orientation occurs in birds and some beach amphipods show lunar orientation. Fish do migrate at night, but the evidence for stellar or lunar orientation in fish is slim. Soles swim on the surface at night, during the period that tidal currents set in the right direction for their migration. This behaviour is disrupted by cloud cover but not affected by the moon, suggesting stellar orientation. In another study, the orientation of sockeye fry toward nursery lakes was most accurate on moonlight nights when the moon was not obscured by clouds, suggesting some form of lunar orientation.

Chapter 3

Chemical Information

3.1 Characteristics and Importance of Chemical Information in an Aqueous Environment

Water can dissolve a wide variety of compounds and both animate and inanimate objects continually release chemicals into the surrounding water. There is, therefore, a chemical "landscape," which reflects the physicochemical and biological aspects of the environment, including the ion and gas content of the water, for example salinity and oxygen concentration, as well as the many secretions and excretions of plants and animals. This wealth of information, coupled with the presence of acutely sensitive chemoreceptors in fishes (Kleerekoper 1969; Hara 1982), is a potential basis for effective orientation and timing.

Chemical information differs from visual information. It will be available throughout the diel cycle, where visual information may be unavailable at night, or during the day in the case of lunar or stellar orientation. There are diel, and longer cycles in some chemical stimuli, dissolved oxygen, for example, but other concentrations remain relatively constant and the information channel remains open. Similarly, depth and turbidity do not interfere with chemically mediated responses to the same degree as they do with vision. The persistence of chemicals in water when combined with water currents can carry information over long distances, far beyond the range of visual or mechanical stimuli. The isotopes from nuclear reactors on the Columbia River allow its waters to be detected by human instruments 115 km to seaward of the estuary (Gross et al. 1965) and the plume of low salinity can be detected 500 km offshore at times (Favorite 1969 a, b). Fish with their ability to detect a few molecules of biologically relevant compounds in contact with the olfactory epithelium (Teichmann 1959; Kleerekoper 1969) should be able to detect the outflow of major rivers over similar distances.

Researchers working with aerial pheromones have coined the term "active space" to describe the volume within which the concentration of a chemical exceeds the threshold sensitivity of the subject animal. This concept is equally applicable to aquatic organisms, and in the case of distinctive chemical stimuli from rivers or ocean currents, the active space may be enormous.

3.2 Fish Chemoreception

Vertebrate chemoreception has traditionally been divided into two main components, taste and smell (or gustation and olfaction), based on the familiar mammalian division of low-sensitivity chemoreceptors sampling dissolved substances in the mouth, and high-sensitivity receptors responding to aerosols in the nose. In fish the situation is more complex. Chemoreceptors require moist surfaces and in an aquatic organism the whole body is bathed in an aqueous medium, allowing chemoreceptors over the whole body surface. Bardach and Villars (1974) suggest that fish have as many as five distinguishable chemosensory systems: (1) olfaction; (2) taste buds, made up of several receptor cells and inervated by cranial nerves; (3) single taste cells inervated by spinal nerves; (4) free nerve endings, and (5) the lateral line system. The last three of these probably comprise the "common chemical sense" suggested by Parker (1912).

Each of these possible sensory systems has different capabilities and each has been studied with different intensity.

3.2.1 Olfaction

Olfaction refers to chemoreception received by the brain through the first cranial nerve and originating from primary chemoreceptor cells, cells derived from nerve tissue, confined to specialized epithelia in the nasal region. The receptor cells are usually equipped with cilia (Bannister 1965; Vinnikov 1965; Wilson and Westerman 1967; Yamamoto 1982) which probably serve to bring the sensory membranes into contact with water from outside the boundary layer of water near the cell. There are several types of receptor cells in fish olfactory epithelia (e.g., Ichikawa and Ueda 1977), but the roles of the various receptor cell types have not yet been elucidated. The mechanism of reception of chemical stimuli at the cellular level is beyond the scope of this volume (Brown and Hara 1982).

The olfactory epithelium must be exposed to new water frequently in order to detect changes in the chemical composition of the medium, and yet it is obviously delicate enough to require some encapsulation for protection. Usually the olfactory epithelium, composed of the sensory cells, supporting cells and glandular cells, is arranged in a folded or convoluted pattern, which increases the surface area in the nasal cavity. This folded structure is referred to as the olfactory rosette. The number and complexity of folds corresponds roughly to the degree of emphasis on olfaction in the biology of a species (Kleerekoper 1969), ranging from none in the sandlance, *Ammodytes,* and pipefish, *Syngnathus,* (Liermann 1933) to as many as 230 lamellae in the striped pargo, *Hoplopagnus guentheri,* (Pfeiffer 1964).

In addition to mechanisms that increase the surface area of the olfactory chamber, there are a variety of mechanisms to increase the flow of water through the chamber, These include ciliary movement and chambers or olfactory sacs, attached to the nasal chamber. The sacs are squeezed by the movement of respiratory or jaw musculature and pump water back and forth (Kleerekoper 1969). The nares (nostrils) are often arranged so that the forward movement of the fish forces water over the olfactory epithelium.

The fish olfactory system is capable of exquisite sensitivity (Table 3.1). There is, of course, tremendous variation from species to species in the degree of sensitivity, as one would expect from the morphological variation in the sensory struc-

Table 3.1. Threshold values of various odours for some species of fish (Kleerekoper 1969)

Ref.	Species	Substance	Threshold	Observations
Bull (1930)	*Blennius gattorugine* *B. pholis*	Whole *Nereis diversicolor* Ground *Mytilus edulis* and *Patella vulgate*	0.00375–0.00075% of weight of living food substance in seawater	
Neurath (1949)	*Phoxinus phoxinus* (L)	Eugenol	$1{:}17\times10^{6}$ or 2.3×10^{14} molecules cc^{-1}	
	Phoxinus phoxinus (*P. laevis*)	Phenyl ethyl alcohol	$1{:}23.3\times10^{6}$ or 2.2×10^{4} molecules cc^{-1}	
Hasler and Wisby (1950)	*Pimephales notatus* (*Hyborhynchus notatus*)	p-Chlorophenol	<0.0005 ppm	Conditioned animals
		Phenol	<0.0005 ppm	
van Weel (1952)	*Thunnus albacares* (*Neothunnus macropterus*)	Tuna flesh (three species)	5 g in 3 l of seawater introduced into tank of unknown volume	
Wisby (1952)	*Oncorhynchus kisutch*	Morpholine	0.000001 ppm	Conditioned animals
Alderdice et al. (1954)	*Oncorhynchus kisutch* *O. tshawytscha*	Alarm substance in skin from hair seal and sea lion	$1{:}80\times10^{9}$	Assumed is that substance is not more than 0.1% of wet weight of skin
Schutz (1956)	*Phoxinus phoxinus* (*Phoxinus laevis*)	Alarm substance in skin of *Phoxinus laevis*	$1{:}5\times10^{10}$ or 1/100 mm^2 of skin in 14 l of water	Assumed is that alarm substance is 0.1% of weight of skin
Tavolga (1956)	*Bathygobius soporator* ♂	Water in which a gravid ♀ of same species had been held for 2 min	0.5 ml	Siphoned into tank holding ♂
	Bathygobius soporator ♂	Ovarian fluid of gravid ♀♀	1:45,000	
Teichmann (1959)	*Anguilla anguilla*	β-Phenyl ethyl alcohol	$1{:}3\times10^{18}$ or 1770 molecules cc^{-1} of water or 1 molecule in nasal chamber	Conditioned animals
	Salmo gairdneri	α-Ionone	$1{:}3\times10^{4}$	
	(*Salmo irideus*)	α-Ionone	$1{:}10^{19}$ of 5×10^{11} molecules cc^{-1}	
	Anguilla anguilla	Macerated *Tubifex*	25 mg in 6.67×10^{15} cc of water	
	Phoxinus phoxinus	β-Phenyl ethyl alcohol	$1{:}67\times10^{6}$ ot 7×10^{13} molecules cc^{-1}	
		1-Methanol	1.93×10^{3} molecules cc^{-1}	
		Citral	2.13×10^{3} molecules cc^{-1}	
		Terpineol	2.75×10^{3} molecules cc^{-1}	
		Eugenol	2.97×10^{5} molecules cc^{-1}	

Table 3.1 (continued)

Ref.	Species	Substance	Threshold	Observations
Marcström (1959)	*Rutilus rutilus* (*Leuciscus rutilus* L)	2,4,6-Trinitrophenol	4×10^{-5} M	Conditioned animals
		Benzene	20.1×10^{-6} M	
		1,3-Denitrobenzene	7.05×10^{-6} M	
		Mononitrobenzene	7.3×10^{-6} M	
		1,3,5-Trinitrobenzene	1.44×10^{-6} M	
		Phenol	9.48×10^{-6} M	
		Resorcinol	54.2×10^{-6} M	
		Phloroglucinol	76.0×10^{-6} M	
Hasler and Wisby (1959)		Potassium phenylacetat	$< 1:10^{13}$	
Pfeiffer (1960)	*Tribolodon hakonensis*	Alarm substance from skin of same species		
McBride et al. (1962)	*Oncorhynchus nerka*	Extract of brine shrimp or zooplankton	2.5 mg of wet weight in 200 liters of water or 0.01 ppm	
Tarrant (1966)	*Oncorhynchus nerka* Walbaum	Eugenol	0.18 ppm	Conditioned animals

tures. There are also seasonal variations within species. The eel, for example, is 60 times more sensitive to b-phenylethyl alcohol in late winter and in midsummer than in late fall and early winter (Teichmann 1959). There are probably variations related to developmental changes and to sex differences as well.

3.2.2 Gustation

Taste information is received by taste buds. These clusters of sensory cells, derived from the epidermis rather than the nervous system, are innervated by the cranial nerves VII, IX, and X, that transmit the information to the dorsal medulla (Bardach and Atema 1971; Bardach and Villars 1974). The taste buds are abundant in the mouth, lips, and branchial arches but in teleosts may also be spread widely over the body. These external taste buds are often innervated by the facial nerve, (VII) leading to elaborate extensions of that nerve. The presence of taste buds on the fins and on other extensions of the body such as barbels and on modified fins such as the feelers of anabantids should allow simultaneous sampling at different points in a chemical gradient. In orientation this may give gustation an advantage over olfaction, since the nostrils of fish are usually placed fairly close together. Orientation will be facilitated by comparison of widely separated receptors.

There are cases of taste receptors, or at least cells with similar histological and sensory capabilities, that are innervated by spinal nerves. The sea robins (Triglidae) have such cells in their modified fins and on the body, served by large spinal ganglia (Scharrer et al. 1947; Bardach and Case 1965). Whitear (1965) found lone cells in fish epidermis that resemble taste cells in their structure and are innervated by spinal nerves.

Table 3.2. Taste thresholds of the minnow *Phoxinus* and of two species of ictalurid catfish

Substance	Capabilities in mol l^{-1}.
	Minnow[a]
Raffinose	1/245,760
Sucrose	1/81,920
Lactose	1/2,560
Saccharin	1/153,600
Quinine hydrochloride	1/24,576,000
Sodium chloride	1/20,480
Acetic acid	1/204,800
	Anosmic bullheads[b]
Quinine	1/10,000
Cysteine HCl	1/1,000,000
	Channel catfish[c]
L-alanine	1/1,000,000,000

[a] Glaser (1966)
[b] Bardach and Villars (1974)
[c] Caprio (1974)

The sensitivity of taste receptors is generally lower than the sensitivity of the olfactory system, although the two systems overlap in the chemicals to which they respond (Table 3.2), and the fish gustatory sense is still very sensitive when compared to human thresholds. For example, the minnow threshold for sucrose is 1/91 of the human threshold (Glaser 1966).

Amino acids are particularly effective gustatory stimulants (Caprio 1974, 1982), but a variety of other compounds such as salts and sugars are also detected and taste receptors are also capable of detecting carbon dioxide (Hidaka 1970). Bardach and Villars (1974) argue that the spinally innervated taste receptors of the sea robins respond to a more restricted range of chemicals than the cranially innervated receptors.

3.2.3 Other Chemoreceptors

Both taste and odour receptors have been demonstrated to be capable of orienting the movement of fishes in space (Kleerekoper 1969, 1972; Bardach et al. 1967), but the other possible chemoreceptor systems have not. The behavioural effects of stimulating the single cells and free nerve endings reported by Whitear (1965, 1971 a, b) are unknown. The lateral line system is capable of responding to chemical stimuli, particularly monovalent cations (Katsuki et al. 1969, 1971; Katsuki 1973; Katsuki and Yanagisawa 1982), but the role of this sensitivity in orientation is unknown. There are examples of migratory fish responding to ion gradients such as salinity differences (see Sect. 3.4.2), but ions can also be detected by both the olfactory and gustatory systems.

3.3 Timing by Chemical Stimuli

3.3.1 Diel Timing

Chemical information may not be well suited to synchronizing behaviour with cyclic changes in the environment. Daily and seasonal changes in light and temperature lead only indirectly to cycles of chemical change in the water. It may, therefore, be more adaptive for animals to cue on light cycles or secondarily on temperature change. Nevertheless, Krogius (1954) has argued that the upstream migration of adult sockeye and the downstream movements of the young may be timed by diel changes in pH and dissolved gases, which occur in a cyclic manner in response to biological productivity in aquatic systems.

3.3.2 Synchronization with Flood Conditions

Chemical timing may be more appropriate where the exact timing of an event cannot be predicted from changes in illumination. Floods or freshets, for example, are broadly seasonal in occurrence but their exact timing, to the day or hour, is important and cannot be determined from the photoperiod. Evolution may therefore have favoured fish that respond to chemical, thermal or mechanical stimuli emanating from such changes in flow or water level.

Entry of migrant fish into streams is frequently associated with freshets, or increases in river discharge. Hayes (1953) found that Atlantic salmon entered the La Have River, Nova Scotia, in response to "artificial" freshets caused by releasing water from dams. Natural freshets were more effective than the artificial version, however. Similarly the pre-spawning migration of rainbow trout into a creek draining into Lake Huron corresponded to freshets (Dodge and MacCrimmon 1971). The role of chemical stimuli was not tested in these studies, but the effectiveness of natural over artificial freshets in Hayes' (1953) study may indicate a chemical effect. Dodge and MacCrimmon noted that temperature changes accompanied the freshets, but Hayes had found temperature ineffective in inducing salmon migration.

This effect may also occur in downstream migrants. Catches of downstream migrant silver eels at fishing weirs were greater during periods of high water and turbidity (Frost 1950). Again, chemical cues were not examined experimentally as timing mechanisms.

Chinook fry left simulated incubation channels during rainy periods even when the stream water was filtered through sand before entering the incubation channel (Thomas 1975). The effect may have been due to temperature, flow or chemical changes.

3.3.3 Synchronization with Tidal Flow

The synchronization of migratory movements with tidal currents also falls into the chemical timing category. Although tides are ultimately timed by the positions of the sun and moon, the actual time and direction of ebb and flood in a coastal area is strongly affected by local topography and cannot be predicted easily from celestial events.

European elvers, in their onshore migration, tend to swim actively during the flood tide, which will carry them shoreward, and to hold position near the bottom on the ebb. Creutzberg (1959, 1963), using circular orientation tanks, found that ebb tide water from estuaries induced positive rheotaxis and that the elvers clung to the bottom. Flood tide water induced negative rheotaxis and movement up into the water column. These responses are controlled by detection of chemical differences between the flood tide water and ebb tide water. Salinity differences alone do not release the behavioural change. Some chemical present in natural freshwater, and hence in the ebb tide water, triggers the positive rheotaxis. When the freshwater is filtered through charcoal, it loses its ability to release positive rheotaxis, indicating that organic compounds are involved. The bottom-hugging response, as distinct from positive rheotaxis alone, occurs in response to high velocities (greater than 36 cm s^{-1}) associated with freshwater odour Creutzberg (1963). Flow is slowest near the bottom, so this response probably prevents elvers from being swept away from shore by strong currents.

American elvers show similar behaviour as they migrate into the estuary of the Penobscot River, Maine, although the controlling stimuli have not been determined (McCleave and Kleckner 1982). In this estuarine situation the strongest shoreward flow is found beneath the halocline, where a tongue of saltwater moves inshore on the flood tide beneath the freshwater outflow of the river. The elvers move up into the water column on the flood tide, but tend to remain below the halocline and hence in the water that is moving inshore (upstream) most rapidly. McCleave and Kleckner (1982) speculate that turbulence, electric currents or circa-tidal rhythms might be involved in timing the entry and departure from the water column.

3.3.4 Pheromones as Timing Cues

Synchronization of the movement of individuals in a population is one aspect of timing which may be well suited to the characteristics of chemical information. Although I do not know of any direct evidence for migration-inducing pheromones in fish, there are cases of surprising convergence in the initiation of migratory behaviour. Lowe (1952) and Boetius (1967) reported that captive eels escaped from laboratory tanks more often on nights when large natural migrations of eels were occurring in nearby rivers. It is not clear whether the water supply used in the eel studies was interconnected with the river or not. If there was a connection then a pheromone might be suspected.

3.3.5 Emergency Responses

Negative chemical stimuli sometimes override normal patterns of behaviour or development. Salmonid alevins in the gravel are vulnerable to low oxygen conditions. High CO_2 levels serve as indicators of hypoxia and Bams (1969) reported that high CO_2 levels induced early emergence from the gravel. He did not indicate which chemoreceptors were used in this response. Pink salmon reared in gravel beds in a hatchery migrate later in time and development than creek fry, perhaps because of lower silt levels in hatchery (Bams 1972). Mason (1969) subjected eggs from a single female coho to three different concentrations of dissolved oxygen; $11\ mg\ l^{-1}$, $5\ mg\ l^{-1}$, and $3\ mg\ l^{-1}$. Low oxygen groups received compensatory temperature increases in order to keep the development rates equal. The smaller, low-oxygen fish emigrated from their stream tank sooner. In such cases the chemical stimuli serve as indicators of unusual events and initiate behaviour that is adaptive under the changed conditions.

3.4 Direction and Distance

The chemical landscape may influence fish distribution and movement by presenting boundaries of physiological tolerance. Natural components of the environment such as O_2, CO_2 or salinity may be too high or too low to allow survival of a particular species, thus limiting its distribution or movements. In the case of these natural substances the fish is usually sensitive to their concentration levels and capable of avoiding unsuitable areas, selecting zones within the tolerance levels of its species. Substances added to the environment by man may also exceed the physiological tolerance of fish and limit their distribution, but the fish may or may not be able to detect such substances and react appropriately. There has not been a long enough period of natural selection favouring behavioural avoidance of pollutants.

Within the limits of physiological tolerance, the spatial pattern of chemical stimuli may guide migrants. Coastal salinity gradients, lake odours and similar chemical variables have been proposed as directing factors in fish migration. There are also minor chemical components of natural waters which will vary from location to location allowing recognition of a specific location – chemical piloting – and home recognition. Substances produced by conspecifics are one important class of minor components in natural waters. Their presence in water would indicate that an area is inhabited by one's own species. Within a local area, short-range orientation may occur in response to chemicals emitted by predators, food, mates or rivals.

3.4.1 Oxygen and Carbon Dioxide

The oxygen levels of natural waters vary from supersaturation to complete absence, high temperature and salinity both decreasing the saturation level for oxygen. Biological and chemical oxygen demand can reduce oxygen levels to below saturation, particularly if the water is not aerated at the surface or if light levels are too low to allow photosynthetic oxygen production. Carbon dioxide levels also vary widely in nature, although in nature low oxygen may be limiting more often than is high carbon dioxide (Fry 1971).

In experimental conditions fish can detect CO_2 through their taste receptors (Hidaka 1970) and respond behaviourally to CO_2 concentration. Atlantic salmon, for example, avoid high CO_2 and low pH (Hoglund and Hardig 1969). Fish are capable of behavioural responses to the oxygen concentration of water. Flounder postlarvae, *Paralichthys lethostigma,* for example, make directed movements to escape from water of low oxygen concentration (Deubler and Posner 1963) and minnows, *Phoxinus phoxinus,* avoid areas of low oxygen concentration (1 mg l^{-1}) in an experimental stream channel (Stott and Buckley 1979). Low oxygen levels occur naturally in stagnant water, under ice and in several other situations. These responses indicate that migratory fish could be capable of orienting to oxygen gradients.

Zones of low oxygen or high carbon dioxide can act as barriers to migrating fish. Adult chinook salmon, tracked with sonic tags, were blocked in their upstream migration by oxygen concentrations below 5 ppm, or by temperatures above 19 °C (Hallock et al. 1970). Carbon dioxide can repel migrating fish as well. Powers and Clark (1943) tested the response of brook trout, *Salvelinus fontinalis,* and rainbow trout, *Salmo gairdneri,* to carbon dioxide and found avoidance responses. They suggested, on the basis of nerve sectioning, that the lateral line was involved in sensing the carbon dioxide. Collins (1952) added gases and chemicals to a flume in a natural stream and found that upstream migrant adult alewives, *Alosa pseudoharengus,* and blueback herring, *A. aestivalis,* avoided high carbon dioxide levels but not altered pH or low oxygen.

The ability of fish to tolerate oxygen deficiency varies widely. Even complete lack of oxygen is not a barrier to those species capable of sustaining themselves through aerial respiration or anaerobic respiration (Hochachka 1980). The Curaçao Basin, near the Netherlands Antilles, becomes deoxygenated when cold bottom water is trapped in this valley in the continental shelf. It then becomes a death trap for some species, but codlets, *Bregmaceros* sp., are able to penetrate the anoxic layer and survive, presumably using anaerobic energy pathways (Richards 1975).

3.4.2 Salinity

The salinity of natural waters varies from near that of distilled water in mountain streams to salinities near saturation in evaporation basins (salt lakes). The salinity of the open oceans is near 35.5. ppm. Coastal waters usually have lower salinity. Fishes have evolved varying degrees of salinity tolerance, but only a minority are capable of moving freely between freshwater and saltwater. For other species, water of a salinity that differs substantially from their normal habitat can form a physiological barrier to migration or dispersal.

3.4.2.1 Oceanic Migration of Salmon

Sockeye salmon from Bristol Bay, Alaska, spend their years of growth in the ocean near the Aleutian Islands. Within this area their movements seem to be bounded on the south by the warmer, more saline water of the transitional and subarctic zones (Favorite 1969 b) (Fig. 3.1). The water in this part of the North Pacific is relatively low in salinity for ocean water, 32 to 34 ppm. To the north of the subarctic boundary water the young sockeye remain primarily in the Oyashio Extension and Subarctic Current (Fig. 3.2) until they start to mature (French and McAlister 1970).

With the beginning of sexual maturation the sockeye congregate in the Alaskan stream and ridge areas. These regions are characterized by lower temperatures and salinities. From this area near the northern edge of the feeding zone, their movement further northwest toward Bristol Bay is blocked in May and June by cold, high-salinity water in the Aleutian passes, which results from strong upwelling (Fujii 1975). In early June this barrier disappears as a result of spring tides and wave action. Water of higher temperature and lower salinity flows through the eastern passes. The sockeye migrate through and into the Bering Sea where they move within a low-salinity zone bounded by the 33.1 and 11.2 ppm haloclines. This mass of water from the Alaskan stream moves up into the Bering sea toward Bristol Bay (Fujii 1975; Leggett 1977). Near shore, plumes of low salinity from coastal rivers may play a role in attracting fish (Favorite 1969 a), but the natural odours of the rivers are probably also important.

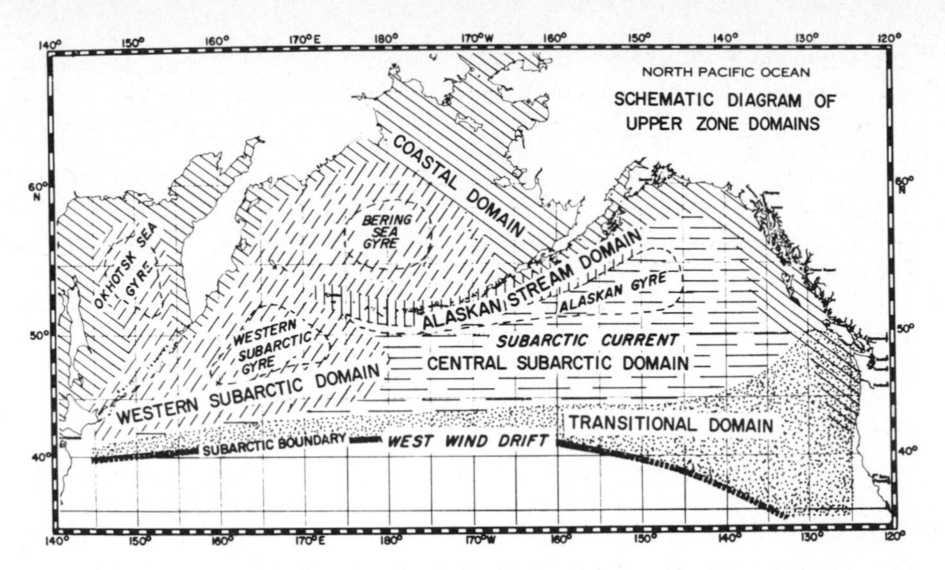

Fig. 3.1. Schematic diagram of upper zone domains in the North Pacific Ocean (Favorite 1969b)

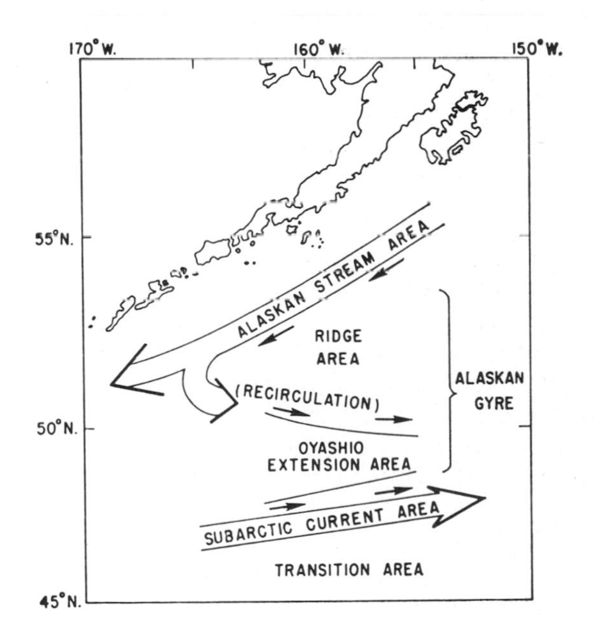

Fig. 3.2. Schematic representation of ocean currents and water masses in the northeastern Pacific Ocean, January to March, 1967 (French and McAlister 1970)

These studies indicate a correspondence between migration routes and salinity, but do not establish salinity as the crucial variable. Temperature or some other unmeasured variable may be more important. The fish may be orienting by celestial or magnetic stimuli in an appropriate compass direction that happens to coincide with oceanographic variables, or the ocean characteristics may be part of a hierarchical system of orientation mechanisms. Experimentation is difficult in such a system.

3.4.2.2 Coastal Migration

The salinity gradient is steepest near the coast and in estuaries, and hence most easily used there as an orientation cue. This has not been lost on researchers. There have been a number of studies of salinity preference of fish as they migrate through estuaries or coastal waters. There are reports of several species of fish responding to the salinity change near estuaries. Juvenile croakers, *Micropogon undulatus,* move up estuaries in the wedge of saline water near the bottom (Haven 1957). Upstream-migrant adults had showed extensive meandering when they reached the saltwater–freshwater interface (there was a 5°–10 °C temperature difference as well as the salinity difference) (Dodson et al. 1972). Most experimental work on this zone has concentrated on three groups, stickleback, Pacific salmon, and eels.

Stickleback. The salinity preference of adult three-spined stickleback (*Gasterosteus aculeatus*) was initially studied by Baggerman (1957) using a two-choice box. She found that salinity preference of sticklebacks changed from seawater to freshwater at about the time they would begin natural upstream migration and that increased thyroid activity was associated with this change in preference (see Chap. 2.3.2).

Pacific Salmon. Houston (1957) used essentially the same apparatus to test the responses of chum, pink, and coho fry to salt and fresh-water, and the response of coho smolts to hypertonic seawater. He found that all three species preferred the seawater solutions to the freshwater. Baggerman (1960a, b) examined chum, pink, coho, and sockeye with similar apparatus. She found that a period of freshwater preference preceded the migratory season in all four species and that coho and sockeye reverted to a freshwater preference if they were held beyond the migratory season in freshwater. Chum and pink fry tended to die if kept in freshwater. She suggested again that the change in salinity preference was associated with high thyroid activity although this time, in contrast to the sticklebacks, thyroid activity was associated with increasing preference for salt water.

McInerney (1964), again using the same type of two-choice apparatus, tested all five eastern Pacific salmon species and found a shift toward saltwater preference with the progress of the migratory season, followed in some species by reversion to freshwater preference if the fish were retained in freshwater. He argued that the gradient of salinities found in coastal areas could serve as a guiding mechanism in the seaward migration. The shift in salinity preference of the juvenile salmon would gradually lead them toward the open ocean.

A subsequent study (Otto and McInerney 1970), using a long trough with a continuous gradient, found bimodal preferences for coho fry when freshwater was included in the gradient and unimodal preferences if the gradient was started at 4 ppm. Since the preference for saltwater developed well before the beginning of migration and the preferred level never reached open-sea salinity levels, Otto and McInerney concluded that salinity preference was neither a releasing factor nor a directing factor in migration. They suggested instead that as the osmoregulatory capacity of the fish developed (Otto 1971) they became indifferent to salinity while other factors initiated and directed the migration. Presumably the sort

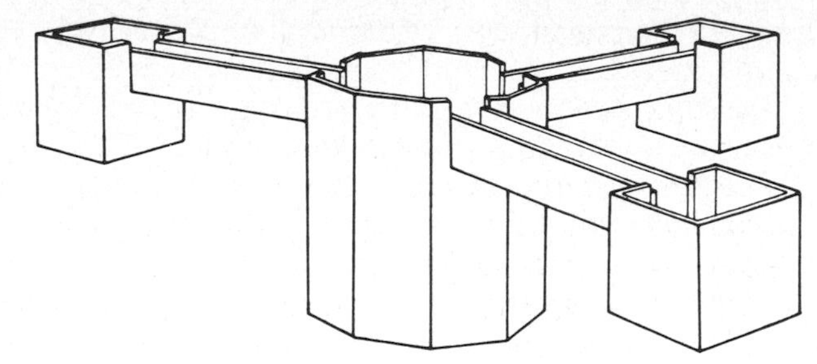

Fig. 3.3. Three-armed choice apparatus used by Hain (1975) to compare responses of eels to two inflowing streams of water and one outflow (Hain 1975)

of orientation reported by Healey (1967) for pink fry (see p. 57) would account for movement through the coastal region. The change in salinity preference might reduce the chance of the fish inadvertently returning to freshwater.

The salinity preference of returning adult salmon has apparently not been examined.

Eels. The upstream migration of elvers has also attracted attention. Creutzberg's (1961) work on the effect of natural freshwater odours on tidal timing has already been mentioned (see Sect. 3.3). Additionally, newly arrived European elvers are initially repelled by low-salinity water and gradually change preference as they wait to enter freshwater streams (Deelder 1958), eventually preferring freshwater to seawater of 18 ppm or 36 ppm when tested in a preference apparatus (Schulz 1975). American elvers also delay entering freshwater on their inshore migration (McCleave and Kleckner 1982). A tentative sequence of events would be: initial approach to the coast through use of odour components of natural freshwater to select onshore tidal currents, a delay close to shore while freshwater preference and perhaps physiological capabilities developed, then movement into freshwater in response to chemical and mechanical stimuli. Neither Deelder nor Schulz performed experiments that would clearly distinguish the effects of odours from the effects of salinity per se, nor did they test the sensory mechanisms in use.

The return migration of eels to the sea would require another change in salinity. Young European yellow eels preferred 18 ppm in a choice apparatus even though they would presumably not yet be migratory (Schulz 1975). Hain (1975), working with American eels, used a three-arm preference apparatus, which allowed upstream and downstream responses as well as a choice of chemical stimuli (Fig. 3.3). He tested 30–50 eels at a time, outside with a view of the sky. Maturing silver (bronze) eels preferred natural seawater over streamwater, yellow eels showed no preference. The seawater preference of silver eels was significantly reduced by cauterizing the olfactory epithelium. Blinded eels and those with just one cauterized nostril retained their strong seawater preference. The eels were not attracted when other eels were placed in one of the inlet traps, eliminating the possibility of an attraction pheromone but not of a migration-releasing (timing) pheromone.

3.4.2.3 Summary

Salinity tolerance and preference seem roughly matched to requirements, but there is little evidence that salinity is a guiding mechanism. The role of specific organic compounds deserves more study. Creutzberg (1959, 1961; see Sect. 3.3.3) found that elvers cued on charcoal-filtrable components of natural freshwater rather than salinity in their shoreward migration. Similarly, young *Mugil liza* are attracted to water from hypersaline bays near Curaçao in the Netherlands Antilles by chemical components in the water that can be removed by charcoal filteration, indicating that an organic compound is involved rather than salinity (Kristensen 1963).

3.4.3 Pollution

As well as the chemicals found naturally in water there are often additional chemicals added by human activities. There is an almost infinite variety of such chemicals and an extensive literature on their characteristics. Many are toxic to fish and obviously a lethal concentration of a chemical will block passage of migrant fish by killing them or deterring their migration. In some cases the fish may be able to avoid death by behavioural avoidance of a pollutant, but many pollutants are new, and fish may lack receptor mechanisms and appropriate behavioural responses. We find a rather variable series of responses to different pollutants, depending on whether or not the fish can detect the substance and whether or not they show an appropriate response to it (Jones 1964). I shall discuss two cases of pollution that have been studied in terms of their effects on migration: gas bubble disease and heavy metal pollution.

3.4.3.1 Gas Bubble Disease

Gas bubble disease deserves mention because of its importance to migrating salmon. Fish tissues become supersaturated with gases, usually nitrogen, and when the gas comes out of solution in the body it forms bubbles in the blood and other tissues which cause injuries to the fish (Wolke et al. 1975; Special tissue: Trans. Am. Fish. Soc. 109, 1980). Although supersaturation can occur in nature it is most widely reported in association with hydroelectric dams, where it can be a serious cause of fish mortality. Air entrapment in cascading waters is one source of nitrogen supersaturation. Air is trapped in the water in bubbles, then carried to a depth where it is dissolved in the water (Wolke et al. 1975). When that water rises near the surface it becomes supersaturated. Below the John Day dam on the Columbia River, for example, water has been recorded at 123–143% of saturation and values as high as 130% have been reported at Astoria, Oregon, over 160 km downstream from the Bonneville Dam (Beiningen and Ebel 1970). Supersaturation can also be induced by an increase in the temperature of saturated water.

The behavioural response of fish to elevated nitrogen levels was studied by Shelford and Allee (1913), but they found no avoidance responses. An unpublished report by Meldrim, Gift and Petrosky, cited by Wolke et al. (1975), indicates some avoidance of nitrogen-supersaturated water by golden shiners and satinfin shiners, but little responsiveness by perch. Fish may sometimes alleviate the effects of gas bubble disease by moving to deeper water, thus increasing the pressure and reducing the size of the bubbles (Ebel et al. 1971). Adult chinook salmon fitted with pressure-sensitive radio transmitters swam deeper when nitrogen-supersaturated water was present than when it was absent (Gray and Haynes 1977). They spent about 89% of their time below the critical supersaturation zone. Deflectors have been added to the spillways of some dams to reduce nitrogen entrainment (Collins 1976).

3.4.3.2 Heavy Metals

Heavy metals from mining and metal processing are often major pollutants of natural waters (Jones 1964). In the most extreme cases the fish populations of entire rivers may be killed. Non-lethal concentrations release behavioural avoidance in migratory fish. In the northwest Miramichi River in New Brunswick, Canada, for example, copper and zinc entering the river led to an increase in the numbers of adult Atlantic salmon that moved downriver without spawning (Saunders and Sprague 1967). When the heavy metal concentration reached 80% of the LD50 (dose causing 50% mortality), the upstream migration was completely blocked until concentrations dropped.

Rainbow trout and Atlantic salmon avoid $ZnSO_4$ at 0.005 mg l^{-1} of Zn, one-hundredth of the lethal threshold found in laboratory studies (Sprague 1968) but for Atlantic salmon, field thresholds for abnormal downstream movements are higher than the laboratory avoidance thresholds (Sprague et al. 1965). Fish in the river would face a conflict between the migratory stimuli and the avoidance response while the fish in the laboratory would have only the avoidance stimulus to which to respond.

3.4.3.3 Other Compounds

Many other chemical substances are added to the water along the migration routes of fishes, including chlorine used as an algicide in cooling water for power plants (Morgan and Carpenter 1978) and agricultural chemicals such as insecticides and herbicides. Laboratory tests of the effects of these chemicals may give inconsistent results: Atlantic salmon, for example, avoid chlorine (Sprague and Drury 1969), while blacknose dace are attracted to low levels of free chlorine (Fava and Tsai 1975, 1976). The relationship between such laboratory tests and field effects on migrant fish is open to question.

3.4.4 Distinctive Chemical Components of Natural Waters

Each water body has a unique mixture of chemical stimuli which could be used to identify a particular location. A fish returning to an area of which it has previous experience could use this to recognize the "home area." At a broader level, classes of water body may share chemical attributes which would allow fish to select an appropriate habitat even if they have never been in a region before. We have already seen examples of this type of response. European elvers respond to charcoal-filtrable components of natural freshwaters (Creutzberg 1963), and juvenile *Mugil liza* are attracted to charcoal-filtrable components of water from hypersaline bays (Kristensen 1963). Other examples of this response to a class of water body occur in the freshwater migration of elvers and the migration of sockeye fry to their nursery lakes.

3.4.4.1 Stream Selection by Elvers

American eel elvers migrate into freshwater for a period of growth. Elvers were captured by Miles (1968) in four Nova Scotia streams and tested in an apparatus which allowed them to choose between water from the stream of capture and other natural waters (Figs. 3.4, 3.5). The elvers were held in seawater prior to testing, but still preferred the natural fresh waters of the streams. There were also strong differences in the attractiveness of the four natural stream waters, and these differences remained the same in several tests, overriding the response to stream of capture in some cases.

Activated charcoal filtration of the freshwater samples reduced their attractiveness, whether the pH was adjusted to compensate for the effects of the charcoal filtration or not. Elvers also preferred charcoal-filtered Sackville water to which 10% unfiltered water had been added over seawater, by a ratio of 41:20, indicating that the results were probably not due to repellent materials from the charcoal.

Sackville water was the most preferred and also had the highest pH; the least-preferred water, Ketch Harbour, was the most acid. In four-choice tests with the pH of each water adjusted to match either Sackville or Ketch Harbour, Sackville

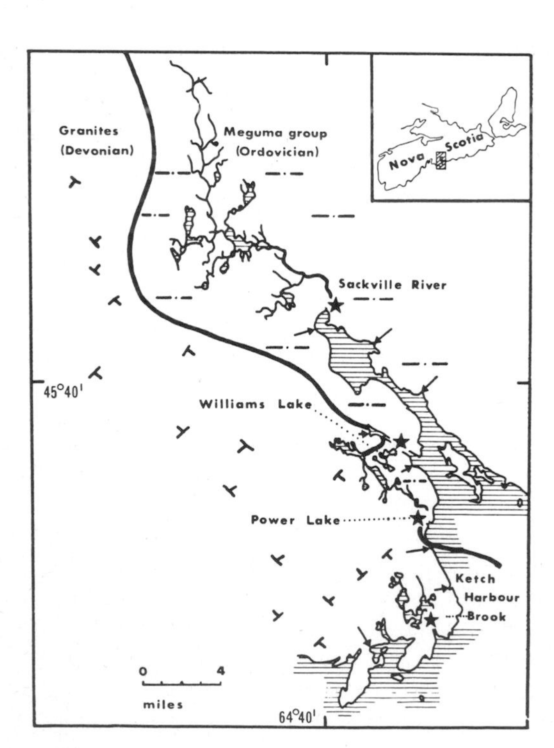

Fig. 3.4. Four streams from which elvers and water samples were collected. The collection points are shown by *stars* and points at which other streams enter the sea are shown by *arrows*. To the right of the wide line running roughly northwest to southwest the rocks are of the meguma group, to the left they are granites (Miles 1968)

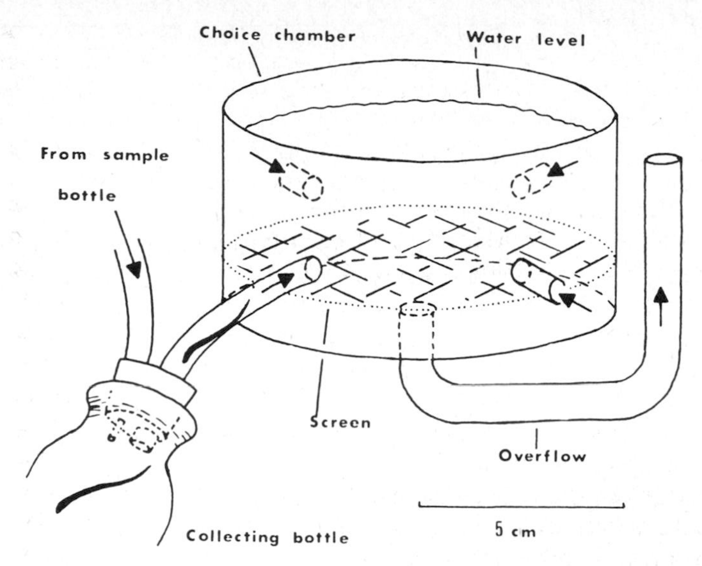

Fig. 3.5. Apparatus used to test the rheotaxis of elvers in response to water samples from the natural streams illustrated in Fig. 3.4 (Miles 1968)

water remained the most attractive, although pH obviously affected the response.

Heating the water to 120 °C did not reduce its attractiveness, but aging of unsterilized water led to a decline in attractiveness over 3 days. Sterilized water did not lose its attractiveness with age even after 2 weeks, indicating that the attractiveness of the water could be altered by bacterial action. Bacterial action did not remove all attractive components, however, because even after 7 weeks of storage the attractiveness of unsterilized water could be further reduced by charcoal filtration. Aeration for 24 h did not reduce the attractiveness, indicating that the attractive components were not volatile. Filtration with a 0.45 μ Millipore filter reduced attractiveness, but did not change the order of preference among the water bodies.

Adult eels in the water increased its attractiveness (64:48) but the order of preference for the water sources remained the same. Elvers in the test water reduced its attractiveness. Elvers migrate in large masses in nature, so this effect cannot be of overwhelming importance during most of the migration. It might play a role in their final dispersion in the new habitat. European elvers, in contrast, seem to be attracted by water which has contained elvers (Pesaro et al. 1981).

In summary, Miles (1968) found that elvers are attracted to natural freshwaters, by non-volatile, heat-stable components, which can be partially eliminated by charcoal filtration, 0.45 μ Millipore filtration or by bacterial action. The natural waters differed consistently in their attractiveness and elvers did not prefer the water from their stream of capture. Eels, but not elvers, in the water increased its attractiveness. Miles did not perform any experiments to determine which chemoreceptor system(s) govern the responses.

3.4.4.2 Lakeward Movement of Sockeye Fry

If lakewater differs chemically from streamwater in predictable ways, then this difference could be used by sockeye fry to solve the "inlet/outlet problem" which faces fish migrating from spawning streams into lakes (Brannon 1972). The geography of the lakes and the streams studied by Brannon was presented in Sect. 2.4.1. Sockeye fry initially move downstream to the nursery lake if they hatch in a tributary stream (Weaver and Seven Mile Creek), or upstream if they hatch in the lake outlet (Chilko, Stellako, and Adams). Lake spawning stocks, in contrast, only need to disperse in the nursery lake (Cultus Lake).

Brannon (1972) used an artificial "migration channel" to test the response of sockeye fry to flowing water of different natural origins (Fig. 3.6). One hundred fry were placed in the central screen chamber. After a 20-min acclimation period the flow in the apparatus was started and 10 min later the screens were removed. The numbers caught in the upstream and downstream traps were then recorded. The duration of the experiment varied with the age of the fry, since young fry took longer to negotiate the apparatus than older fry. When several different genetic stocks of fish were incubated in Cultus Lake water and then tested in Cultus Lake water, they tended to move upstream but when they were tested in water from nearby Hatchery Creek they tended to move to the downstream trap (Table 3.3). Chilko fry were also tested in the field with Chilko Lake water versus Madison Creek water (Table 3.4). Again there was a strong upstream response in lake (and incubation) water and downstream movement in stream water.

When fry of Chilko stock were incubated in Cultus Lake water and in Hatchery Creek water, the results indicated a positive rheotactic response in lakewater regardless of incubation water. But the incubation water also had an effect, tending to increase the upstream response in creek water for fish incubated in creekwater in this outlet spawning stock.

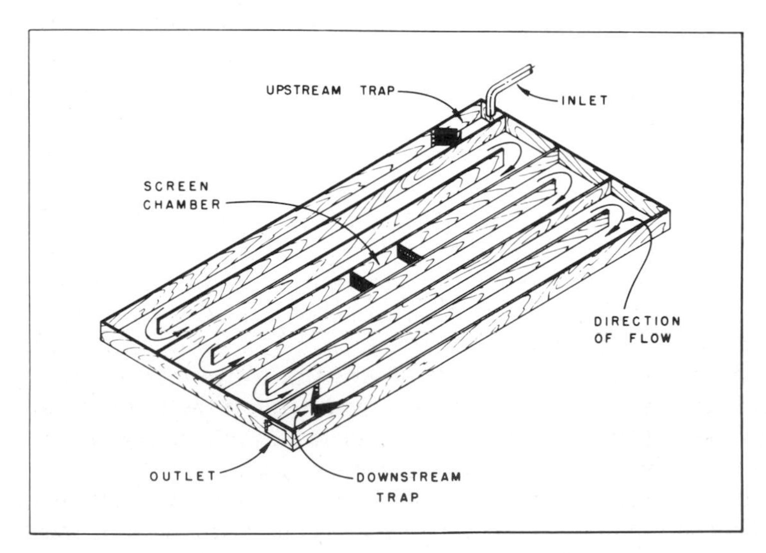

Fig. 3.6. Migration channel for testing response of sockeye fry to water current (Brannon 1972)

Table 3.3. Total number and percent response of five experimental stocks of sockeye fry responding upstream (U) and downstream (D) to Cultus Lake and Hatchery Creek water. These results could be due to differential response to lakewater versus streamwater or a response to familiar incubation water over unfamiliar water (Brannon 1972)

Stock	No. of tests	Water	Number of fry		Percent	
			U	D	U	D
Chilko	30	Lake	2,750	173	91.7	5.8 cm:
	20	Creek	277	1,723	13.9	86.1
Cultus	8	Lake	514	286	64.3	35.7
	8	Creek	322	479	40.3	59.7
Adams	7	Lake	296	404	42.3	57.7
	7	Creek	61	639	8.7	91.3
7-Mile	9	Lake	527	373	58.6	41.4
	9	Creek	241	659	26.8	73.2
Weaver	11	Lake	762	338	69.3	30.7
	11	Creek	415	685	37.7	62.3

Table 3.4. Total number and percent response of native Chilko fry responding upstream (U) and downstream (D) in Chilko Lake and Madison Creek water (Brannon 1972)

Water source	No. of tests	No. of fry		Percent	
		U	D	U	D
Lake	10	842	158	84.2	15.8
Creek	10	47	953	4.7	95.3

To offer the fry a choice of two water sources at once, Brannon (1972) used a Y-maze. Chilko fry, captured in the wild and offered a choice between Chilko Lake (incubation) water and Madison Creek water, preferred the lakewater. The choice was usually made "unhesitatingly" as the fry approached the area where the two water sources mixed. Even fry captured from groups holding in the mouth of Madison Creek preferred the lakewater.

Chilko stock were incubated in two different water sources, Cultus Lake and Hatchery Creek, then tested in the Y-maze at two developmental stages, alevin and fry. Some groups of fish were transferred from one medium to another halfway through yolk absorption, about 2 weeks before alevins were tested. Alevins preferred the water they were in prior to the time of transfer, although in the transferred groups the second water type attracted about a third of the alevins. By the fry stage, each of the groups preferred lake water. There was no significant difference between the four groups; these 90-day-old Chilko fry therefore preferred lakewater regardless of incubation and rearing water. For the group transfered from creekwater to lakewater and the group raised entirely in creekwater, this represented a change in preference after the alevin stage.

One group of alevins was tested 2 weeks after the main tests, about 7 days before emergence. They showed a 66.2% lake preference against a 15.2% creek preference, so the shift to lakewater preference had occurred prior to emergence.

Post-incubation experience did not alter the lakewater preference of Chilko stock in the fingerling stage. When migratory Chilko fry were captured in the wild then held for 90 days in Cultus Lake or Hatchery Creek water until they reached the fingerling stage, they still showed a strong lakewater preference, 86% for fish reared in lakewater and 96% for fish reared in creekwater.

Brannon (1972) summarized the situation: alevins, fry, and fingerlings all show preference responses. Alevins prior to emergence prefer incubation water and incorporate their early incubation experience into their later preference behaviour. Sometime prior to emergence, preference for lakewater develops, regardless of incubation water source, and persists even if the fish are reared in creekwater. Lakewater diluted with creekwater is still preferred even in a choice between full creekwater and a mixture of 95% creekwater with 5% lakewater.

A further study of this sockeye fry migration problem by Bodznick (1978 a) yielded similar results. He used a Y-maze and experimental techniques similar to those of Brannon (1972). Stocks from a population of sockeye that spawn in the Cedar River, the main inlet stream of Lake Washington, in Washington State, were used. When eggs from this stock were fertilized and reared in Lake Washington water, the resulting fry strongly preferred Lake Washington water in the Y-maze but when they were fertilized and raised in well water they showed no consistent preference for either type of water. These results support the hypothesis that lakewater is especially attractive to sockeye but that incubation in other water can override the effect.

To test the response of fry that had never experienced either of the test media, Cedar River eggs were reared in Cedar River Water and in dechlorinated Seattle

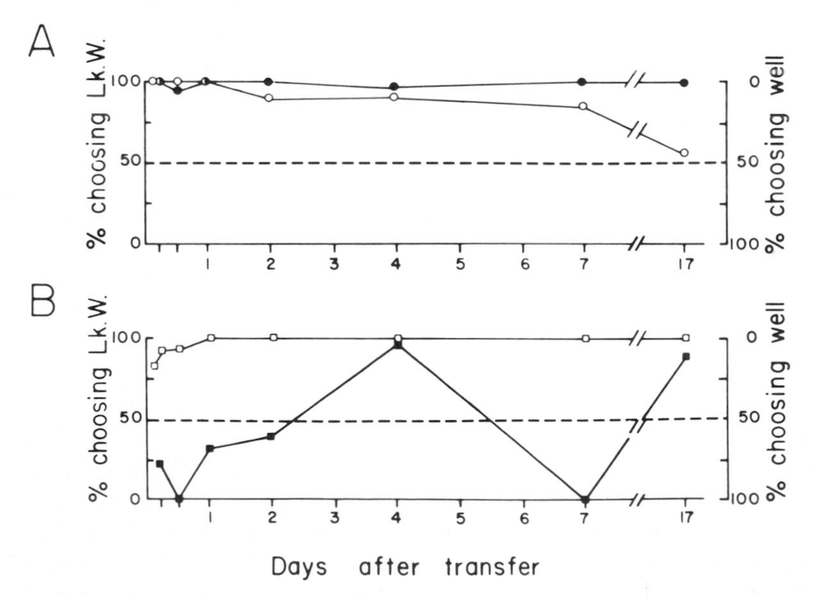

Fig. 3.7. The effect of recent experience on water source preference of sockeye fry. *A* Fish reared in lakewater and transferred to well water (*open circles*) and controls that were not transferred (*closed circles*). *B* Fish reared in well water and transferred to lakewater (*open squares*) and controls reared and retained in well water (*closed squares*) (Bodznick 1978 a)

city tap water. Then they were tested with a choice between well water and Lake Washington water. Chilko fry and Lake Washington fingerlings raised in city water were offered the same choice. In all but one case these fish preferred Lake Washington over well water although they had no previous experience with either. Bodznick (1978 a) was unable to explain the preference for well water shown by Lake Washington fry reared in tap water.

To test the effects of recent experience, fry reared in Lake Washington water were transferred to well water about 40 days after the normal migration time and well water-reared fry were transferred to Lake Washington water. The lakewater-reared fish gradually lost their lake preference over about 17 days. The well water-reared fish acquired a strong lake preference within 1 day, while the controls kept in well water continued to be inconsistent in their responses (Fig. 3.7). Again, the fishes' recent experience affects their response, but a strong tendency to prefer lakewater is evident.

3.4.4.3 Perception of Differences Between Lake Water and Creek Water

The receptors used in distinguishing the differences between lake and creek water were then examined. The olfactory epithelium of the embryo 3 weeks prior to hatching appeared as well developed, on the basis of light microscopy, as that of the adult. Alevins preserved 1 week before yolk absorption had a similar epithelium, though of greater surface area (Brannon 1972). When Chilko fry, with their nostrils plugged with petroleum jelly, were tested in the Y-maze, they showed no preference. Lakewater therefore seems to contain some odour component that serves as an attractant for sockeye fry. Often the anosmic fry chose one side of the maze and persisted in that choice even as the water types were switched from side to side, suggesting use of visual or other cues when the olfactory sense was occluded.

Olfactory occlusion with petroleum jelly also eliminated the lakewater preference of Bodznick's (1978 a) Lake Washington fry. Taken together, the studies by Brannon (1972) and Bodznick (1978 a) provide strong evidence for lake odours as a directing mechanism in the lakeward migration of sockeye fry. Positive rheotaxis in the presence of lake odour or negative rheotaxis in its absence will each carry the fry toward a lake in most sockeye spawning habitats.

The response to lake odours was also examined at the cellular level. Bodznick (1978 b) tested the EEG response of single olfactory bulb units to natural waters and amino acids. Fifteen- to 30-g sockeye reared in either Lake Washington water or in well water were used. Units that responded specifically to particular water sources were found (Fig. 3.8) and two cells clearly showed different electrophysiological responses to lakewater as a class compared to non-lakewaters and amino acids (Fig. 3.9). These lakewater units were capable of responding to dilutions as low as 1 part lakewater to 100 parts distilled water.

Although the fish used were reared in two different water types, this was not reflected in any unique set of cells responding to incubation water. The established ability to distinguish incubation water from other waters may be dependent on higher levels of the central nervous system.

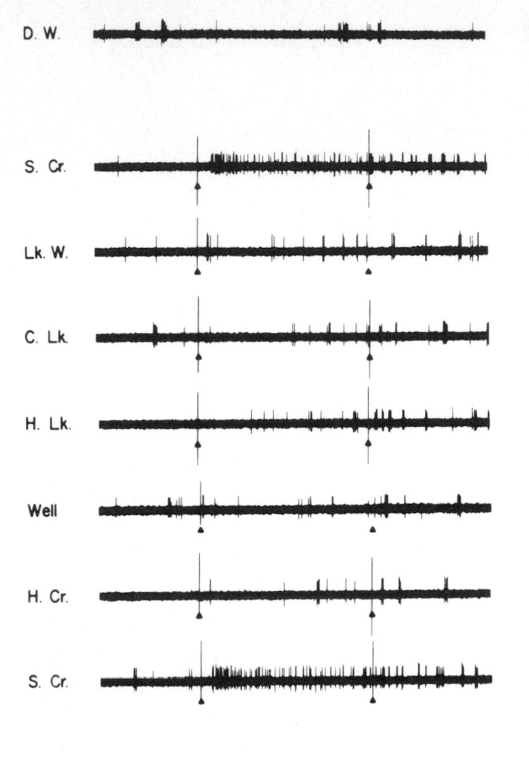

Fig. 3.8. Olfactory bulb unit demonstrating specificity for one natural water. In this case the unit responded to water from Sekwi Creek (*S. Cr.*) and not to other natural waters such as Lake Washington (*Lk. W.*), Cultus Lake (*C. Lk.*), Harrison Lake (*H. Lk.*), well water, and Hatchery Creek (*H. Cr.*). The fish had been reared in well water (Bodznick 1978 b)

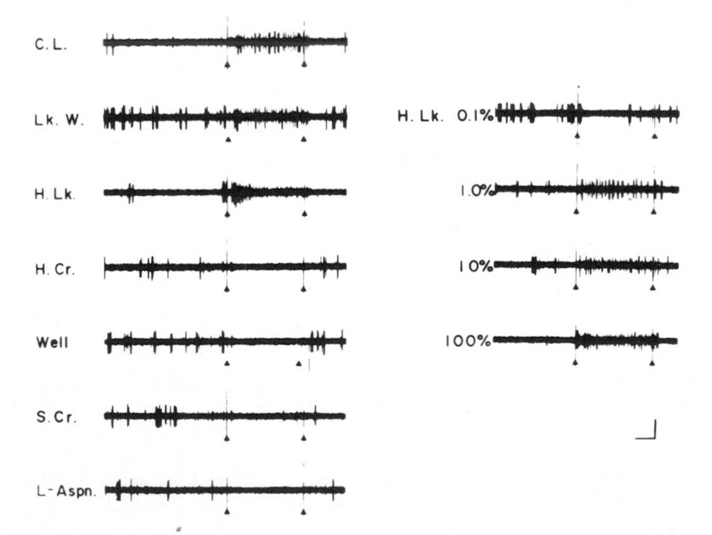

Fig. 3.9. Olfactory bulb unit demonstrating discrimination between lake and non-lake natural waters, recorded from the granule layer of the olfactory bulb of a fish reared in well water. The responses of the same cell to serial dilution of Harrison Lake water with distilled water are also shown. Abbreviations as in Fig. 3.8 (Bodznick 1978 b)

Table 3.5. Concentrations of CaCl found in some natural water bodies used in studies of sockeye lake-odour responses (Bodznick 1978 b)

Water body	Concentration
Lake Washington	$1.2\text{–}2.2 \times 10^{-4}$ M
Cultus Lake	$3.1\text{–}3.2 \times 10^{-4}$ M
Harrison Lake	$0.9\text{–}1.2 \times 10^{-4}$ M
Well	$3.6\text{–}6.0 \times 10^{-4}$ M
Hatchery Creek	$1.4\text{–}2.1 \times 10^{-4}$ M
Sekwi Creek	$1.1\text{–}1.3 \times 10^{-4}$ M

Natural waters would contain a complex variety of odorous compounds. Bodznick (1978 b) suggests that one such compound may be $CaCl_2$, which gave olfactory unit responses resembling the responses to natural waters at some concentrations and which varied in concentration from one natural water body to another (Table 3.5).

Gross potential (olfactory EEG) responses from the olfactory bulb of 9- to 20-month-old Lake Washington sockeye supported the hypothesis that $CaCl_2$ might be an effective odorant (Bodznick 1978 c). The EEG response to $CaCl_2$ at concentrations similar to those found in natural lakes qualitatively resembled the response to natural lakewater but did not resemble the responses to single amino acids. Of the four cations (Ca, Mg, Na, and K) and three anions (HCO_3, SO_4, and Cl) that make up the major part of the ionic composition of natural freshwaters, only calcium provided a strong, graded EEG response over the whole range of concentrations found in eight representative natural sockeye waters, including four lakes and four rivers. Magnesium gave a response at concentrations near the maximum found in the natural waters. As well as the bulbar EEG responses, single cells were found that responded to $CaCl_2$ at concentrations of 10^{-4} or lower. It is not clear that Ca^{2+}, which affects so many aspects of receptor physiology, is an appropriate stimulus for separating lakewater from streamwater, but it was nevertheless investigated by Bodznick (1978 c).

Perfusion of the olfactory epithelium with $CaCl_2$ solutions tended to cause adaptation of the calcium response, but did not decrease the response to amino acids. This effect was specific and reversible. This adaptation to calcium in perfusion water may be responsible for some of the differences in response to natural waters found by other researchers (Bodznick 1978 c). Some used distilled water as a rinse between stimuli (e.g., Oshima et al. 1969 a, b; Ueda et al. 1971; Bodznick 1975), while others used tap water, which may have significant levels of calcium (e.g., Ueda et al. 1967; Dizon et al. 1973; Cooper and Hasler 1976). This difference between the two rinses could have affected the results.

Tests of behavioural preference using Lake Washington sockeye fry (70 to 90 days after yolk absorption) indicated that calcium could influence the response of fry to natural waters. However, it was not the only factor as the fry could still distinguish between lakewater and well water when the two stimuli had equal $CaCl_2$ concentrations (Bodznick 1978 c). Organic compounds would seem more likely indicators of lakewater than inorganic ions.

3.4.4.4 Chemical Stimuli in Homing

Situations in which chemical components are used to distinguish between classes of water bodies, such as between lakes and streams or freshwater and seawater, have just been considered. Even within a class, natural bodies of water will differ in their chemical components, as seen in both the elver study of Miles (1968) and the work of Brannon (1972) on sockeye fry. Both those studies dealt with first-time movements to an area that the fish had never experienced before. However, the unique chemical signature of a region might be most valuable to fish returning to a known location.

The possibility of olfactory discrimination of natural waters was established by Hasler and Wisby (1951), who trained bluntnose minnows, *Pimephales notatus,* to distinguish between odours of natural streams using negative reinforcement (electric shock) or positive reinforcement (food). The role of such odour-discrimination ability in the biology of bluntnose minnows is unknown, but the minnows discriminated between natural waters quite readily and retained the response for up to 15 weeks. This is a reasonably long period for retention of a response of unknown function in the 2-year life cycle of the minnow. The active chemical components were volatile and heat-labile. Cautery of the olfactory epithelium eliminated the conditioned response, indicating that the minnows were responding to odour rather than taste or some other sense.

Studies on several species of fish have indicated that homing may be dependent on intact olfactory receptors. Homing has been defined by Gerking (1959) as return to a place formerly occupied instead of going to other equally probable places. This definition would include the return migration of fish like salmon and shad to their natal areas, but those species will be covered separately because of the extensive literature dealing with return migrations of these commercially important fishes.

The return to a home area, where it may have an established territory, local knowledge of feeding and hiding places or other benefits, such as a place in a dominance hierarchy, is important to a displaced fish. This is indicated by the frequency with which homing is found in fishes (e.g., Gerking 1953, 1959; Parker and Hasler 1959; Harden Jones 1968, Chap. 12).

Olfaction is often important in fish homing. Longear sunfish, *Lepomis megalotis,* for example, have a restricted home range, covering about 60 linear m of stream (Gerking 1953; Gunning 1959, 1963, 1965). When longears are displaced 60 to 80 m from their home range, they home even when blinded but interfering with olfaction by cauterizing the nares or plugging them with latex significantly reduced homing success (Gunning 1959, 1963), although 40%–50% still homed successfully. Similarly, the small, rocky shore sculpin, *Oligocottus maculosus,* which homes to its original tide pool when displaced 30 to 120 m, can home well when blinded (Khoo 1974). Cauterization of the nares, however, reduces homing ability. White suckers home as accurately when blinded as when unimpaired, but plugging the nostrils decreases homing accuracy. Werner (1979) displaced suckers from inlet stream spawning sites at either end of Wolf Lake, N.Y. Fish blinded by lens removal homed as accurately as controls, although the total number of recoveries of tagged fish was lower. However, of fish with the nostrils plugged with cotton plugs, only 52% of the recovered fish were in the home stream, compared with 92% of the controls.

The radiated shanny, *Ulvaria subbifurcata,* is a benthic stichaeid that is able to return to a specific (3 m^2) home site after displacement of 270 m and to orient toward the home site, even under ice at night (Green and Fisher 1977). When shannys were blinded (lens removal) or rendered anosmic (cautery of the nares) then released 19 m from their home site, they homed poorly compared with unimpaired controls (Goff and Green 1978). These results indicate that both vision and olfaction are important in shanny homing. The shannys also possess the interesting ability to orient toward home, even when blind. Anosmic fish also showed non-random orientation but in the opposite direction, 180° away from home (see p. 190). Thus cues other than vision and olfaction may be involved in shanny homing.

3.4.4.5 Home Stream Odour

Salmonids and clupeids which return to their rearing site or "home stream" make use of the chemical differences between natural water bodies (Hasler and Scholz 1978, 1983; Hasler et al. 1978; Cooper and Hirsch 1982). The fidelity of salmon returning to their home-stream has been known for years and has been repeatedly confirmed by marking experiments. The suggestion that responses to home-stream odour might be involved in this return has a long history. Buckland (1880) is often listed as the first to have set out the home-stream odour hypothesis, although Kleerekoper (1969) suggests that Treviranus (1822) has precedence. Buckland's explanation of home-stream recognition remains a delightful presentation of the concept: "When the salmon is coming in from the sea he smells about till he scents the water of his own river. This guides him in the right direction, and he has only to follow up the scent, in other words to 'follow his nose,' to get up into freshwater, i.e., if he is in a travelling humour. Thus a salmon coming up from the sea into the Bristol Channel would get a smell of water meeting him. 'I am a Wye salmon,' he would say to himself. 'This is not the Wye water: it's the wrong tap, its the Usk. I must go a few miles further on,' and he gets up steam again" (Buckland 1880, p. 320; from Harden Jones 1968).

As Harden Jones (1968) points out, the elements of the modern home-stream odour hypothesis are presented in Buckland's description. The hypothesis deals with migration near shore and in the rivers, "when the salmon is coming in from the sea." It is usually felt that other mechanisms bring the fish from distant feeding grounds to the coastal region near its home stream. The fish must also know its home stream from among all the other salmon waters which may be nearby "I am a Wye salmon" and it must be able to select the appropriate stream from among several.

These elements have been more formally stated by Hasler and his co-workers in their many studies of the home-stream odour hypothesis. "(1) because of local differences in soil and vegetation of the drainage basin, each stream has a unique chemical composition and, thus, a distinctive odor; (2) before juvenile salmon migrate to the sea they become imprinted to the distinctive odor of their home stream; and (3) adult salmon use this information as a cue for homing when they migrate through the home-stream network to the home tributary." (Hasler et al. 1978). Extensive studies have been performed or inspired by Hasler and his colleagues. These studies are reviewed in detail by Hasler and Scholz (1983).

The term "imprinting" is used frequently in discussions of the home-stream odour hypothesis. This term is used by ethologists to describe situations in which an animal learns a set of stimuli during a short "critical period," often early in development, and then retains the ability to respond to these stimuli until a later stage of the life cycle, even if a period intervenes during which response to the stimulus is not reinforced. Mature male mallard ducks, for example, respond sexually to objects that resemble the object they followed as a duckling even if the object is a person or other biologically inappropriate stimulus. If the home-stream odour response of some salmonids, such as coho salmon, shares these attributes of a short critical period and long retention of the response, then the term imprinting will be appropriate.

3.4.4.5.1 Coho Salmon and the Home-Stream Odour Hypothesis

Duration of the Critical Period. Coho emerge from the gravel of their freshwater incubation site in the spring and spend a year feeding in streams prior to smolting and seaward migration. Only 1 to 3% of the smolts return as adults, but those that do return with considerable accuracy to their home stream. Transfer experiments have shown that the "home stream" is the stream in which they underwent the smolt transformation. When fry were transferred from their natal rivers to new rivers they returned as adults to the foster river (e.g., Rounsefell and Kelez 1938; Donaldson and Allen 1957; Vreeland et al. 1975).

The amount of time required for the acquisition of this home-stream response is apparently quite short, if the fish are at the correct developmental stage. Coho fingerlings were raised to smolt size at the Leavenworth Hatchery on the Wenatchee River, then transported to a fish-handling facility downstream from the hatchery (Jensen and Duncan 1971). There they were held in spring water for 36 to 48 h (in March, April, and May of 1967) while being heat-branded for identification. They were then released about 1 km upstream of the holding facility to test the effect of the turbines at Ice Harbor Dam on downstream migrants. In September, 1967, jacks (precocial males) were seen congregating near the handling facility outlet, something that had not been observed before. A floating trap was brought to the location, the outflow from the facility was pumped through the trap, and a number of marked fish from the previous release were recaptured. To test the possibility that the fish were attracted to the flow alone, river water was pumped through the trap and compared with a similar volume of holding facility water. No fish were caught when the river water was flowing through the trap and 399 were captured when the holding facility water, from the spring, was flowing through the trap. None of the marked fish returned to the hatchery. In 1968 more fingerlings were released in the same way and again the jacks returned in the fall of the same year, along with the adults of the 1967 release. None of the marked fish returned to the Leavenworth Hatchery, although fish from other releases did return there. The apparent short learning period supports the use of the term "imprinting" in this situation, as does the report by Mighell (1975) cited by

Hasler et al. (1978) that coho returned to a site after 4 h of experience at the location. The response to a small volume of water (less than $800 \, l \, min^{-1}$) entering a major river, indicated a very low threshold of response to the active components.

The sensitive period for learning home-stream odour is controlled by endogenous hormone levels, probably thyroid hormones (Scholz 1980; Hasler and

Table 3.6. Responses of juvenile coho tested in a natural stream after various imprinting and hormonal treatments. The stream had a fork allowing morpholine to be metered into one side and PEA into the other side. The behavioural responses of the fish were divided into three categories: Down, downstream movement; None, keeping position near the release site and Up, upstream movement. The fish and been subjected to one of six endocrine treatments at the time exposure to the artifical odours: Smolt, natural onset of smoltification; TSH+ACTH, injection of TSH and ACTH; TSH, injection of TSH only; ACTH, injection of ACTH only; Saline, control injection of saline and Uninjected. In each treatment group some fish were exposed to morpholine and others to PEA. At the time of testing some fish were treated with gonadotropins, G+, while others were given control saline injections, Saline+. Upstream migration occurred primarily in fish that had elevated TSH levels, natural or induced, during exposure to the odour and that were treated with gonadotropins during testing (Scholz 1980)

	Treatment		n*	Odour absent (n × 1 trial)			Odour present (n × 2 trials)			Odour discrimination	
				Down	None	Up	Down	None	Up	Mor	PEA
G+	Smolt	Mor	10	9	1		1	1	18	18	
		PEA	8	8			1		15	1	14
	TSH+ACTH	Mor	13	11	1	1	2		24	21	3
		PEA	12	11		1	2		24	2	20
	TSH	Mor	8	8			1		15	14	1
		PEA	8	6	1	1	1	1	14	2	13
	ACTH	Mor	8	7	1		15		1		
		PEA	4	4			7		1		
	Saline	Mor	13	12		1	27	1	1		
		PEA	11	9	1	1	20		2		
	Uninjected	Mor	6	6		1	11		1		
		PEA	4	3			8				
Saline+	Smolt	Mor	6	1	4	1		10	2		
		PEA	4		3	1		7	1		
	TSH+ACTH	Mor	6		5	1	1	9	2		
		PEA	4		4		1	6	1		
	TSH	Mor	6		5	1	1	9	2		
		PEA	3	1	1	1		3			
	ACTH	Mor	4		4			7	1		
		PEA	0								
	Saline	Mor	6	1	5		2	8	2		
		PEA	4		2	2	2	5	1		
	Uninjected	Mor	4		3	1	1	6	1		
		PEA	0								

Scholz 1983). Juvenile coho which were injected with thyroid-stimulating hormone (TSH) and exposed to an artificial odour, either morpholine or phenylethyl alcohol (PEA), later moved upstream in a natural river system in response to the imprinted odour (Table 3.6). Fish exposed to the same odours but without TSH injections moved downstream. When the two artificial odours were metered into different branches of the stream, the imprinted fish preferentially entered the branch containing their odour stimulus. The fish were tested prior to natural sexual maturity and the upstream response occurred primarily in fish which had been injected with gonadotropin, artificially inducing an endocrine state resembling natural maturity (Scholz 1980; Hasler and Scholz 1983).

Coho held in estuarine or saltwater during the learning period return to the general area in which they were held and enter streams draining into the area (Mahnken and Joyner 1973; Heard and Crane 1976) or even salt lagoons near the imprinting site (Novotny 1980). For example, in 1973 four groups of 5000 yearling coho from Issaquah hatchery were branded, then each was subjected to one of four treatments: (1) trucked from Issaquah to Beaver Creek, Washington and released immediately in the pond closest to tidewater (2) reared in net-pens in saltwater in Clam Bay for 3 weeks then released in Beaver Creek where it enters Clam Bay (3) reared in Clam Bay for 3 weeks then released directly into the bay (4) released directly into Clam Bay upon arrival from Issaquah (Novotny 1980). None of these fish returned to Issaquah. At Beaver Creek a salmon trap recovered 2.2% of the fish released in treatment (1), 1.6% from treatment (2), 1.4% from treatment (3), and 0.2% from treatment (4).

If the fish are held in estuarine conditions until after smoltification and after the normal time of migration, then they tend to stay in the waters near the holding site instead of migrating to their normal oceanic feeding grounds. This "delayed release" technique is being used in combination with saltwater imprinting to enhance sports fisheries and may be useful in salmon ranching (Mahnken and Joyner 1973; Novotny 1980).

The Role of Olfaction. Evidence that return to the home stream was based on olfaction came initially from a study by Wisby and Hasler (1954) performed at the junction of Issaquah Creek and East Fork. Adult coho that had homed to each branch of the system were transported back downstream and released. Some had their nostrils plugged with cotton, (sometimes mixed with vaseline or benzocaine), others were left as controls. None of the controls from Issaquah Creek was caught in other streams while 12 (23%) of the 39 plugged Issaquah fish that were recovered entered the smaller East Fork and 27 (77%) returned to Issaquah. Nineteen (71%) of the recaptured East Fork controls were in the East Fork while 16 (84%) of the 19 plugged East Fork fish were recaptured in Issaquah Creek. Assuming that the larger Issaquah Creek was more easily entered, these results suggested the use of olfaction in the return migration.

Imprinting on Artificial Chemical Stimuli. Another approach to the home-stream odour problem has been to "imprint" coho on artificial stream odours. If the fish learn distinctive local odours during their sensitive period, then perhaps one could add an odorous compound to their water at smolting time, then decoy them by adding the compound to a stream when they were returning to the area as adults.

This would provide strong evidence for olfactory home-stream recognition based on learned information. The compound chosen for the majority of such studies has been morpholine, a heterocyclic amine. Studies reported by Hasler (1966) and Hasler et al. (1978) showed that unconditioned coho respond to concentrations as low as 1×10^{-6} mg l^{-1} and that morpholine is neither a strong attractant nor a strong repellent for coho. Subsequently, the use of morpholine as a neutral odorous compound has been questioned by Hara and MacDonald (1975) and Hara and Brown (1979), who suggest that morpholine is a non-specific irritant of the olfactory epithelium and that behavioural responses to morpholine involve non-olfactory chemoreceptor systems such as taste or general chemical sense. On the other hand, Hirsch (1977) successfully conditioned the heart rate of coho to change in response to the association of electric shock with morpholine or phenylethyl alcohol (PEA), another compound used in imprinting studies. Fish with plugged nostrils did not become conditioned to these compounds while fish with open nostrils did, indicating an olfactory response.

The primary study area for artificial imprinting studies has been the west shore of Lake Michigan, using stocks of Pacific salmon that had been introduced to the Great Lakes (Fig. 3.10) (Hasler et al. 1978; Hasler and Scholz 1978, 1983).

In 1971, 16,000 coho fingerlings, all of the same genetic stock and rearing history, were held in water pumped from Lake Michigan. Morpholine at 1×10^{-4} or 1×50^{-5} mg l^{-1} was metered into a tank containing half of the fish, the other

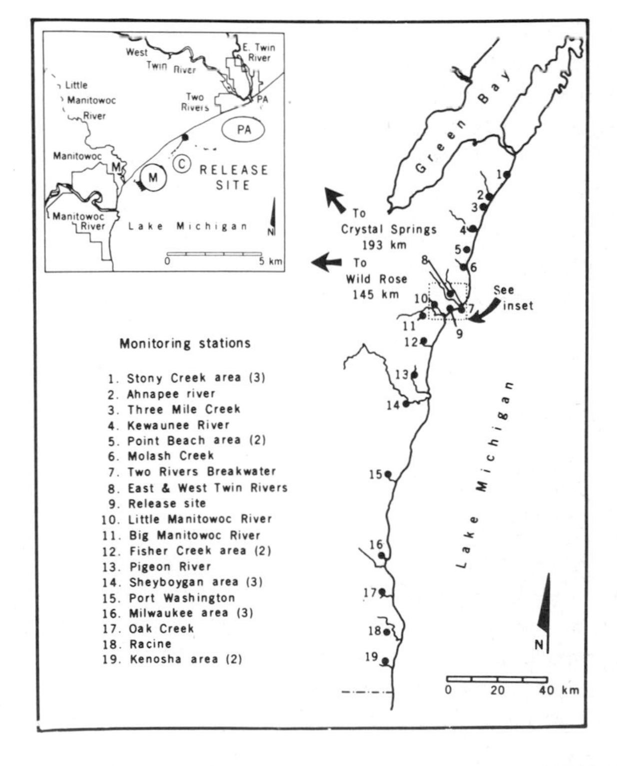

Fig. 3.10. Research area used for studies of imprinting of salmon on artificial home stream odours. *Numbers in parentheses* the number of streams that were surveyed in the general area of the monitoring station. *Insert* release site used in a study comparing the attractiveness of the morpholine-scented (*M*) Little Manitowoc River and the phenethyl alcohol-scented (*PA*) breakwater area at Two Rivers (Scholz et al. 1976). Copyright 1976 by the AAAS

Table 3.7. Record of salmon caught at Oak Creek in 1971–74. Experiment 1; imprinted and control fish released at Oak Creek, morpholine added to creek in 1971 and 1972, experiment 2; fish released 13 km north of Oak Creek, experiment 3; no morpholine added to Oak Creek in 1974. Probabilities based on Chi square (Cooper et al. 1976)

Experiment	Treatment	Morpholine concn.	No. released	No. jacks recovered and %	No. adults recovered and %	Total no. recovered and %	P
				1971	1972		
1	Imprinted	1×10^{-4}	4,000	4 (0.10)	103 (2.58)	107 (2.68)	
	Imprinted	1×10^{-5}	4,000	27 (0.68)	82 (2.02)	109 (2.70)	<.001
	Control	None	8,000	3 (0.04)	25 (0.31)	28 (0.35)	
				1972	1973		
2a	Imprinted	5×10^{-5}	5,000	4 (0.08)	433 (8.66)	437 (8.74)	<.001
	Control	None	5,000	3 (0.06)	46 (0.89)	49 (0.95)	
2b	Imprinted	5×10^{-5}	8,200	20 (0.24)	627 (7.65)	647 (7.89)	<.001
	Control	None	10,000	1 (0.01)	64 (0.64)	65 (0.65)	
2c	Imprinted	5×10^{-5}	5,000	16 (0.32)	423 (8.46)	439 (8.78)	<.001
	Control	None	5,000	2 (0.04)	53 (1.06)	55 (1.10)	
					1974		
3a	Imprinted	5×10^{-5}	5,000		51 (1.02)	51 (1.02)	<.10
	Control	None	5,000		55 (1.10)	55 (1.10)	

half were kept as controls. The fish were then fin-clipped to identify the two groups and released directly into Lake Michigan, 0.5 km south of the mouth of Oak Creek (creek 17 in Fig. 3.10). In the fall of 1971 and 1972 when the fish would be returning, as jacks or adults respectively, morpholine was metered into Oak Creek at 5×10^{-5} mg l^{-1}. The actual concentration in the stream would vary inversely with the flow rate. Stream census showed 2.7% return by morpholine-imprinted fish compared to 0.35% by controls, setting the stage for a series of similar experiments (Table 3.7) (Scholz et al. 1975; Cooper et al. 1976).

Repetitions of the experiment yielded similar results, even when the fish were released 13 km north of Oak Creek rather than at the creek mouth. When no morpholine was metered into Oak Creek in 1974, as a control, the number of morpholine-treated fish captured was the same as the number of control fish. The previous year, when morpholine was in the stream, the jacks of this same group had responded (Table 3.7). The full range of effective morpholine concentrations was not determined but levels as low as 1×10^{-5} and as high as 1×10^{-2} seemed equally effective (Cooper et al. 1976). Exposure time to the morpholine was usually from 3 weeks before until 2 weeks after the first signs of smolting, but exposure times of as little as 2 days were still effective.

The returning fish, whether morpholine-imprinted or control, and regardless of the hatchery where the imprinting was carried out or their release location, tended to return within a predictable period (Cooper et al. 1976). This indicates that the various treatments did not have much effect on the time it took for the fish to return to Oak Creek from their feeding areas in the lake.

Table 3.8. The total number of morpholine-exposed, M, phenylethyl alcohol-exposed, PEA, and control, C, coho salmon recovered at the monitoring stations shown in Fig. 3.10. Figures in parentheses are the number of streams surveyed in the general area of the monitoring station. Streams 7 and 8 were scented with PEA and stream 10 was scented with morpholine (Scholz et al. 1976)

Location	1974						1975					
	Number recovered			Number of trips			Number recovered			Number of trips		
	M	PA	C	Creel census	Gill net	Electro-fishing	M	PA	C	Creel census	Gill net	Electro-fishing
1 Stony Creek (3)	1		4	90	13	13			12	40		
2 Annapee River		2	7	138	3	5	6	1	37	224	4	14
3 Three Mile Creek	2	1	1	27	5	5			2	26		
4 Kewaunee River				71		5				9		
5 Nuclear Power Plants (2)	1		4	123					2	3		
6 Molash Creek			2	8						1		
7 Two Rivers Breakwater	3	118	15	184	3	1	3	192	12	126	14	1
8 East & West Twin Rivers		15	7	123		9	3	8	21	17		14
9 Stocking Site	1		7	90	1				1	30		
10 Little Manitowoc River	207	6	24	189		8	452	14	52	135		
11 Big Manitowoc River	2	3	31	44		5		1	26	7		
12 Fisher Creek (2)			3	44					1	2		
13 Pigeon River				23								
14 Sheboygan River (3)	1		3	75		1			3	10		
15 Port Washington				38								
16 Milwaukee area (3)				65								
17 Oak Creek		1	7	306		5				180		
18 Racine			1	11								
19 Kenosha (2)				14								

In a more elaborate variation of this type of experiment, coho smolts held in artesian well water were exposed to 5×120^{-5} mg l^{-1} morpholine, 5×10^{-3} mg l^{-1} PEA or control conditions (Scholz et al. 1976). In 1973, 5000 fish received each treatment and in 1974, 10,000 fish. During the fall of 1974 and 1975 morpholine was metered into Little Manitowoc Creek (number 10 in Fig. 3.10) and PEA was monitored into the water at Two Rivers Breakwater (number 7 in Fig. 3.10). The two treated areas and 17 other locations were then sampled for marked fish. The results strongly supported the hypothesis that coho could be imprinted on artificial odours such as morpholine and PEA, since morpholine-imprinted fish returned primarily to Little Manitowoc River and PEA-imprinted fish returned primarily to the Two Rivers Breakwater area (Table 3.8). The PEA fish tended to concentrate near the breakwater, where the PEA source was located, which was reflected in good sports fishing in that area.

Tracking Migrants Through Morpholine Plumes. One of the most convincing demonstrations of the behavioural response of the returning coho to morpholine was provided by tracking studies (Madison et al. 1973; Scholz et al. 1973; Hasler and Scholz 1978). Migrants were captured as they returned to Oak Creek, in 1971 to 1973, fitted with ultrasonic transmitters then transported 32 km north along the Lake Michigan coast, released and tracked with hydrophones. As they approached a small stream, the "decoy area," north of Oak Creek (Fig. 3.11; Scholz et al. 1973) morpholine was dispersed into the water across their path.

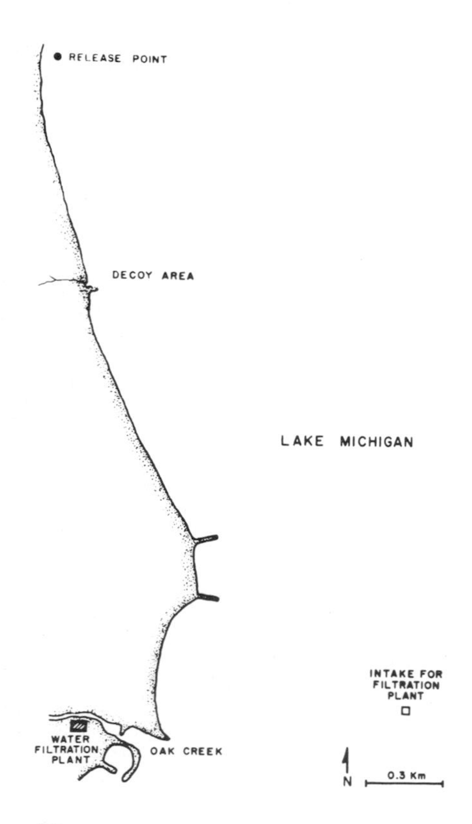

Fig. 3.11. Research area: South Milwaukee, Wisconsin (Scholz et al. 1973)

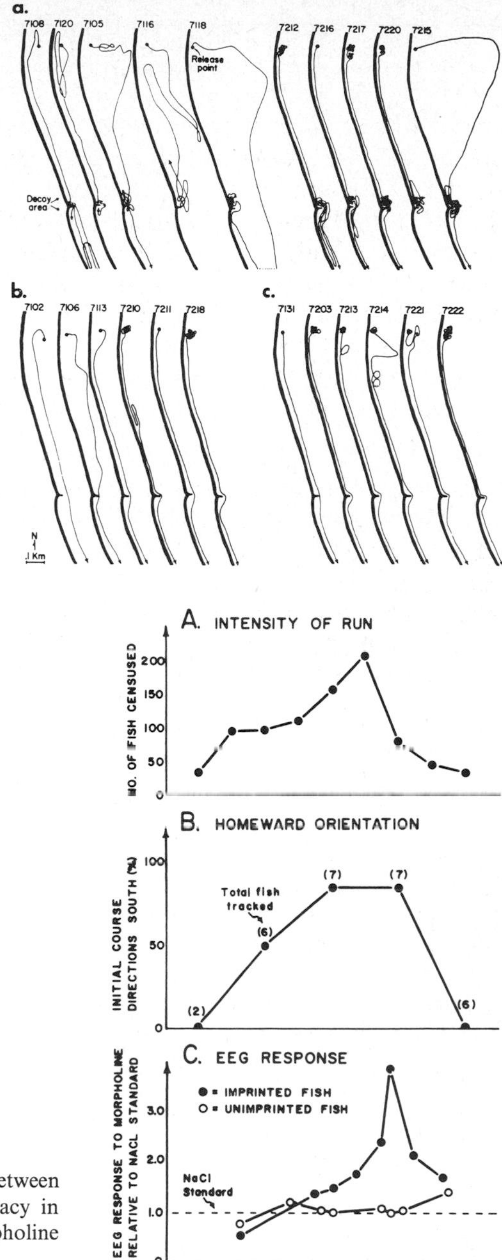

Fig. 3.12 a–c. Tracks of coho salmon captured at Oak Creek, displaced to the North and tracked as they moved south past the decoy area. *a* imprinted salmon when morpholine was present in the decoy area; *b* imprinted salmon when morpholine was absent from the decoy area, non-imprinted salmon when morpholine was present in the decoy area (Scholz et al. 1973)

Fig. 3.13 A–C. Temporal relationships between number of fish returning, orientation accuracy in tracking studies and EEG responses to morpholine (Scholz et al. 1973)

In the 20 cases where morpholine was dispersed in front of morpholine-imprinted fish, they stopped migrating and swam in a convoluted pattern in the morpholine cloud until the morpholine dispersed (Fig. 3.13). When the same experiment was performed on unimprinted fish they swam through the morpholine-scented decoy area without delay. Similarly, morpholine-imprinted fish swam

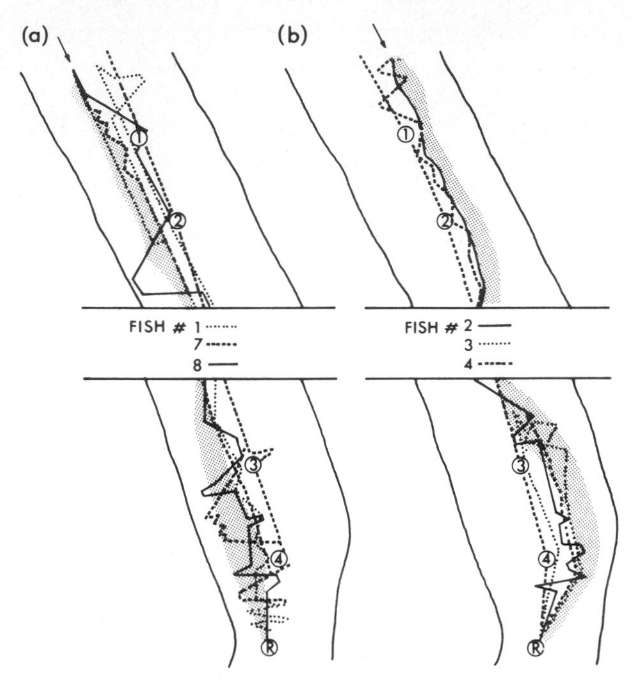

Fig. 3.14 a, b. Paths of morpholine-imprinted fish. *a* When morpholine was introduced in the left half of the stream, fish confined their movements to that side. Similarly, when morpholine was introduced in the right half of the streams as in *b* fish confined their movements to that side. With permission from Johnsen PB, Hasler AD (1980) The use of chemical cues in the upstream migration of coho salmon *Oncorhynchus kisutch* Walbaum. J Fish Biol 17:67–73. Copyright: Academic Press Inc. (London) Ltd.

through directly when no morpholine was added to the water, controlling for any effect of the small creek. When PEA or N-B-hydroxyethyl morpholine were dispersed in the water in the same way as morpholine, the morpholine-imprinted fish also migrated directly through the area. The results of these tracking experiments indicate a strong behavioural response to the imprinted chemical, a response that suggests a searching for the odour source. All of the fish, whether imprinted or not, showed remarkable orientation toward their capture stream, Oak Creek. The accuracy of this orientation tended to peak at the same time as the peak in number of returns (Fig. 3.14).

When morpholine was introduced to one side of West Twin River, "imprinted" fish tended to confine themselves to that half of the stream on their upstream migration (Fig. 3.14) (Johnsen and Hasler 1980). Of 19 fish that had been imprinted on PEA or hatchery water alone, 15 moved downstream when released in West Twin River during the morpholine release, presumably because their imprinted stimulus was absent. Of the 18 morpholine-imprinted fish tested, 13 moved upstream, 2 held position, and 3 moved downstream. Several of the fish moved along the edge of the odour plume with occasional forays into the unscented water. This behaviour may reduce sensory adaptation to the imprinted odour.

Hara and MacDonald (1975) and Hara and Brown (1979) argue that there is no direct evidence that these tracked fish were responding to olfactory information. They suggest that the response may be gustatory or mediated through the general chemical sense and suggest that fish with plugged nostrils should be tracked through a morpholine plume. The suitability of morpholine as an olfactory stimulus has been vigorously discussed (Cooper 1982; Hara 1982).

Separation of Open Water and Stream Orientation. The artificial imprinting studies on coho salmon in Lake Michigan support the hypothesis that coho can imprint to artificial odours, specifically morpholine and phenylethyl alcohol (PEA) and home to streams where these odours are metered into the water. The fish probably reach the area near the scented stream by using an "open water" orientation technique that is independent of the imprinted odour (see Sect. 3.4.4.5.8). Fish released at the Milwaukee Breakwater 13 km north of Oak Creek were captured in Oak Creek when it was scented with morpholine. In 1974, however, when Oak Creek was not scented and Little Manitowoc River about 110 km to the north was scented with morpholine, there were no reports of any of the 5000 morpholine-imprinted fish released at Oak Creek in 1973 being caught at Little Manitowoc River. This suggests that the released fish either return to a point in the general area of release or that they do not move very far from the release point. In apparent contradiction to these findings, seven control fish from the 1973 Manitowoc-Two Rivers release were recovered in Oak Creek and four in Stoney Creek, 58 km to the north, in total 116 of the 5000 control fish were recovered (Scholz et al. 1976). The control fish scattered widely.

Scholz (1980, cited p. 357) by Cooper and Hirsch 1982) recaptured morpholine-imprinted fish and control fish that had been stocked at Algoma (location 2, Fig. 3.10) and had subsequently returned there. He transported them 60 km south to Manitowoc. There he fitted the fish with ultrasonic tags to allow tracking. The fish were released south of the morpholine-scented Little Manitowoc River and tracked by boats with directional hydrophones. In this situation, if the fish moved north toward Algoma they would pass through the scent plume of the Little Manitowoc River.

When the eight morpholine-imprinted fish were released they moved northward to the Little Manitowoc River mouth where they spent several hours, then continued moving north. The six control fish moved through the morpholine plume without stopping.

The imprinted fish responded behaviourally to the morpholine, but then continued on toward their home location. This suggests an open-water orientation system that is "dominant" over the stream-selection mechanism (Cooper and Hirsch 1982) at least at this stage of migration. Stream survey results suggest that the open-water mechanism brings the fish back to within about 20 km of their home stream (Scholz 1980).

Electroencephalograph Studies of Migrating Coho. Electroencephalograph recording from the olfactory bulb has been used as a technique for investigation of the role of olfaction in adult coho migration. The initial use of the technique depended on the assumption that the magnitude of the electrical activity of the olfactory bulb reflected the importance of an odour stimulus to the animal and that, in adults on their spawning migration, the strongest EEG response would indicate the imprinted odour. Early results appeared promising. Hara et al. (1965) and Ueda et al. (1967) found that coho and chinook salmon captured on the spawning ground showed a larger EEG response to water from the capture stream (presumably the home stream) than to other natural waters. There was also an indication that water from areas along the freshwater migration route produced lesser responses than the home stream, but greater response than waters not on the migration route.

However, Oshima et al. (1969 a) using coho and chinook homing to streams in the Columbia River, found that some non-home natural waters gave bulbar EEG responses equal to or greater than the response to home-stream water. For example, fish homing to Skykomish Creek or to the University of Washington Hatchery responded as strongly to Issaquah Creek water as they did to home-stream water. The effects of Ca^{2+} in the tap water rinse used in this study may have influenced the results (Bodznick 1978 c).

Oshima et al. (1969 a, b) used fish that they considered to be in better physical condition, on the basis of appearance, than those used in the earlier studies, and argued that their results were, therefore, more likely to be accurate. This view may be incorrect. Migration is a seasonal phenomenon and orientational accuracy and EEG response to imprinted odours peak near the time of migration (Fig. 3.13). The response may be restricted to periods of high circulating gonadotropins or gonadal hormones (Table 3.6). In Pacific salmon arrival at the home stream often coincides with deterioration in physical condition. So the "healthier" fish used by Oshima et al. (1969 a, b) may have been more removed from peak response than the fish used by Hara et al. (1965) and Ueda et al. (1967), and hence less likely to discriminate correctly. Hara et al. (1965) had noted that there was very little electrical activity in the brains of their test fish except for the olfactory bulb and the posterior cerebellum. The results of Oshima et al. (1969 b) did confirm the potential olfactory discrimination of natural waters, since the responses to each pair of water stimuli were sufficiently different to indicate that the fish might perceive the difference between them, even if the home-stream water gave a lower response than other waters along the route. The assumption that preferred water would induce the greatest EEG response was probably incorrect.

The fish tested in these experiments were captured in their presumed home stream and therefore had their most recent experience with "home-stream" water. This may have reinforced the response to the home stream (Brett and Groot 1963). Oshima et al. (1969 a), for example, found that chinook salmon that initially did not respond to University of Washington hatchery water developed a strong response after a 67-day period of captivity in the hatchery.

Bulbar EEG's were also recorded from some of the fish that returned in the artificial imprinting studies in Lake Michigan. In each case reported, the morpholine-imprinted fish showed a strong bulbar EEG response to morpholine, while the unimprinted fish did not (Scholz et al. 1973; Cooper and Hasler 1974, 1976) even when 10–18 months had intervened between the initial exposure to morpholine and the EEG recording (Dizon et al. 1973). Fish did not respond strongly to water of their capture stream alone, indicating that they are not just responding to the stimuli most recently experienced. (The fish had originally been released far

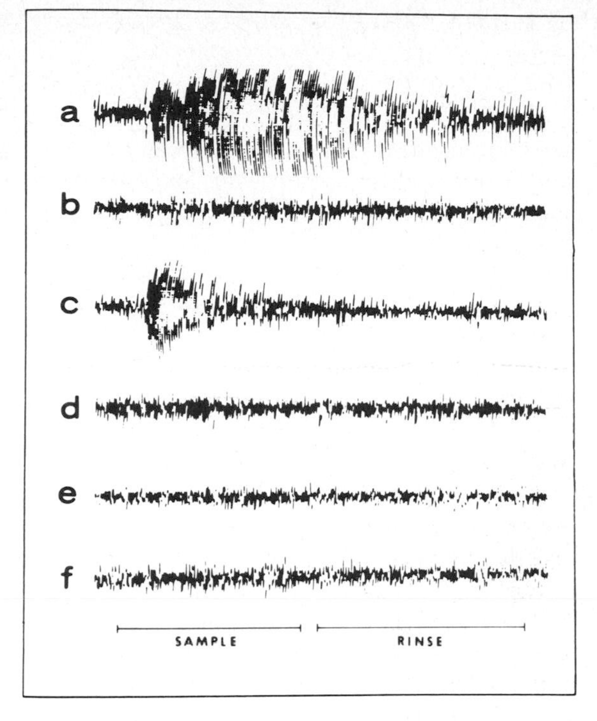

Fig. 3.15. EEG responses of *a* imprinted salmon to morpholine, *b* nonimprinted salmon to morpholine, *c* imprinted salmon to Oak Creek water containing morpholine, *d* imprinted salmon to Oak Creek water alone, *e* imprinted salmon to phenethyl alcohol (PEA), *f* imprinted salmon to N-b-hydroxyethyl-morpholine (Hasler and Scholz 1978)

enough away from the test stream that they would not have imprinted on the stream's natural odours.)

Despite the apparently convincing results, some caveats have been lodged. Hara's views on the suitability of morpholine as a stimulus have been noted. The fact that the Lake Michigan studies did not show an EEG response to morpholine at concentrations below 1% argues in favour of Hara's suggestion that morpholine acts as an irritant. However, physiological thresholds are often much higher than behavioural thresholds and one would expect an irritant to affect unimprinted fish and imprinted fish in the same way, in contrast to the results actually obtained. Hara and MacDonald (1975) and Cooper and Hasler (1974, 1976) point out that the EEG response to 1% morpholine differs qualitatively from response to natural water or amino acids by being slower to develop and more prolonged. They found that perfusion with 1% morpholine inhibited the response to the amino acid L-serine, suggesting a general inhibitory effect. In contrast, Cooper and Hasler (1976) found no morpholine inhibition of the response to a different amino acid, L-methionine. Interpretation of qualitative differences in integrated EEG responses is not a highly developed field.

On a more general level, it is important to remember that gross bulbar EEG activity may not give an accurate indication of the behavioural response of the animal. A strong attractant and a strong repellent might both be expected to give significant increases in brain electrical activity. Nevertheless, in spite of the uncertainties and objections, the EEG work seems to reinforce the conclusions of the tag returns and tracking studies: that coho salmon learn the odour of the region

Table 3.9. The five most common types of evidence bearing on the home-stream odour hypothesis are: (A) an experimentally demonstrated preference for home-stream water; (B) evidence that olfaction is an important element in homing successs, usually through releases of fish with plugged nostrils; (C) differential olfactory bulb EEG responses to different natural waters; (D) differential olfactory bulb EEG responses to artificial odours in fish exposed to the odours as smolts and (E) attraction of fish to streams scented with odours they were exposed to as smolts. The response categories are: +, response reported; (+), inconclusive reports of response; −, no response when tested and N, no reported tests. The lower case letters refer to the reference listed below the table. The list includes only one reference per point although there may be other tests that reached the same conclusion

Species	A Prefer home odour	B Olf. role in homing	C Diff. EEG to waters	D Diff. EEG to imprint	E Home to imprint
Coho salmon	+a	+b	+c	+d	+e
Chinook salmon	+f	+g	+c	N	N
Sockeye salmon	N	(+)h	+i	N	N
Chum salmon	N	(+)j	N	N	N
Pink salmon	N	N	N	N	N
Atlantic salmon	+k	+l	N	N	N
Rainbow trout	N	N	N	+m	+n
Cutthroat trout	N	(+)o	N	N	N
Brown trout	+p	(−)q	N	N	+r

a, Jensen and Duncan (1971); b, Wisby and Hasler (1954); c, Hara et al. (1965); d, Scholz et al. (1973); e, Scholz et al. (1975); f, Whitman et al. (1982); g, Groves et al. (1968); Lorz and Northcote (1965); i, Bodznick (1975); j Hiyama et al. (1967); k, Sutterlin and Gray (1973); l, Bertmar and Toft (1969); m, Cooper and Hasler (1976); n, Cooper and Scholz (1976); o, LaBar (1971); Stuart (pers. commun.), in reference r; q, Stuart (1957); r, Scholz et al. (1978a)

in which they undergo the smolt transformation and that they use that odour in recognizing their home stream when they return.

The home-stream odour hypothesis has been tested primarily on coho. Evidence for home-stream odour responses in other salmonid species is less complete but, where it has been tested, it tends to fit the coho picture (Table 3.9). Some studies on other salmonid species also require specific discussion.

3.4.4.5.2 Sockeye Salmon EEG Responses Prior to Reaching Their Home Stream

Bodznick (1975) captured adult migrant sockeye near the mouth of the Fraser River and at various points along the river. He then used body proportions, scale measurements and the timing of their migration in an attempt to determine which local race they belonged to, and hence the home area. Fish assigned by this method to the Horsefly river stock or the Adams River Stock did not show their strongest bulbar EEG response to their putative home areas. They did, however, show different responses to the waters at each choice point tested on the migration route, again suggesting a basis for discrimination. An example is the response of fish at the point where the Thompson River enters the Fraser. Horsefly fish should take the Fraser Fork and Adams River fish the Thompson Fork (Fig.

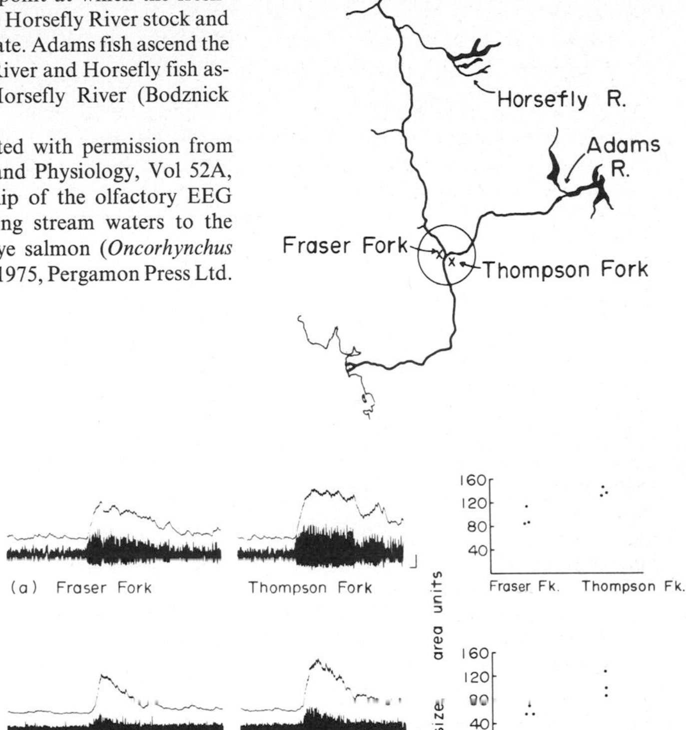

Fig. 3.16. Map showing the point at which the freshwater migratory routes of the Horsefly River stock and the Adams River stock separate. Adams fish ascend the Thompson Fork to Adams River and Horsefly fish ascend the Fraser Fork to Horsefly River (Bodznick 1975)

Figs. 3.16 and 3.17: Reprinted with permission from Comparative Biochemistry and Physiology, Vol 52A, Bodznick D, The relationship of the olfactory EEG evoked by naturally-occurring stream waters to the homing behaviour of sockeye salmon (*Oncorhynchus nerka* Walbaum). Copyright 1975, Pergamon Press Ltd.

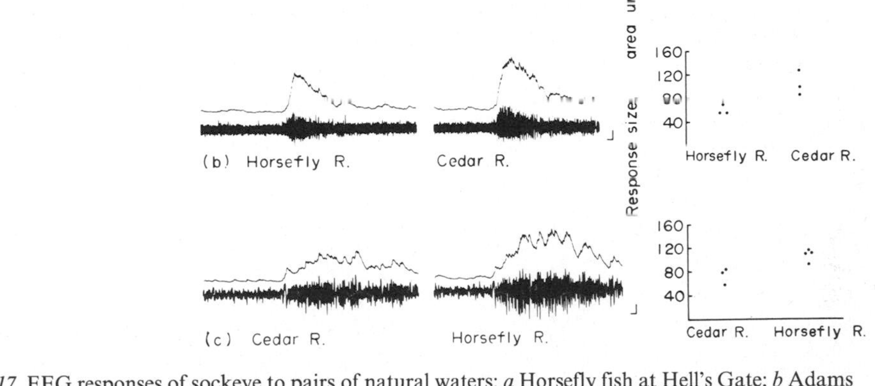

Fig. 3.17. EEG responses of sockeye to pairs of natural waters: *a* Horsefly fish at Hell's Gate; *b* Adams fish at Hell's gate; *c* Adams fish at Adams River. *Points on graphs* at *right* represent all of the responses of the fish to the two waters and demonstrate that in each trial one water consistently elicits a larger response (Bodznick 1975)

3.17). The supposed Horsefly fish showed a stronger response to Thompson Fork water than to Fraser Fork water while the response of Adams fish was quite variable. Similarly, Horsefly fish caught at Hells Gate in the Fraser showed a greater response to water from Cedar River, a sockeye spawning stream near Seattle, Wash. (not on the Fraser) than to water from the Horsefly River (Fig. 3.18), while Adams River fish captured at Adams River showed a stronger response to Horsefly water than to Cedar River water. It may be that a strong EEG response does not necessarily indicate preference. A person seeking a drink might turn toward the odour of a brewery rather than a sewer treatment plant, even if the sewer plant evoked a stronger EEG response.

3.4.4.5.3 Water Preference in Atlantic Salmon

In a study on the St. John River in New Brunswick, Sutterlin and Gray (1973) allowed marked, hatchery-reared salmon smolts to migrate out of the Mactaquac Hatchery voluntarily. They trapped returning adults at the hatchery and at a power station dam upstream from the hatchery. The ratio of marked hatchery

Table 3.10. Distribution of adult hatchery-reared and wild fish during the fall runs in 1971 and 1972 (Sutterlin and Gray 1973)

Year	Hatchery-reared fish		Wild fish	
	Recovered at hatchery	Recovered at dam	Recovered at hatchery	Recovered at dam
1971	104 (71%)	42 (29%)	25 (2%)	1029 (98%)
1972	151 (65%)	82 (35%)	25 (2%)	1289 (98%)
Total	255 (67%)	124 (33%)	50 (2%)	2318 (98%)

Table 3.11. Selection of various waters by hatchery-reared and wild Atlantic salmon) Sutterlin and Gray 1973)

A. Hatchery-reared fish (river water both sides)

No. of fish	Right	Left
6	3	2
6	1	2
10	4	3
10	3	4
14	6	4
—	—	—
46	17	15

B. Hatchery-reared fish

No. of fish	River water	River water plus hatchery effluent
7	0	3 (R_1)[a]
7	0	4 (L_1)
10	1	4 (R_2)
10	1	4 (L_2)
11	1	6 (R_3)
11	0	5 (L_3)
12	1	5 (R_4)
12	1	4 (L_4)
—	—	—
80	5	35

C. Wild fish

No. of fish	River water	River water plus hatchery effluent
9	2	3 (R_1)[a]
9	3	1 (L_1)
11	3	1 (R_2)
11	4	2 (L_2)
10	2	1 (R_3)
10	3	3 (L_3)
—	—	—
60	17	11

D. Hatchery-reared fish

No. of fish	River water	Well water
6	4	0 (R_1)[a]
6	3	0 (L_1)
8	4	0 (R_2)
8	3	0 (L_2)
11	6	0 (R_3)
—	—	—
39	20	0

E. Wild fish

No. of fish	River water	Well water
10	4	0 (R_1)[a]
10	5	0 (L_1)
—	—	—
20	9	0

F. Hatchery-reared fish

No. of fish	River water	River water plus hatchery effluent plus Cu^{2+}
10	4	0 (R_1)[a]
10	4	1 (L_1)
10	2	1 (R_2)
10	3	0 (L_2)
—	—	—
40	13	2

[a] R and L indicate the side (right or left) of the apparatus in which the river water was modified or replaced. Subscripts denote the experiment group of fish used for two tests

fish to unmarked fish from rearing grounds above the dam indicated some home-stream fidelity (Table 3.10).

Returning fish were then tested for water preference in a choice tank. Control tests indicated that the two sides of the apparatus were similar in attractiveness. Since the hatchery water constitutes only one one-thousandth of the river flow, the fish were then offered a choice between river water and river water with one ppm hatchery effluent. Hatchery-reared fish preferred the diluted hatchery water, while wild fish were indifferent (Table 3.11). Response to water from the well that supplies 50% of the hatchery water was negative for both groups. The other half of the hatchery water supply comes from the river, so the distinctive stimuli must originate within the hatchery.

In another study Atlantic salmon smolts from a distant hatchery were held at a marine site 12 km from their parent stream for 2–3 weeks before being released. They returned to the marine site 13 to 25 months after release and stayed in the region of the release site for 2 months, during which time many were captured (Sutterlin et al. 1982).

There is no direct evidence from the work of Sutterlin and Gray (1973) or Sutterlin et al. (1982) that the return of the salmon was based on olfactory discrimination. It may have been based on other chemoreceptor systems on non-chemical attributes of the water or on some "open-water" orientation mechanism (see Sect. 3.4.4.5.8). It does provide evidence for ability to discriminate between the water from the rearing area and water from other nearby areas.

3.4.4.5.4 Rainbow Trout in Lake Michigan

Rainbow trout, *Salmo gairdneri,* have been introduced widely to eastern North American waters, including Lake Michigan, where they were tested along with coho salmon in artificial imprinting studies. In 1972, Cooper and Scholz (1976) released three groups of 3000 rainbows derived from sea-run "steelhead" stock. One group, exposed to 5×10^{-5} mg l^{-1} morpholine during the time of normal seaward migration and a control group were released 13 km north of Oak Creek, while the second control group was released at Oak Creek (Location 17, Fig. 3.10). In the fall and spring of 1973 and the spring of 1974 morpholine was metered into Oak Creek at approximately 5×10^{-5} mg l^{-1}. Steelhead often return in the spring. The experiment was repeated in 1973 with 3900 controls and 2000 morpholine fish released 13 km to the North of Oak Creek, at Milwaukee Breakwater (Location 16, Fig. 3.10).

A higher percentage of morpholine-exposed fish than controls returned to Oak Creek and more of the Oak Creek-released controls returned than the Milwaukee Breakwater controls (Table 3.12). Despite the spring release of morpholine at Oak Creek, the majority of the fish returned in the fall, all three groups at about the same time. Cooper and Hasler (1976) recorded olfactory bulb EEG's from the returned rainbows and found that morpholine exposed fish responded more strongly to 1% morpholine than did non-exposed fish, indicating that imprinting had occurred.

Scholz et al. (1978 b) reported on returns of rainbows released in September, 1973 near Little Manitowoc River on the shore of Lake Michigan (Fig. 3.10). The

Table 3.12. Census of morpholine-imprinted and non-imprinted (control) fish at Oak Creek, 1972–1974. MB = Milwaukee Breakwater, OC = Oak Creek (Cooper and Scholz 1976)

Exptl. group	No. released	Release site	No. captured fall 1972	No. captured spring 1973	No. captured fall 1973	No. captured spring 1974	Total no. captured	% captured of no. released
1972								
Imprinted	3000	MB	12	19	67	76	174	5.8
Control	3000	MB	0	1	4	11	16	.53
Control	3000	OC	21	15	14	8	58	1.9
Strays			80	70	123	–	–	–
1973								
Imprinted	2000	MB	–	–	18	48	66	3.3
Control	3900	MB	–	–	1	7	8	.21

fish were held in ponds or in the hatchery until September 1973, then released. One hatchery group was exposed to 5×10^{-5} mg l^{-1} morpholine, the other only to the control hatchery water during the smolt transformation. In the fall of 1974 and 1975, morpholine at approximately 5×10^{-5} mg l^{-1} was metered into the Little Manitowoc River. Morpholine-imprinted fish returned in greater numbers than the other two groups and showed less straying. These results are similar to those found with coho in the same experimental system. Although they do not prove that olfaction was used, they do indicate that rainbow trout may learn chemical stimuli and subsequently recognize their home stream. As further evidence Scholz et al. (1978 b) report that the returning fish tended to congregate near the morpholine outlet.

In contradiction to the studies reported above, Hara and Brown (1979) found that rainbow trout exposed to 5×10^{-5} mg l^{-1} of morpholine during smolting did not show stronger bulbar EEG responses to morpholine than controls when tested immediately and 12 months after exposure. The trout tested by Hara and Brown may not have been in a migratory phase when tested and may not have even belonged to a migratory stock (Cooper 1982). The results of Scholz et al. (1973) (Fig. 3.13) with coho indicate that the EEG response to imprinted morpholine reaches a peak near the peak of the behavioural migration.

3.4.4.5.5 Brown Trout

The homing of brown trout, *Salmo trutta*,to their natal streams has been established by the experiments of Stuart (1957) at Dunalastair Reservoir in Scotland. In 1955 he displaced 113 spawners from their home stream to another stream across the reservoir. Fifty-eight of the fish had their nostrils plugged with cotton wool or cotton wool impregnated with vaseline or chloroxyphenol, while 55 were controls. Eighteen control fish returned to the natal stream, while 14 experimentals, ten with intact nose plugs, also returned. The results do not support the home stream odour hypothesis. On the other hand, Scholz et al. (1978 a) report a personal communication from Stuart, stating that when the water from a brown trout-spawning stream was changed to a new channel after the fish had left, the returning adults preferred the new channel containing the water from "their" stream over the old channel which was still flowing with water from other tributaries. This suggests that chemical stimuli, although not required, may override the spatial location or visual characteristics of the channel used on the outward migration.

Introduced brown trout were also tested in the Lake Michigan artificial imprinting experiments (Scholz et al. 1978a). Three hundred fish, 18 months old, were exposed to 5×10^{-5} mg l^{-1} of morpholine for 34 days. They were then released, along with 300 control fish, 13 km North of Oak Creek Wisconsin at the Milwaukee Breakwater. Another 300 controls were released at Oak Creek. Morpholine was then metered into Oak Creek during 1972 and 1973 as reported above. A total of 53 (17.7%) of the morpholine fish were captured in Oak Creek while 3 (1%) of the Milwaukee controls and 14 (4.7%) of the Oak Creek controls came to Oak Creek.

These results suggest that brown trout use learned chemical attributes of the home stream, and possibly other stimuli, in their return migration.

3.4.4.5.6 Shad and Alewife Responses to Home-Stream Odour

Anadromous shad and alewives have also been examined from the viewpoint of the home-stream odour hypothesis. Thunberg (1971) captured returning adult migrant alewife, *Alosa pseudoharengus,* and tested their response to natural waters in a two-choice selection tank. They preferred the water of their home stream over other natural waters, even if the other water contained an alewife population. In a choice between two non-home waters, one with a spawning alewife population and one without, the alewives did not prefer the water with the alewife population, suggesting that the home stream response is either not due to a pheromone or is based on a population-specific pheromone. Thunberg considered a population-specific pheromone to be unlikely since the populations under study were artificially introduced to the region. His conclusion that the response is olfactory is based on the results of tests with fish whose nostrils were plugged with cotton. They no longer showed home-stream preference. Atema et al. (1973) tested alewives in a y-maze. Adult migrants captured in the Bourne River preferred their home river over three of four other natural waters. Various fractions of the home-stream water were tested on groups of alewives in the maze and the preference response was elicited by a heat-stable, non-volatile, polar component of molecular weight less than 1000.

American shad, *Alosa sapidissima,* also migrate into rivers prior to spawning and may also make non-reproductive intrusions into freshwater (Gabriel et al. 1976). Shad, tracked by ultrasonic tags as they entered their home river, tended to swim against the ebb tide and hold position on the flood (Leggett 1973), a behaviour pattern consistent with attraction to, or positive rheotaxis in, water of home-stream origin. The attempt of Dodson and Leggett (1974) to test the effect of blinding and olfactory occlusion on the migration of shad toward the home stream was inconclusive.

3.4.4.5.7 Summary of the State of the Home-Stream Odour Hypothesis

The three main points of the home-stream odour hypothesis are: (1) that local differences in water chemistry allow recognition of a home stream (2) that fish learn, "imprint to," the home-stream odour during a short critical period, probably constrained by development or endocrinology (3) that this learned information is used in selecting a spawning stream when the fish return as mature adults.

Evidence bearing on the first point includes the behavioural preference for home-stream water that has been demonstrated in tests offering the fish a choice between home and non-home waters. Such a preference has been found in the four salmonids that have been tested (Table 3.9). Additionally, studies in which the olfactory apparatus is blocked indicate that chemosensory information is important in stream selection in six of the seven salmonid species tested. In the seventh species, brown trout, the fish may have made the migration one or more times without blocked nostrils and have learned other cues. Studies of the electrophysiological response (EEG) of the olfactory system to natural waters all indicate differential response to waters from different natural sources. This should allow the fish to distinguish between natural waters, even though the home-stream water may not necessarily elicit the strongest EEG response.

Evidence for imprinting is provided by the responses of adult fish to artificial odours such as morpholine and PEA that had been presented to them during smoltification. Both coho and rainbow trout, when tested as adults, show stronger EEG responses to these odours if they have been exposed to them as smolts than they do if they were not exposed as smolts (Fig. 3.15). If morpholine were a simple irritant, as suggested by Hara and MacDonald (1975) and Hara and Brown (1979, 1982), it should affect both groups equally. Similarly, the differential return of adults to streams scented with artificial odours supports the imprinting portion of the home-stream-odour hypothesis (Tables 3.8, 3.9). Fish exposed to morpholine return primarily to morpholine-scented streams. PEA-imprinted fish return primarily to PEA-scented streams, while fish exposed to neither odour scatter among nearby streams. Early exposure to artificial odours also affects the response of fish tracked through odour plumes (Fig. 3.12). Fish which have been exposed to an odour as smolts circle in a plume of the odour when they encounter it in the lake, while fish which were not exposed continue directly through the plume without turning. Thus three different types of evidence – EEG studies, return to scented streams, and tracking through odour plumes – support the hypothesis that fish respond differently as adults to odours present during a short period near smoltification. Use of the term "imprinting" is justified.

The third element in the hypothesis, use of the imprinted odour in home-stream selection, is supported by the differential return of migrant fish to streams scented with the odour to which they were imprinted (Table 3.8). The behavioural response of tracked fish to scent plumes provides indirect support of this element of the hypothesis, as does the selection of home-stream water in preference tests. The repeated finding that olfactory blocking interferes with home-stream selection (Table 3.9) also indicates that olfaction is important in stream selection.

The home-stream hypothesis is well supported by data available for the coho salmon and by most of the information available for other species that have been tested. However, it may not be the only mechanism at work; fish did return in spite of olfactory blocks in several of the experiments reported above, so other senses or random scatter may be sufficient for returning to the home stream in some cases. No study reported has combined artificial imprinting with blocking of the olfactory channel. Hara's proposal that morpholine may also be detected by returning fish through some other chemoreceptor system is thus untested.

3.4.4.5.8 Separation of Home-Stream Selection and Open-Water Orientation

A separation between open-water orientation bringing the fish back to the general area of the home stream and near-shore orientation governing river entry has often been proposed (e.g., Bertmar and Toft 1969; Peters 1971; Scholz 1980), and there is some evidence for such a separation. First, on theoretical grounds, the dilution of home-stream water would seem to be impossibly low to serve as a directing stimulus for oceanic migrations, from the North Pacific to the Fraser River, for example. Second, in the artificial imprinting studies the fish seemed to return to the general area offshore from the release point, whether imprinted or not. In the tracking studies, both imprinted and unimprinted fish moved in a direct manner toward the capture stream.

In apparent contradiction to the separation of offshore orientation and river entry there was wide scattering of anosmic fish in a study of Atlantic salmon by Bertmar and Toft (1969) and the scattering of unimprinted controls in the coho and trout releases near Two Rivers, Wisconsin (Scholz et al. 1976). However, fish that return to the appropriate near-shore area but then fail to recognize a home stream may wander widely after the initial return. This would be difficult to detect.

The timing of maximum EEG responses to home-stream water or artificial odours supports the separation of open water and river entry mechanisms. Hara et al. (1965) and Ueda et al. (1967), using fish near the end of their spawning migration, found a maximum response to home-stream odour. Oshima et al. (1969b), using "healthier" fish, did not find the same degree of home-stream response, nor did Bodznick (1975), who tested sockeye captured downstream from their home river. These results agree with those of Cooper and Hasler (1974) (Fig. 3.13), who found a peak EEG response to morpholine in imprinted fish at about the time of peak migration into Oak Creek. In other words, the home-stream odour response may only be activated for a short time as the fish near their objective, possible by gonadal hormones (Scholz 1980). This could explain the results of Hara and Brown (1979), who tested rainbow trout that may not have been in the migratory phase (Cooper 1982) and may not have belonged to a migratory genetic stock.

The Sequential Odour Hypothesis. Harden Jones (1968) proposed a sequential odour hypothesis for salmon homing: that the fish learn the sequence of different odour stimuli in the streams and rivers on their downstream migration, then on returning, respond to each of these odours in turn, in reverse order. They would thus follow a string of imprinted odours back to the home stream. This hypothesis recognizes some of the objections to the home-stream odour hypothesis based on the extreme dilution of small streams by large river systems. Arguing on the basis of an estimated threshold of 10 ppm home water in river water, Harden-Jones suggested that sockeye in the Fraser River would not be able to react to Cultus Lake water. On the other hand, the results of Sutterlin and Gray (1973) on Atlantic salmon indicate a threshold below 1 ppm. Nevertheless, dilution must eventually reduce the stimulus to below-threshold levels and the sequential learning of odour stimuli along the route would greatly extend the range of olfactory orientation. There is no direct proof or disproof of the sequential hypothesis. However, coho trucked around Columbia River reservoirs (and hence prevented from forming the correct series of imprints) return to their final release point rather than to the hatchery. This would suggest that a complete sequence is important (Vreeland et al. 1975). In contrast, chinooks trucked 400 km downstream did return to their hatchery (Ebel 1980). EEG studies by Oshima et al. (1969 a, b) and Bodznick (1975) failed to find a neat sequence of EEG responses, but the weaknesses of EEG studies have already been discussed and there might be a temporal sequence in response which would not be easily detected in conventional EEG studies.

Mechanism of Response to Home Stream Water. It remains a major open question whether the returning salmon use the chemical stimuli from the home stream as directing stimuli, by, for example homing in on the concentration gradient of home-stream odour or use it as a switching stimulus, initiating positive rheotaxis. If the chemical gradient is to serve as a directing influence, it must be sufficiently steep for some form of kinesis or taxis (Fraenkel and Gunn 1940, 1961) to be effective. Kinesis orientation involving simple changes in the rate of swimming or turning in relation to the concentration of the home odour would seem, as Harden-Jones (1968) suggests, too inefficient to account for salmon homing. Taxes or directed orientation to the gradient would face two problems, the gradual nature of the gradient and the adaptation of the receptors to the stimulus. The finding of Kleerekoper (1972) that the olfactory response of male catfish to the sex pheromone of females had a very low or non-existent adaptation rate indicates that some fish olfactory orientation responses have low adaptation rates. Fish nostrils are very close together for comparison of concentration differences in a gradient; but if the thresholds for discriminating concentration differences are sufficiently low, a taxis, probably tropotaxis, remains possible. Taste receptors with wider separation on the body would seem better suited for gradient orientation and they possess low response thresholds for some classes of compounds, particularly amino acids (Caprio 1974, 1982).

A more likely response would be a simple rheotactic response in the presence of the stimulus odour (Harden Jones 1968). If the fish swam upstream in the presence of home-stream odour and drifted or searched in its absence, they would move upriver until they passed the odour source then drift downstream until they detected it again. If a fish took the wrong fork in a river it would lose the scent and drift down to the junction again for another try. In a simple tidal current near the home-stream mouth the fish would swim "upstream" against the scented ebb tide then drift inshore with the unscented flood. If the rheotaxis was stronger with increasing scent concentrations it would facilitate the process.

One would expect to obtain evidence of the response mechanism from tracking individual fish with ultrasonic or radio tags. Madison et al. (1972) found no evidence of different tidal responses in their tracking of adult sockeye. Scholz et al. (1972), tracking chinook salmon in Lake Michigan, reported that during seiches (tide-like changes in lake level) the fish swam against the "ebb" of the seiche and would only enter the home stream on the ebb. During the "flood" they made local movements. Stasko et al. (1973), tracking pink salmon 70 km south of their presumed home stream, the Fraser River, found two categories of fish. Passive fish drifted with the tidal currents. Active fish swam directly toward the Fraser without regard to the direction of tidal currents. Similarly Groot et al. (1975) found that 15 of 18 adult sockeye tracked in the Skeena River estuary were passive drifters, two fish swam actively upriver regardless of tide and one swam against both ebb and flow. The fish in some of these studies were tested in estuarine areas and the approaches to rivers. Their orientation may have been based on an "open-water orientation mechanism" rather than the "home-stream odour mechanism" making the results meaningless in terms of response to the odour of tidal streams.

The evidence from tracking studies is thus mixed, perhaps in balance offering little support of the odour hypothesis as a sole mechanism or dominant mechanism in a hierarchy in the close offshore area. The number of passive drifting fish reported may indicate problems with the tracking technique. Zimmerman (1980) found that internal or external tracking tags induced rainbow trout to show reduced activity.

3.4.4.6 Pheromones

The term "pheromone" was coined by Karlson and Luscher (1959) to refer to chemical signals between conspecifics. There are examples of chemical warning signals, mating signals, parental care signals, aggregation signals and dominance signals in fish. Water is an excellent solvent, fish have good chemosensory ability and, like any animal, will "leak" a variety of chemicals into the surrounding water. It is not surprising, therefore, that the chemical senses have come to serve social communication (Barnett 1977; Liley 1982).

Atlantic Salmon. White (1934, 1936) may have been the first to suggest that odours from conspecifics served to attract migrating fish to suitable spawning areas. The Apple River in Nova Scotia has East and West branches. The West Branch supported an Atlantic salmon population, but the East branch population had been wiped out by a dam 60 years before White's study. Although the two branches joined at low tide, the East Branch population had not re-established itself after the removal of the dam. In 1932 Atlantic salmon fry were transplanted to the East Branch, to which they returned as grilse in 1935, supporting the imprinting hypothesis of home-stream recognition. The interesting result was the presence of "unusual numbers" of "stray" adult fish in the East Branch in 1932, 1933, and 1934 when seining prior to 1932 had not yielded any salmon. This suggested to White that young salmon in the system released odours that attracted adults into the river.

A similar observation and conclusion was reported by Solomon (1973), working in the Bristol Channel region. Atlantic salmon returning to spawn in the Usk, Wye, and Severn often paused in the estuary of the River Parrett and supported a small fishery there, despite the absence of a spawning population in the Parrett itself. From 1953 to 1956 salmon eggs were planted in the River Tone, a tributary of the Parrett. The catch of the fishery at the river mouth increased in 1953 before any of these planted fish could have returned and the catch remained relatively high until the mid 1960's. The last salmon was reported in the Tone in 1964 and in 1966 water levels were very low, probably eliminating the population. Solomon (1973) felt that the increased fishery was due to fish from other runs being lured into the estuary by the odour of the fry and parr in the system. This theory was supported by the recovery of tagged fish from the other salmon rivers in the system and by the absence of corresponding changes in rainfall or of the salmon populations of the other rivers that might have accounted for the increased numbers in the Parrett estuary.

The finding of White (1934, 1936) and Solomon (1973) indicate that some Atlantic salmon are attracted to non-home waters by the presence of conspecifics in the water system. They do not establish that the response is olfactory, nor do they account for the well-established tendency of Atlantic salmon to return to their natal stream, unless one postulates a population-specific chemical stimulus.

Sea Run Trout and Char. The pheromone hypothesis of migration has been developed in a more elaborate form by Nordeng (1971, 1977) in an attempt to account for the entire landward migration from the oceanic feeding grounds rather than just entry into the river. He suggested that the seaward migrating smolts could lay down odour trails that guide the adults returning to the home stream. The evidence presented was, however, quite tentative.

Sea run char, *Salmo alpinus = Salvelinus alpinus,* in the Salangen River system in northern Norway go to sea for about a month each summer beginning when they are 3 to 4 years old. They become sexually mature after their second or third trip to sea. When Salangen fish were reared in southern Norway at Voss then returned to the Salangen region and released at the mouth of the Loksbotn River, 21 km from the Salangen, 10 of 31 recaptured fish entered the Loksbotn and 21 returned to the Salangen. Nordeng (1971) interpreted this result as indicating a population specific attractive pheromone.

The life cycles of char, Atlantic salmon and sea run brown trout (*Salmo trutta*) are compatible with Nordeng's (1977) pheromone homing hypothesis. In each of these species the seaward migration of smolts overlaps with the homeward migration of the adults and, in the case of the char, with the migration of the returning smolts as well. In the case of the Atlantic salmon the departure of the smolts was followed in 15 days by the first return of the adults to the river. The arrival of the smolts at the offshore feeding grounds may trigger the return migration of the adults, which then follow the odours of the outward migration of smolts of their own population. The migration times for the outward trip of the smolts and the return trip of the adults are compatible with published migration speeds.

There is some evidence for the release of attractive substances by these salmonids. Hoglund and Astrand (1973) tested juvenile char in a fluvarium choice apparatus and found that they showed a preference for water in which conspecifics were caged upstream and that this preference was lost when the nares were cauterized.

Electrophysiological examination of 107 cells in the olfactory lobes of the char, revealed 93 that responded to fish odours (Døving et al. 1974). Of the 45 cells studied in detail, 39 gave differential EEG responses to different classes of char. For example, fish of the migratory Salangen population could be differentiated from fish of the non-migratory Salangen population by 47.4% of the responsive single cell units. Populations from two lakes, Lake Vangsraten and Lake Lona, which had been separated by a waterfall for at least 5000 years, were also differentiated, as were old and young fish within a population.

Mucus from the stimulus fish gave results similar to those with tank water, but the mucus-collecting technique may have contaminated the mucus with other substances (Selset and Døving 1980). Using a better method of mucus collection, Selset and Døving found that anadromous char responded more strongly to intestinal contents, particularly bile, than they did to mucus.

Bile acids may be suitable candidates for migration pheromones because some are quite stable and tend to be absorbed by organic materials in the water and they have been shown to elicit bulbar EEG responses in char and grayling, *Thymallus thymallus,* at very low concentrations (Døving et al. 1980). Bulbar EEG's of Atlantic salmon also indicate that they were more sensitive to intestinal contents than to skin mucus (Fisknes and Døving 1982). Mucus and intestinal contents from Atlantic salmon from three different Norwegian rivers differed from one another chromatographically, indicating chemical differences between stocks, which could serve as a basis for population-specific pheromones (Stabell et al. 1982), although three salmon tested by Fisknes and Døving (1982) did not distinguish between odours of donor fish from their own stock and of different stocks.

Whatever they were responding to, the char smolts showed a preference for water from smolts of their own population (Hammerfest) in a choice apparatus (Selset and Døving 1980). In one case, however, the Hammerfest smolts showed attraction to water that had contained young mature char from Tinnsjø. The Tinnsjø char had been kept with Hammerfest fish in the holding facility for one and a half years before the experiment, suggesting the possibility of odour exchange or learning the odour of tankmates.

In summary the pheromone hypothesis proposed to account for the return migration of Atlantic salmon, anadromous char and anadromous brown trout is supported by appropriate life cycles and by evidence of attraction to water conditioned by conspecifics, possibly specifically by bile acids. In the case of Atlantic salmon, there is also the evidence in the field for attraction of "stray" adults to streams containing conspecifics. This is far from proving that fish use conspecific chemicals as a migratory cue, and there is at present no direct evidence to support Nordeng's (1977) suggestion that the returning fish follow odour trails left in the ocean by the outward migrating smolts.

Pacific Salmon. Attempts to apply the pheromone hypothesis to Pacific salmon encounter the problem of incompatible life cycles (Nordeng 1977). Pink salmon fry and chum fry migrate out of the spawning river long before the adults return. Sockeye might be lured back to their home lakes by pheromones from the fry and smolts, but at the time of spawning the spawning streams have been vacated by the fry for several months so the established fidelity to the home stream must depend on other factors. Of the Pacific salmon, the coho is probably the best candidate for pheromone-based homing, since the juveniles may remain in the natal area long enough to provide a stimulus for returning adults.

In EEG studies, hints of pheromonal responses have been found. For example, 2-year-old chinook salmon gave a higher bulbar EEG response to water that had contained other chinooks than they did to water from the same creek that had contained coho (Oshima et al. 1969a). Dizon et al. (1973) also found an increased bulbar EEG response by chinooks to water conditioned by conspecifics. These responses may, of course, be related to social interactions and not to migration. Mature rainbow trout, for example, were attracted to the water from spawning pairs and males were attracted to female water, while females were attracted to water from males and females (Newcombe and Hartman 1973). These results indicate a role for chemoreception in the spawning behaviour of salmonids.

Adult coho have been tested in a Y-maze for their response to water conditioned by conspecific fry (Quinn et al. 1983). Spawners returning to the University of Washington hatchery were given a choice between "home water" from the hatchery, which does not normally contain juvenile coho, and city water. They preferred the home water (69%), as we would expect. City water on both sides of the Y-maze led to few (25%) of the fish moving upstream in the apparatus. When they were given a choice between city water containing juvenile coho from their population and city water with no juvenile coho, more fish moved upstream than with city water alone (57%) and of those that chose, 79% preferred water conditioned by juvenile coho. In a choice between city water conditioned by their own population and the same water conditioned by an unrelated population, there was no significant difference in preference. Other species of salmonids were not tested as odour sources. As Quinn et al. (1983) point out, these results do not account for the homing behaviour of coho salmon. Returning adults normally pass many streams containing conspecific juveniles along their homeward route, and they may correctly return to home-water sources such as the University of Washington hatchery, that lack juvenile populations. Perhaps the response would allow stray fish to locate suitable spawning streams if they are unable to find their home stream for some reason.

There is no reason to believe that salmon would depend exclusively on one orientation or home-recognition mechanism. The postulated pheromone response may be present and supplement other mechanisms such as home-stream odour imprinting. The apparent attraction of mature fish to non-home waters found by White and Solomon suggests that a percentage of Atlantic salmon may be drawn to other salmon waters despite the normal home-stream fidelity of the species.

Non-Salmonids. The possible role of pheromones in the migration of non-salmonids is virtually unexamined. Miles (1968) found that American eel elvers were attracted to water that had contained adult eels but were repelled by water that had contained elvers (see Sect. 3.4.4.1). Pesaro et al. (1981) found that European elvers and juvenile eels were attracted to water which had contained elvers or juveniles. The difference in results may be due to species differences or to methodological differences between researchers. In the spawning migration of eels or in the location of mates on the spawning grounds chemosensory responses would seem appropriate for these macrosmic fishes. Edel (1975a) noted that eels that had been brought into sexual maturity by hormone injections had a unique pungent odour to the skin and mucus, which he termed an "odour of ripeness." He speculated that the odour might be a mating pheromone. Pheromones might also maintain the social organization of migrating fish. Schooling jacks (*Caranx hippos*) were able to maintain school formation better when separated by screen barriers than when separated by glass, suggesting the use of chemical or mechanical information as well as vision in their schooling (Shaw 1969).

One might speculate that fish of nocturnal habits or those living in deep or silty water where vision is limited would be most likely to use chemical signals in migration, but these types of fish are seldom used in experimental studies.

3.4.4.7 Predator Odours

Fish, like many other prey animals, respond to chemical stimuli from their predators. In one case this type of response has been examined as a means of directing migrating fish away from dangerous areas. Pacific salmon, on their upstream migration, respond to water that has been in contact with mammalian skin by stopping their migration and moving downstream (Brett and MacKinnon 1952, 1954). L-serine, an amino acid, is one of the active components in the mammalian skin rinse (Idler et al. 1956, 1961). Brett and Alderdice (1958) mixed human hand rinse water with odours of two predatory fish, Dolly Varden (*Salvelinus malma*) and cutthroat trout, then allowed the mixture to seep into one side of an experimental flume in an attempt to deflect sockeye smolts migrating downstream. Neither the mixture of predator odours nor a dilute formalin solution was effective in diverting the sockeye. It is not clear that salmon smolts respond to human hand rinse or to the odour of predatory fish under any circumstances. Downstream migrants would also be difficult to divert which chemical stimuli. Forrester (1961) was able to divert adult salmon into a net by use of bear paws dipped in the water and several mammalian skin rinses repelled adult coho and chinook salmon (Alderdice et al. 1954). A well-known alarm pheromone (Schreckstoff) occurs in cypriniform fishes such as minnows and carp (Pfeiffer 1982; Smith 1982) but it has apparently not been tried as a means of directing these fish away from hazards to migration.

3.4.4.8 Food

Food as a chemical component of the environment may affect the distribution of fish during portions of the life cycle devoted to feeding and growth. For example, the level of food in a stream may determine the degree of out-migration by subordinate trout or salmon during the period of stream life (Chapman and Bjornn 1969; Symons 1971; Mason 1976). Fish are often attracted by chemical stimuli emanating from food, and may use taste or smell to home in on prey. Bardach et al. (1967) found that ictalurid catfishes (*Ictalurus natalis* and *I. nebulosus*) use the taste receptors on their barbles and bodies to locate food. Amino acids are particularly effective stimuli and are released by many potential prey items. Similarly, lemon sharks (*Negaprion brevirostris*) orient upstream when they detect glutamic acid or trimethylamine oxide and nurse sharks (*Ginglymostoma cirratum*) show a true gradient response to such chemical stimuli (Hodgson and Mathewson 1971). Sutterlin (1975) has been successful in attracting a number of marine species to food odours, indicating the importance of chemical stimuli in their foraging.

The role of food stimuli in determining migration routes is unknown. At the very least, effects on stream distribution would affect the starting point of seaward migration in juvenile salmonids. In an evolutionary sense, the productivity of various regions must have had an important role in determining the present form of migratory life cycles as fish moved from regions suitable for reproduction to other areas with higher food availability that were less suitable for their type of reproduction. The short-term local orientation of fish may often be determined by food odours, as in the foraging sharks and catfish, but again the role of such factors in migration is not clear. Fish that feed actively during migration might conceivably be deflected or influenced in their routes by artificial placement of food or food odours. Pollution from food processing might alter fish distribution through release of chemicals that serve as feeding stimuli.

At least one case of food levels determining the end-point of migration has been decribed. Stage 4 and 5 plaice larvae move toward their inshore feeding grounds by swimming in midwater on the flood tide and sinking to the bottom and holding position during the ebb (Creutzberg et al. 1978). The mechanism by which they recognize the tide is unknown. Chemical stimuli from food play a role in terminating the inshore migration. When food is abundant, the larvae remain on the bottom in experimental tanks and eventually metamorphose into O-group plaice. However, if food is not readily available, they resume their excursions into midwater and would, in nature, be carried to another area. Creutzberg et al. (1978) argue that this response accounts for the retention of plaice larvae in the Wadden Sea nursery area, once they reach the area.

3.5 Summary

Water is a good medium for the transport of chemical information. It dissolves many substances, and water currents carry dissolved chemicals over long distances. Thus, information about the physical and biological environment, a "chemical landscape," is available to fish with their sensitive chemoreceptors.

At least five separate types of chemoreceptors have been described in fish, but olfaction and gustation are the best understood of the chemical senses. Both taste and smell are "long-distance" senses in fish, which have taste receptors on the outer surface of the body. Both senses are also capable of great sensitivity, particularly to organic compounds.

Chemical stimuli apparently have a limited role as timing stimuli in migration. Light stimuli, such as photoperiod or the diel cycle of light and dark, are more commonly used by migrating fish. However, chemical stimuli will be reliable timing stimuli for some types of behaviour. Irregular events such as floods or tidal currents may be characterized by predictable chemical changes that fish can respond to in the timing of their behaviour.

Chemical social signals or pheromones could synchronize social aspects of migration. Fish may, for example, benefit by traveling together as a group and synchronizing pheromones might allow the group to remain together. This function is, however, speculative.

Major chemical gradients such as salinity differences between ocean zones or oxygen gradients associated with depth may limit the distribution of fish, but there is little evidence to indicate that they guide migrations. Pollution may also limit fish distribution through lethal effects or behavioural avoidance of polluted areas.

Natural waters differ in the concentration and presence or absence of many chemicals. The mix of these distinctive components can serve to identify general types of habitats, or even specific unique locations. Sockeye salmon fry, for example, migrating toward lakes, show a preference for lakewater in general over streamwater. EEG studies indicate that the olfactory system is able to distinguish between lakewaters and streamwaters even if the fish has no previous experience with the particular body of water.

The unique chemical mix from each location may allow a fish to detect its home from some distance downstream. This recognition of home odours is particularly well developed in coho salmon. Young salmon learn home-stream odours during a critical period that coincides with the smolt transformation. The critical period may be timed by thyroxine levels. Later, when the fish are sexually mature and returning to the spawning streams, they preferentially move into water bearing the scent that they implinted on as smolts. Fish imprinted on artificial odours can be lured into streams near the original release point by adding the imprinted odour to streams the fish have never been in before. They will also show increased EEG responses to the imprinted odour and will respond with circling behaviour when they encounter a plume of the odour in open water. Fish that were not exposed to the odour during the critical period do not show these responses when they are exposed to it as adults. The response is restricted in time to a period near sexual maturity, perhaps controlled by reproductive hormones. There is evidence that several other salmonid species and some anadromous clupeids show similar responses to home-stream water.

The home-stream odour hypothesis may account for home stream recognition and upstream movement within a river system, but there is evidence that other mechanisms are involved in offshore orientation.

Conspecifics may be one source of odours in a stream. Adult Atlantic salmon seem more likely to enter non-home rivers if there are juvenile salmon in the river than if the river has no resident salmon. Population-specific odours would allow fish to select streams occupied by their own genetic population. There is evidence of preferential response to odours from members of their own population in the genera *Salmo* and *Salvelinus*. This sort of pheromone-mediated homing is not compatible with the life cycles of some of the Pacific salmon of the genus *Oncorhynchus*.

Chemical stimuli from predators and prey may influence the route of migration or affect the levels of emigration from feeding areas, but there is little evidence that they are important guiding stimuli in migration. In at least one case, that of larval plaice, the presence of adequate food levels is an important stimulus for the termination of the onshore migration.

Chapter 4

Mechanical Stimuli

4.1 Mechanical Stimuli

Mechanoreceptors respond to physical movement in the receptor. This can result from water currents, sound, body movements, pressure or gravity. Each of these types of stimulation is available in water and may be involved in migration and orientation.

Fish mechanoreception is relatively difficult to investigate. The responses of several different types of receptors seem to overlap. Mechanical stimuli are hard to record without interfering with the stimulus itself, particularly stimuli from touch or small turbulent water movements. Unlike vision and olfaction, it is relatively difficult to neutralize the mechanical senses experimentally. Touch and the lateral line sense are widespread on the body and would require nerve cutting, which is more difficult than stuffing cotton into nostrils to eliminate olfaction, for example. The senses of hearing and balance are virtually inextricably linked in fish, so any attempt to deafen the fish will also interfere with its balance and orientation to gravity. Although there are many experiments in which blind or anosmic fish were tested for their ability to home or migrate, no studies that I know of test the effect of deafness or lack of the lateral line on migration or homing.

4.2 Mechanoreceptors

4.2.1 Touch and Proprioception

These two senses have received little recent attention in fishes. The sense of touch is well developed in fishes and seems to be mediated through the myriad of free nerve endings in the skin (Marshall 1966). Touch probably plays a role in courtship and in responses to substrate. At least in some fish, the same receptors are responsive to temperature and touch. It is not clear how the two types of information are separated (Murray 1971). Proprioception, sensory information about the movement and posture of the body, is present as well, and could provide information about the action of water currents on the fins or other parts of the body. I know of no studies on the degree to which proprioception is used by fish in orientation or migration.

4.2.2 Pressure

Fish are able to respond to relatively small changes in water (hydrostatic) pressure (Blaxter and Tytler 1978). In some species this ability is linked with the presence of a gas-filled swim bladder. Deflation of the swim bladder of the saithe, *Pollachius virens,* for example, reduces the sensitivity to pressure changes by seven to ten times, and reinflation of the bladder restores sensitivity (Tytler and Blaxter 1977).

This sensitivity is not dependent on connections with the inner ear. Qutob (1960) found that blind European minnows still showed good bouyancy regulation and responded to pressure changes of 4–7 cm of water at 70 cm min^{-1} even when the Weberian apparatus was removed. The Weberian apparatus links the swim bladder mechanically to the ear in the Ostariophysi, minnows and related fishes. Changes in pressure induced changes in the discharge rate of sensory nerves from the swim bladder while cannulating the swim bladder and changing its volume by adding or subtracting gas also induced changes in nerve discharge (Qutob 1962).

It seems likely that, in fish with swim bladders, stretch receptors in the bladder provide information on pressure changes. The sensitivity achieved is impressive. Tsvetkov (1969) reported the thresholds of response of six teleosts as being between 4 and 20 mm of water. The weather loach, *Misgurnus fossilis,* responds to changes in pressure of 4–14 mm of water (Tsvetkov 1972). McCutcheon (1966) used a tilting tube to test 12 species of fish. As the tube is tilted toward the vertical, the height of the water column increases. He found thresholds as low as 2 mm of water in some cases. The sea bass, *Centropristis striata*, showed a slight response to 0.4 mm of water, a depth change of 0.004%. Blaxter and Tytler (1972) conditioned haddock, *Melanogrammus aeglefinus,* cod and saithe to pressure changes over a range of 1 to 20 atm. Responses followed Weber's Law with thresholds about 0.5% of the adapted value in cod and saithe and 1.2% in hake.

Swim bladders are not essential for pressure sensitivity. Dabs, *Limanda limanda,* which lack a swim bladder, could be conditioned to pressure changes of 1–2% of the adapted value (Blaxter and Tytler 1972). Larval plaice and larval blennies, *Centronotus gunnellus,* which lack swim bladders, responded to changes in water depth of 25 cm. *Blennius pholis,* which does have a swim bladder, showed a lower threshold, 5–10 cm of water (Qasim et al. 1963). The mechanisms involved in pressure detection without a gas space are not clear. The otoliths of some fish are peizoelectric. They produce electric currents in response to mechanical force. But it has not yet been determined whether this plays any role in sensory reception of pressure changes (Morris and Kittleman 1967).

4.2.3 The Lateral Line System

The remaining mechanoreceptor systems: lateral line, hearing and the detection of gravity and acceleration, are all based on the same basic type of receptor, the sensory hair cell (Pumphrey 1950). This receptor cell has stereocilia arranged in a steplike pattern leading up to a larger kinocilium (a true cilium). When the cilia are bent toward the kinocilium by a water current or similar mechanical stimulus, the cell becomes depolarized. Bending in the opposite direction polarizes the cell (Flock 1971). The depolarization increases the spontaneous discharge rate of the cell.

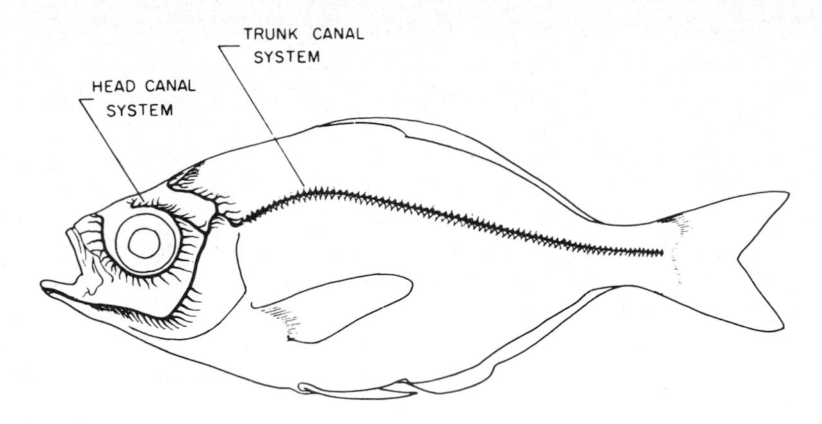

Fig. 4.1. Lateral line system of the walleye surfperch (Walker 1967)

In the fish lateral line system these hair cells are arranged in pairs, facing in opposite directions, and many such pairs are embedded in the base of a gelatinous projection termed a cupula. The cupulas may be encased in canals, opening to the outer surface through pores, or they may be exposed on the outer surface of the fish. The lateral line system gets its name from the prominent line of pores that runs along the side of many fish. However, the sensory system also spreads over the head of the fish and sometimes onto the dorsal surface (Fig. 4.1). Free-standing cupulas may also be scattered over the body.

The lateral line system is primarily sensitive to water displacement. The direction of displacement may be detected by the orientation of the hair cells and perhaps by the pattern of stimulation in the canals, which will be affected by the orientation and anatomy of the canal pores (Cahn 1967; Flock 1971). An object vibrating in water produces displacement waves, which attenuate over short distances (the near field effect) and pressure waves, which travel longer distances (the far field effect). The lateral line organs may be sensitive to the displacement wave, the so-called near field effect (van Bergeijk 1967). In herring and sprat, which have an air chamber associated with a portion of their lateral line system, responses have been detected over a range of 0.01 Hz to 1000 Hz, a range which could include stimuli from depth changes, water movements from nearby fish and low frequency sounds (Denton et al. 1979). Lateral line systems that are not associated with air spaces, e.g., most non-clupeid fishes, may be insensitive to pressure changes associated with depth. In summary, the lateral line will allow the fish to detect small-scale water movements, including surface waves (Schwartz and Hasler 1966; Bleckmann 1980), in its nearby environment.

4.2.4 Hearing

The study of hearing in fishes has a short but vigorous history. Until the early part of the 20th century, fish were thought by many to be deaf, despite the work of Parker (1903). The suggestion that fish could hear had to be defended as re-

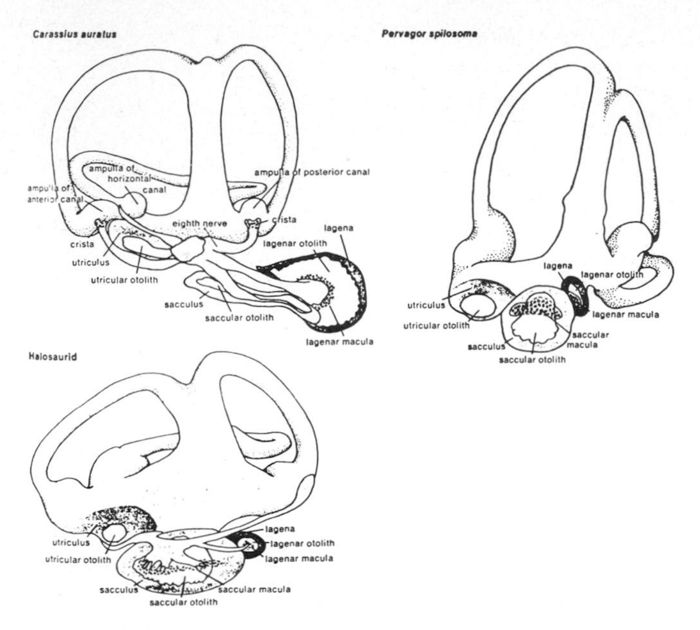

Fig. 4.2. The teleost inner ear consists of three semicircular canals, each ending in a widened area or ampulla, containing sensory epithelial tissue, the cristae and three otolithic organs – the sacculus, lagena, and utriculus. Each otolithic organ contains a single hard otolith of calcium carbonate and an area of sensory epithelium, the maculae. (The utricular maculae and certain other features are not visible in these views.) Some of the variability of these organs in different species is evident here in schematic drawings of the inner ear of goldfish, *Carrassius auratus,* an ostariophysian; a filefish, *Pervagor spilosoma,* a monocanthid; and a deep-sea halosaurid, genus and species unknown (Popper and Coombs 1980)

cently as 1938 (e.g., Parker 1903; von Frisch 1923, 1938: reprinted in Tavolga 1976). Since that time there has been an increasing number of studies published, perhaps stimulated initially by the use of hydrophones for submarine detection during the second world war. The placement of hydrophones near harbour entrances and military installations led to the realization that the sea was noisy. Fish and many invertebrates produced loud, biologically significant sounds. Although the war experience led initially to little more than the identification of the sound-producing organisms, the foundation was laid for extensive studies of sound production and hearing of fish and for the intense modern interest in the subject (Schuijf and Hawkins 1976; Tavolga 1976; Fay 1978; Fay and Popper 1980; Popper and Coombs 1980; Schuijf and Buwalda 1980; Tavolga et al. 1981). In spite of the research activity in this field, definitive information on fish hearing was slow to emerge, perhaps because of the difficulty and complexity of underwater acoustics and the relatively inaccessible location of the sense organs themselves.

The fish ear is made up of two portions, the pars superior composed of the three semicircular canals, and the utriculus and the pars inferior made up of the sacculus and lagena (Fig. 4.2). The pars superior has generally been considered to sense changes in equilibrium and acceleration, the vestibular senses. The pars inferior may be responsible for hearing in most teleosts. Each of the semicircular canals contains a sensitive area, the crista, surmounted by a gelatinous cupula, just as in the lateral line. The crista is placed in a widened area of the canal, the ampulla. The utriculus, sacculus, and lagena each contain a dense bony otolith. In elasmobranchs the otoliths are gelatinous and contain embedded calcium carbonate crystals (Popper and Coombs 1980). Both the cristae and the otolith organs depend ultimately on the same type of sensory hair cells as already described for the lateral line. These hair cells are arranged in patterns so that the movement of the cupula or otolith will bend the cilia and change the rate of discharge of the cells.

In teleosts sound may initiate movement of the otolith through two different routes; by direct action on the otolith or indirectly through the swim bladder. The soft tissues of fish have about the same density as water; hence, sound waves, (pressure waves) will travel through the body without causing much relative displacement of the various body parts. The otoliths, however, are substantially harder than most of the tissues and may vibrate out of phase with the rest of the body and thus move relative to the sensory epithelium, bending the sensory hairs (Dijkgraaf 1960). This route for sound reception will be open to both teleosts and elasmobranchs.

The swim bladder is much less dense than the surrounding water and is compressible. The pressure changes associated with sound will cause the swim bladder to vibrate and these vibrations can then be transmitted to the ear. Several mechanisms to facilitate such transmission are found in fish. The superorder Ostariophysi (e.g., suckers, minnows, catfish, loaches) is characterized by a set of bones, the Weberian ossicles, that transfer vibrations of the swim bladder to a fluid-filled sinus that connects to the sacculi of the two ears. Other taxonomic groups have used different mechanisms to link the swim bladder to the ear. Mooneyes, *Hiodon,* and knifefish, *Notopterus,* for example, have extensions of the swim bladder that run anteriorly and connect with the sacculi. The herring and the sprat, *Sprattus sprattus,* also have anterior extensions of the swim bladder, but they connect with the utriculi, suggesting an auditory role for the utriculus in these clupeids (Blaxter et al. 1979; Denton et al. 1979; Gray and Denton 1979) and perhaps in other fish as well.

The swim bladder may still aid hearing even if there is no direct connection to the ear. Deflation of the cod swim bladder raises the response threshold by 30 dB even though there is no direct connection between the cod swim bladder and the ear (Sand and Enger 1973). There are, however, anterior projections of the gadid swim bladder that terminate on the skull near the ear. Placing a small air-filled balloon next to the head of a swim-bladderless flatfish enhanced its sensitivity to sound (Chapman and Sand 1974). Movements of the swim bladder may induce displacement (near field) movements in body of the fish, which are then detected by the ear, the swim bladder serving as a transformer, converting pressure changes into displacement (Sand 1976).

It is not yet clear how differences in frequency (tone) are decoded by the fish ear. In the cod, different portions of the saccular maculae may respond to different broad frequency ranges (Sand 1974). Popper and Coombs (1980) suggest that the complexly shaped otoliths associated with patterns of orientation and re-

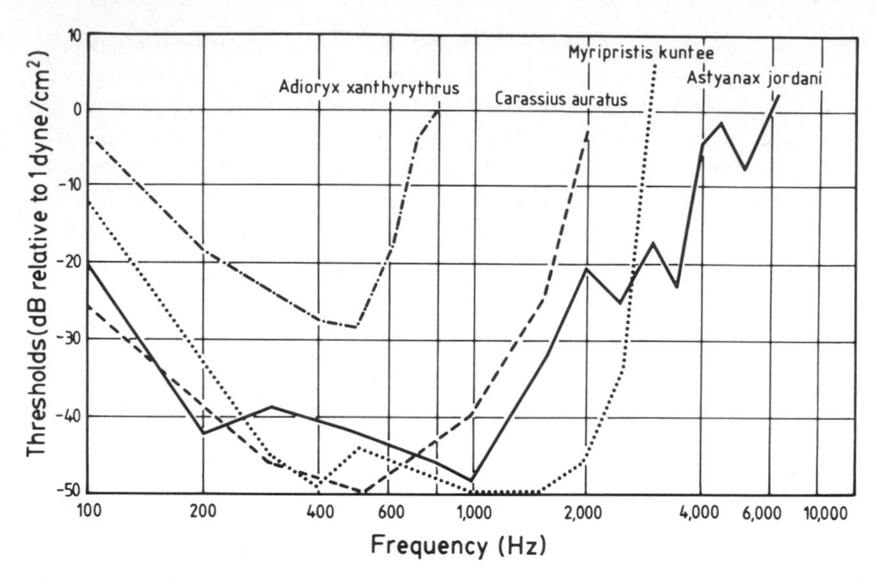

Fig. 4.3. Auditory thresholds for four species of fish determined by conditioning the fish to respond to pure tones ranging from 100 to 7400 Hz. The goldfish and the Mexican blind cave fish, *Astyanax jordani,* are ostariophysians with Weberian ossicles connecting the swim bladder to the ear. The soldierfish, *Myripristis kuntee*, and the squirrelfish, *Adioryx xanthyrythus*, are both in the family Holocentridae, a group of fish that produce sounds as social signals (Popper and Coombs 1980)

sponse threshold in the sensory hairs may allow peripheral decoding of frequency and displacement information, but the exact mechanisms involved are still unresolved.

There are numerous studies of the auditory responses of fish in controlled laboratory settings. These indicate considerable variation in hearing range, with Ostariophysi and fish with similar connections between the ear and swim bladder showing the greatest range (Fig. 4.3). Even in the Ostariophysi, the upper limit of response occurs in the area of 1 to 4 kHz, well below the upper limits of hearing in terrestrial vertebrates. The human upper threshold is 20 kHz, for example. Fish hearing seems to be limited to lower frequencies, but fish do possess intensity discrimination and the ability to distinguish sound in the presence of other masking sounds of similar frequency (Fay and Popper 1980).

The ability to locate the direction of a sound source would be advantageous. Terrestrial vertebrates locate sounds by comparing time of arrival, phase and loudness at the two ears. The use of a single swim bladder for sound reception by both ears could interfere with sound localization in fish, as there are no independent responses by the two ears. The high speed of sound transmission in water would also make the use of time of arrival and phase differences between the two ears more difficult than in air, and the transparency of the fish body to sound in water would reduce the production of sound shadows, which are important in generating loudness differences between the ears of terrestrial vertebrates. These conditions led researchers to conclude that localization of sound by fishes would be poor, at least for the pressure or far-field component of sounds (Van Bergeijk

1964). It is now obvious from behavioural studies that fish can be attracted to sounds from distances well beyond the near field (see Sect. 4.4). Thus a reassessment of the directional hearing ability of fish has occurred (Sand 1976; Fay 1978; Hawkins and Horner 1981; Schuijf 1981), leading to the proposal of several mechanisms including ability to process very short time intervals between the stimulation of the two ears and directional variation in the movement of the otoliths or the sensitivity of the hair cells. Whatever the mechanisms involved, cod can discriminate between two sound stimuli separated by an angle of 22°, an ability that requires two intact labyrinths but not the lateral line (Schuijf 1975).

Fish hearing is apparently extremely variable. As with other sensory systems, some groups have emphasized hearing and developed adaptations that improve the range or sensitivity of their ears. Other groups have presumably depended on other sensory systems with less well-developed hearing abilities. Fish are most sensitive to low frequencies from 100 to about 3000 or 4000 Hz, and can distinguish tone and intensity; some can locate the direction of distant sounds. There is therefore the basic capability present for behavioural responses to biologically significant sounds, at least in the lower frequency range. Combined with tactile and lateral line senses this gives fish a range of response to mechanical stimuli from a few cycles per minute to thousands of cycles per second.

4.2.5 Gravity and Acceleration

As well as detecting the various components of sound, the ear is also sensitive to movements of the fishes' body (Lowenstein 1971). The three semicircular canals each have a widening, the ampulla, which contains a sensitive epithelium of hair cells embedded in a cupula. Movement of the fish around any axis will lead to movement of the canal relative to the fluid it contains, causing a flow that will bend the cupula and affect the hair cells. The spontaneous rate of discharge from these cells is increased by depolarization when the kinocilia are bent away from the stereocilia. The discharge rate decreases with bending in the opposite direction (Lowenstein 1971). Each of the canals is oriented roughly at right angles to the other two, and the hair cells in each canal are polarized so that movement in any direction will be detected. This information is used in controlling eye movements and postural adjustments but has also been proposed as a source of inertial information to keep a fish on course. It is not clear whether fish can detect linear acceleration or not (Arnold 1974).

Gravity acts on the dense material in the otoliths. Changes in the orientation of the fishes' body relative to gravity will change the pattern of bending of the hair cells attached to the otoliths (Lowenstein 1971). Maintenance of normal body orientation does not depend entirely on gravity. Many fish also show a dorsal light response, keeping their dorsal surface toward the source of light, a reaction that will keep them upright in most natural situations. When light and gravity are placed in conflict experimentally, by shining a light on the fish from the side or from below, most fish will compromise, taking an intermediate orientation between the vertical and the direction of light (von Holst 1950; Braemer 1957).

4.3 Mechanoreception and Timing

If mechanoreception has a role in the timing of migration, it may be in relation to social synchronization, tides, weather changes, and periodic flooding. Social signals transmitted by sound could synchronize the activity of migrants, but I do not know of any examples. Each of the other factors, tides, weather, and flooding, is potentially important to migratory fish and could be detected by mechanoreceptors. Thus the potential exists for each to have a role in the timing of migratory behaviour.

4.3.1 Tidal Rhythms

The movement of the moon around the earth takes 24 h and 51 min or 24.8 h. This leads to high tides every 12.4 h (Palmer 1973). Superimposed on the lunar tides are smaller solar tides, with highs 24 h apart. The cycle of changes in water level at a coastal site will be determined by an interaction between the two rhythms, with the most extreme tides occurring when the two cycles are in synchrony. The actual change in water level and the timing of the tides at a specific location will also be determined by the local hydrographic conditions. For example, if the tides have to flow around islands or through narrow channels then the time of high or low tide may be offset from the time predicted by celestial events.

Fish migrating through coastal areas will be subjected to changes in water depth and illumination. Rising water will cut down the light reaching a bottom-dwelling fish, with each tidal cycle. Those living in the intertidal zone itself will experience changes in access the surface area as the shoreline is alternately flooded and exposed, along with changes in temperature and turbulence, as the surf line moves up and down the beach and as the incoming water is warmed or cooled by the land it floods. Some fish, such as juvenile plaice, *Pleuronectes platessa;* flounder, *Platichthyes flesus;* and turbot, *Scophthalmus maximus*, move in and out with the tide, staying at about the same depth as they forage over the covered intertidal zone (Tyler 1971; Gibson 1976). Others wait out low tides in home tide pools and forage at high tide, or, if displaced, perform their homing movements at high tide as in the British Columbian tidepool sculpin, *Oligocottus maculosus* (Green 1971).

Cycles of activity that correspond to the tidal cycle have been found in several fishes when held under constant conditions (Gibson 1978). Green's (1971) tidepool cottid maintained a tidal rhythm for 1 or 2 days when tested under constant conditions in a cage with an activity meter attached to a false bottom. Cottids are negatively buoyant and their weight on the false bottom was detected by the meter. Plaice, flounder, and turbot also maintained tidal activity rhythms as measured by breaking an infrared beam, for 1–3 days under constant conditions (Gibson 1973, 1976). Persistant tidal rhythms occur in *Coyphoblennius galerita* (Gibson 1971) and *Blennius pholis* (Gibson 1965, 1967, 1971) and the hogchoker, *Trinectes maculatus,* maintains peaks of activity corresponding to the times of slack tide for at least six cycles in constant light (O'Connor 1972).

The role of tidal cycles in the timing of migratory behaviour is not clear. Migratory fish sometimes select tidal currents for migration but the selection may be based on chemical (see Sect. 3.3.3) or rheotactic cues rather than depth or pressure changes. Coastal migrants presumably must avoid stranding at low tide, but this point is seldom studied.

4.3.2 Weather and Barometric Pressure Changes

Weather may be important to fish. The idea that storms or other sudden changes in barometric pressure may effect terrestrial animals is familiar, as is the folklore surrounding the ability of such animals to anticipate meteorological changes. This folklore may have some basis in fact. Both the homing pigeon, *Columba livea* (Kreithen and Keeton 1974) and the pocket mouse, *Perognathus longimembris* (Hayden and Lindberg 1969), are capable of responding to changes in barometric pressure of the order of magnitude of the changes that accompany weather disturbances. Humans, of course, have used atmospheric pressure, measured by barometers, as an indicator of weather changes for many years. For shallow-water fish, at least, changes in weather may bring changes in water temperature, salinity, turbidity, and turbulence, higher or lower tides from the effects of onshore or offshore winds and changes in the flow level of streams.

The weather loach, *Misgurnus fossilis*, a common aquarium fish, is named for its supposed ability to signal a change in the weather by increasing its activity. Tsvetkov (1972) did find that the weather loach is sensitive to small changes in pressure, although whether it can use this ability to predict the weather is still not established.

Fish migrating in streams may be strongly affected by rainstorms and thus benefit by anticipating weather changes. Peterson (1972) noted that the number of rainbow trout trapped while entering their spawning area increased during periods of dropping barometric pressure and decreased during rising barometric pressure (Fig. 4.4). The decrease in pressure, rather than an absolute threshold pressure, seemed to stimulate movement. The annual timing and duration of the

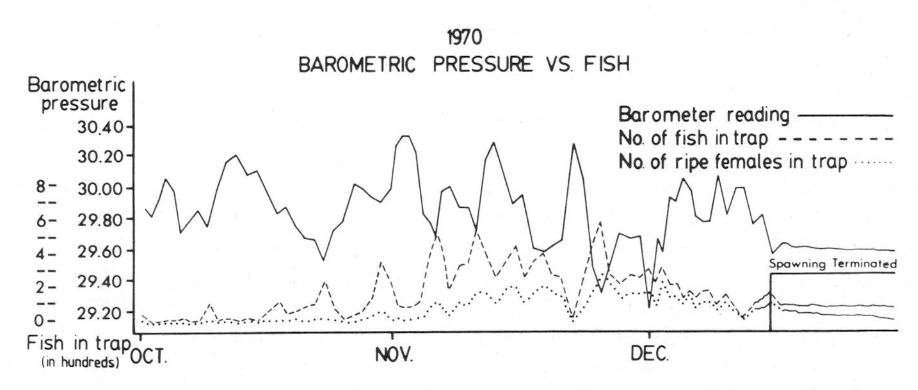

Fig. 4.4 Barometric pressure and the number of rainbow trout entering a trap on their spawning migration in 1970 (Peterson 1972)

Table 4.1. Coefficients of correlation between physical factors in the environment and the number of coho and chinook salmon returning to the University of Washington hatchery during the 1953–1954 migratory season (Allen 1959)

Factor	Correlation coefficient	
	Coho	Chinook
Total radiation	+ 0.06	− 0.19
Air temperature	+ 0.09	+ 0.12
Total daily precipitation	− 0.07	+ 0.30
Day-time precipitation	− 0.17	+ 0.06
Night-time precipitation	+ 0.16	+ 0.63
Barometric pressure	− 0.50	− 0.20
Water temperature	+ 0.03	+ 0.11
Water discharge	+ 0.27	+ 0.20
Number of fish	28	76
Degrees of freedom	26	74
1% level of significance	0.49	0.29

spawning season were probably controlled by temperature or photoperiod with barometric pressure only affecting variations within the spawning season. Falling barometric pressure often signals the passage of a frontal system and could indicate rainfall or other biologically significant changes. Peterson (1972) found some effect of cloud cover but little relationship between temperature and spawning activity.

Chinook and coho salmon returning to the University of Washington hatchery were trapped as they reached the top of a short fish ladder leading into the hatchery (Allen 1959). Entry of chinooks into the trap correlated most strongly with falling barometric pressure while coho entry correlated with natural precipitation (Table 4.1). Falling barometric pressure and precipitation often coincided. Allen (1959) suggests that the species difference is due to a difference in spawning habitats. Chinooks tend to use large rivers where additional rainfall might not be required for entry at the river mouth, although the frontal weather indicated by falling barometric pressure might be expected to provide greater stream depth when the fish reached the headwaters. Coho, on the other hand, tend to spawn in smaller streams where actual precipitation would facilitate entry into the stream.

4.3.3 Microseisms: Small Earthquakes

A unique suggestion for a timing mechanism arises from the seaward migration of silver eels in the Netherlands. Deelder (1954) noted, as have others (see Sect. 2.3.3.2.1), that eels tended to migrate on dark, stormy nights with rain. He also noted, however, that peaks of migration sometimes occurred on nights when these weather conditions did not occur. In a sample of 28 migration peaks, six peaks were not accompanied by bad weather, but in each of the 28 cases the weather of the Netherlands was dominated by a meteorological depression. As well as changes in weather, barometric lows produce microseisms, possibly through the effects of the high waves under the center of the low. In the Netherlands these depression-induced microseisms have an amplitude of 1 to 20 μm that varies with the intensity of the low and they have a period that varies with the location of the centre of low pressure. Each of the 28 cases of peaks of silver eel migration was associated with microseisms of about 3 s in period length, furthermore there were no cases of microseisms of about 3-s period that were not associated with peaks of eel migration. Thus in addition to light and precipitation, eels may be keying on seismic changes or possibly barometric changes. The adaptive significance of responding to stimuli associated with lows might be to synchronize the runs with periods of low illumination and/or high water normally associated with weather fronts. Deelder (1954) suggests that microseisms might account for the synchronization noted by other workers (e.g., Lowe 1952) between the migration of captive eels and peaks of migration by wild populations. Barometric pressure responses might also provide such synchronization.

4.3.4 Changes in Water Level

Freshwater fish, particularly those dwelling in rivers and streams, and fish that migrate through such environments, may be strongly affected by changes in water level, whether flooding or low water. Such changes might be detected through mechanoreceptors, particularly by a pressure sense detecting changes in depth or the lateral line detecting increased turbulence. Changes in water level are also accompanied by other stimuli including changes in temperature, turbidity and chemical composition. There are numerous reports of the movement of anadromous migrants into streams being timed by floods or freshets (Alabaster 1970). Rainbow and brown trout move upstream and spawn in response to freshets (Dodge and MacCrimmon 1971; Munroe and Balmain 1956), as do Atlantic salmon (Huntsman 1948; Hayes 1953) while seaward migrant European eels migrate downstream during flood conditions (Frost 1950). However, it is not clear what sensory stimuli are used in coordinating these movements with the flood conditions.

Rivers and streams provide an environment where the direction of flow will be predictable over geological time spans, and the presence of flow alone may be an adequate cue for adaptive migrations. Young alewives, *Alosa pseudoharengus,* tend to move downstream on days of heavy water flow (Kissil 1974; Richkus 1975a). Although the movements downstream also correlated with decreases in temperature and increased precipitation, experiments with young alewives in a choice apparatus indicated that changes in flow were sufficient to induce downstream movement by themselves (Richkus 1975b). A velocity of 9.4 cm s^{-1} was more effective than the lower rate of 7.1 cm s^{-1}. The downstream movement of rainbow trout fry in a tributary of Sagehen Creek may be governed by a similar mechanism; at least downstream movement takes place during periods of high stream discharge (Erman and Leidy 1975). It is difficult to distinguish downstream displacement by high flow rate from the release of downstream movement by flow acting as a releasing stimulus.

4.4 Mechanical Direction and Distance

Several different types of mechanical stimuli could indicate direction or distance. Water currents will give consistent direction information in rivers and streams and in tidal regions. The physical aspects of the habitat, rocks, coral heads, plants, and substrate type will provide landmarks, which could provide information for pilotage. Pressure will be a reliable indicator of depth. Sounds have been used in attempts to divert migrant fish and inertial guidance through the mechanoreceptors in the ear is a possibility. Each of these subjects is sufficiently distinct to warrant separate discussion.

4.4.1 Water Currents

Water currents are a fundamental attribute of natural water bodies (Arnold 1974). They may act on migrant fish in three main ways. First, velocity may act as a limiting factor, excluding fish from portions of the environment. Second, currents may carry fish toward the goal of their migration, increasing the speed and decreasing the energy requirements of the migrant. Third, currents may act as directional cues, indicating the appropriate swimming direction.

4.4.1.1 Water Velocity as a Limiting Factor

High velocity may limit the access of migratory fish by physically preventing entry to a region of fast water. The threshold for this type of exclusion will depend on the degree to which the fish is morphologically and behaviourally adapted to holding position or moving upstream in high flow rates. Arnold (1974) reviews adaptations to high velocity environments. The drag of the current can be reduced by streamlining. Negative buoyancy makes it easier for the fish to stay in the boundary layer of low velocity water near the bottom and may increase the traction of gripping mechanisms such as fins or suction disks. Loss of buoyancy may be obligate or facultative. Many fast-water and bottom-dwelling fish lack a swim bladder altogether or lose the swim bladder early in development. In others, the swim bladder is retained and its size (and hence the fishes' buoyancy) can be adjusted to the flow conditions (Gee and Machniak 1972; Machniak and Gee 1975), so that the fish becomes heavier in fast water and lighter in slow water. Larger fish are able to hold position better in fast water than are small fish (Arnold 1974). Everest and Chapman (1972) for example, found a correlation of +0.92 between the body length of juvenile chinook salmon and the water velocity at the focal point of their territory, while in juvenile steelhead trout body size showed approximately a +0.75 correlation with focal velocity. These considerations will also apply to migratory fish moving from one flow regime to another.

4.4.1.2 Transport by Water Currents

The use of water currents for transport along a migration route depends on selection of a water mass that is moving in the appropriate direction. For juvenile fish, there are two points in the species life cycle at which the selection may occur. The young may select the water current through their own behavioural responses, or the water mass could be selected behaviourally by the parental generation. Elvers select onshore tidal currents by swimming up when the water is not scented by freshwater odours (Creutzberg 1959, 1961, 1963). Adult salmonids spawning in inlet streams of a nursery lake place their young in a geographical location where the water current will reliably carry them toward the lake, in which case the young need not recognize water direction or identify a water mass but merely swim up off the bottom after hatching.

A fish might reliably enter water currents in the preferred migration direction by swimming up into midwater at specific times. The most obvious example would be transport by tidal currents with the fish timing its excursion to coincide with either the ebb or flood tide. This "modulated drift" does occur (see Sect. 4.4.1.2.2).

The cases of chemical, thermal, electrical or genetic control of drift are discussed under the appropriate headings in other chapters. Here I will deal with cases in which the key stimuli are unknown or are mechanical in nature.

4.4.1.2.1 Passive Drift

Pelagic eggs or very small larvae may drift passively in currents (Harden Jones 1968; Arnold 1974). An example may be the movement of eels from the presumed spawning grounds in the Sargasso Sea to the edge of the continental shelves of Europe and North America. The passive phase of the migration probably ends with metamorphosis of the leptocephalus into the eel-shaped elver or glass eel (Tesch 1977).

Although the migration of young skipjack tuna, *Katsuwonus pelamis,* into the eastern Pacific has been the subject of both active and passive migration models (Seckel 1972; Williams 1972), the question is still unresolved. Downstream migration of newly hatched pike-perch, *Lucioperca lucioperca,* may be an example of passive drift but again not all workers agree (Belyy 1972). LaBar et al. (1978) used ultrasonic transmitters to track nine hatchery-reared Atlantic salmon smolts through the estuary of the Penobscot River in Maine and found that a combination of river currents and tidal flow could account for the movements observed.

4.4.1.2.2 Modulated Drift

The effective use of passive drift still requires appropriate entry into the current and exit from it (Harden Jones 1968), so use of drift may not be as easy as might be naively thought. Use of water currents to assist active migration could certainly be adaptive for many fish. Pettersson (1926) pointed out that fish could conceivably use the various ocean currents much as mariners used the trade winds, selecting currents that were moving in the appropriate direction for their stage of migration. This would be a form of what Arnold (1974) terms modulated drift. Houde and Forney (1970) found that very young, 7-mm walleye, *Stizostedion vitreum,* tended to stay near the bottom and be carried westward by deep currents in Oneida Lake, when they lost their yolk sacs they swam at the surface and were carried back to the east by surface currents. The adaptive significance of these drift patterns is not clear.

Coastal fish may use selective tidal stream transport, selecting those phases of the tidal current that are moving in the direction of their migration (see Sect. 3.3.3). Flatfish are particularly well suited to the use of modulated drift. Their negative buoyancy and streamlined, bottom-hugging shape allow them to ride out adverse currents with little or no displacement, and the assistance of favourable currents will be most useful to such slow-swimming fish. Some North Sea plaice migrate southward in the autumn and spawn in the Southern Bight of the North Sea, then migrate North again during the winter (Greer Walker et al. 1978; Harden Jones et al. 1979). There are also other plaice populations in the North Sea, with other spawning grounds and migratory routes.

Greer Walker et al. (1978) attached acoustic transponding tags to 12 plaice and released them off Lowestoft, U.K., in the region of the migration route. Transponding tags send out a signal when they are "interrogated" by the sonar on the tracking ship (Greer Walker et al. 1971). Their return signal then shows on the sonar display as a distinctive echo. Sector-scanning sonar allows the detection of the range, depth and bearing of the transponder equipped fish. The transponding plaice showed modulated drift, or to use the terminology of Greer Walker et al. (1978), "selective tidal stream transport." The track of plaice number 7 illustrates this method of transport (Fig. 4.5). The fish moved off the bottom during the northward-flowing ebb tides and returned to the bottom again near low water slack tide. During the southward-flowing flood tides it stayed on the bottom, and moved around slowly. This tactic of moving off the bottom during one direction of tidal flow but not the other allowed plaice 7 to cover 43 km in 26 h. In total, 8 of the 12 tagged plaice showed active movement, defined as traveling over 15 km. Of these, six moved North and two moved South and all eight used selective tidal stream transport. (Of the other four, one tag failed and three fish were inactive.)

The tagged plaice had been held in captivity for about 2 months. It was not known if they were in migratory condition or to which spawning stock they belonged. These results therefore indicate that plaice are capable of using selective tidal stream transport, but the movements of the experimental fish cannot be directly related to natural migration times or routes.

The results allow the formulation of a testable hypothesis about the natural migrations in the region. If selective tidal stream transport is used, then during the southward migration toward the Southern Bight in the fall one should find more ripe fish in midwater during the southward-flowing tide than during the northward-flowing tide. During the return winter migration to the north, more plaice should be caught in midwater during northward-flowing tides and they

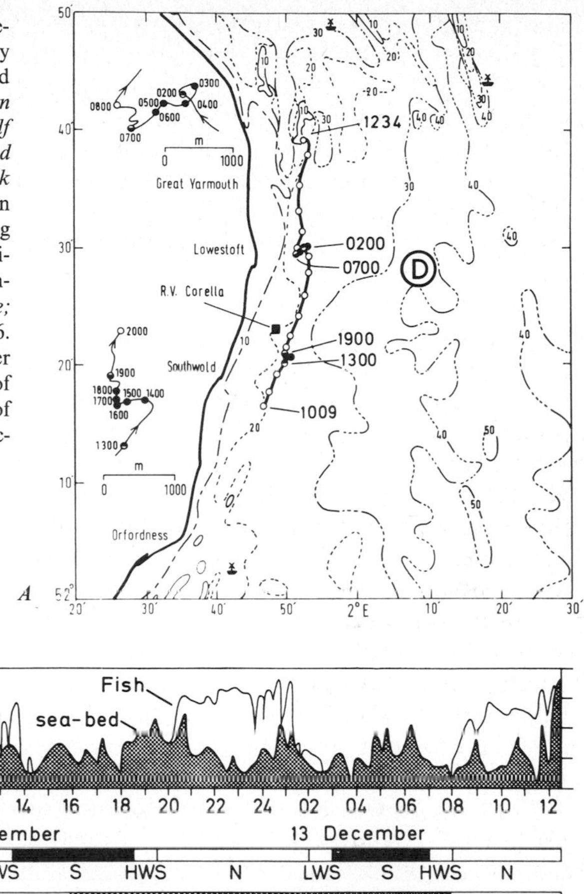

Fig. 4.5. A Track chart of plaice 7, released at 10:09 h, Dec. 12, 1971. Hourly positions of the fish are indicated and the times of slack water are given. *Open circles* north-going tides; *Upper half black circles* low water slack; *closed circles* south-going; *lower half black circles* high water slack. *Insert figures* on the land show details of the track during the south-going tides at the scale indicated. The position of RV Corella's anchor station is shown by a *black square; D* Station D referred to in Fig. 4.6. Depth contours in meters (Greer Walker et al. 1978). *B* The depth of plaice 7 in relation to the direction of the tide and other environmental factors (Greer Walker et al. 1978)

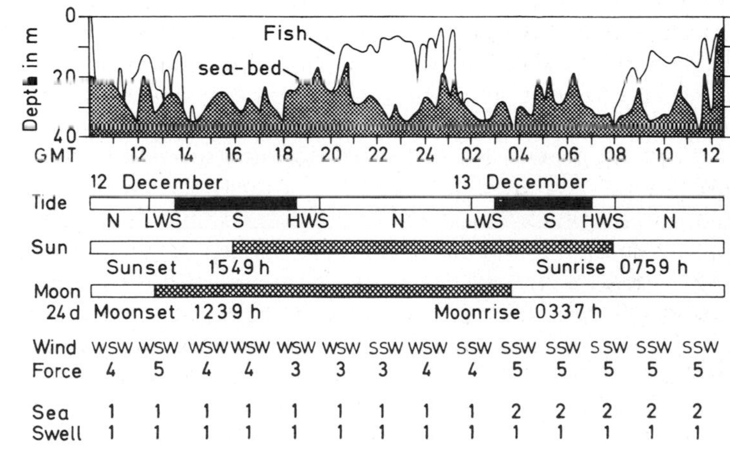

should be spent, post-spawning fish. Harden Jones et al. (1979) tested and confirmed this hypothesis by fishing along the migration route with a midwater trawl. They made a series of paired midwater trawl tows, one tow of each pair during the ebb, the other during the flood. During November and December, more plaice were in midwater on the southerly tide, while during February more were in midwater on the northerly tide (Table 4.2). As further confirmation, ripe fish predominated in November and December catches, while many of the February fish were spent. Several diferent spawning populations occur in the region. Some of the fish caught on the "wrong" tide may have belonged to these other populations.

Table 4.2. A comparison of plaice catches in paired hauls of an Engle trawl on consecutive northerly and southerly tides in the Southern Bight of the North Sea

Months	No. of paired tows	More plaice caught on tide flowing		Catches equal
		Northerly	Southerly	
Nov., Dec.	39	5	33	1
Feb.	21	20	1	0

Another case of selective tidal stream transport in plaice has been reported. Stage-4 and -5 plaice larvae are buoyant and swim upright. They approach the inshore nursery areas by swimming up in the flood tides and staying near the bottom on the ebb (Creutzberg et al. 1978). This brings them inshore to the areas where they settle on the bottom and metamorphose into O-group plaice (see Sects. 4.3.1 and 3.4.4.8). "Artificial tides" in experimental tanks indicated that the behavioural responses to ebb and flood were not due to changes in temperature, salinity or odour. Depth or pressure changes were not tested. Lack of food induced midwater swimming, but on a time scale of days rather than the short-term time scale appropriate for synchronization with the tides.

The elements of selective tidal-stream transport deserve some examination. Ramster and Medler (1978) used self-recording current meters, anchored in the study area off Lowestoft, to record the speed and direction of tidal currents during the tracking study on adult plaice. Current direction shows a sharp reversal at the time of slack water, but current speed changes more gradually with minimum velocities at slack water and maximum velocities near the midpoint of the tide (Fig. 4.6). As expected, the flow is faster near the surface than near the bottom and the time of change differs by a few minutes between the surface and the bottom. Near the time of slack water, current speeds are slow, so there is little to be gained by precise timing of movement off the bottom. As long as the fish is in the water column during the period of rapid flow in the preferred direction and back on the bottom before the current flows strongly in the opposing direction, it will benefit. The directional change at slack water or the velocity minimum might be adequate timing stimuli. The fish would have to distinguish high-water slack from low-water slack by use of its pressure sense or some other indicator in order to make the correct choices. Some of the tracked plaice showed increased activity at slack water (Greer Walker et al. 1978).

The researchers involved do not know what cues are used for selecting the appropriate tides (Greer Walker et al. 1978; Harden Jones et al. 1979). The tracking system that they have developed would seem to lend itself to experiments using sensory deprivation. One could release and track blind or anosmic plaice or plaice with magnets or Helmholtz coils attached to their bodies. Such experiments might indicate which sense receptors were used in sensing the tidal change.

Selective tidal-stream transport has also been the subject of theoretical analysis. Weihs (1978) calculated that energy savings in the order of 40–90% might be achieved through use of the tidal currents instead of constant swimming. Tidal transport will lead to greater energy savings for slow-swimming than for fast-

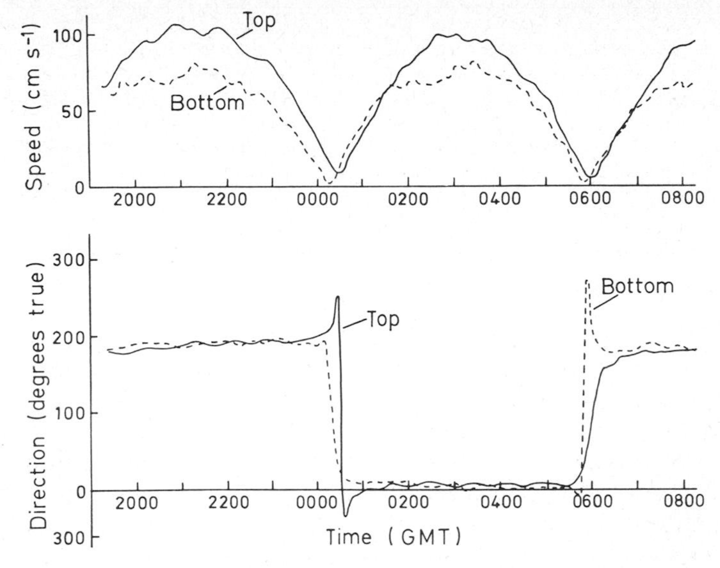

Fig. 4.6. Ten-min values of current speed and direction in the near-surface and near-bottom layers at station D (see Fig. 4.5 A) (Ramster and Medler 1978)

swimming fish, although both can benefit. Fish, e.g., small larvae that cannot swim faster than the tidal velocity will fare better by just drifting than by swimming actively with the current, while fish able to swim faster than the tidal velocity will benefit by actively swimming along in the direction of the current flow. Obviously, swimming with the flow requires more information than just drifting. A fish that must swim actively to maintain position will use more energy between favourable tides than a fish, like the plaice, that can lie on the bottom or even bury itself in the sand.

Weihs (1978) model is based on the assumption that if migratory fish do not use selective tidal transport they would attempt to maintain a constant speed over the ground during both favourable and opposing tides. This assumption may overestimate the energy cost of non-tidal migration, and hence overestimate the savings to be achieved through use of the tide. It seems at least as likely that migrants would regulate their speed through the water and avoid high-speed, and therefore costly, swimming during the opposing tide. Lateral line or proprioceptor stimuli should provide adequate information regarding speed through the water, while visual contact with the bottom or some similar reference point may be required for maintenance of a constant ground speed.

For fish like the plaice it would be advantageous to swim along with the current in the direction of migration. This requires that the fish be able to maintain a compass heading, as well as identify the time and location of favourable water currents. To ascertain whether the tracked plaice were swimming or drifting, the researchers had to separate movement due to water flow from movement due to the fish's activity. For this they needed accurate information on current direction at the depth and location of the tracked fish (Greer Walker et al. 1978; Ramster

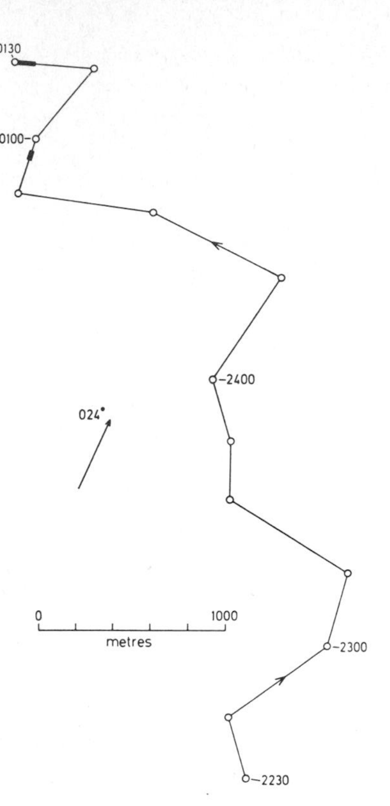

Fig. 4.7. Progressive vector diagram to show the movement of plaice 7 through the water, taking the relative movement of the water currents into account. Positions are shown each 15 min. *Thickened parts of the track* periods when the fish made excursions to the bottom. *Arrow* direction of the tide (024°) (Greer Walker et al. 1978)

and Medler 1978). Current measurements were taken from the tracking vessel, from an accompanying vessel and from automatic recorders anchored in the study area (Fig. 4.5). By comparing the current information with the movements of the tracked fish, they were able to work out the velocity of the fish relative to the water, in other words, the direction in which the fish were swimming (Fig. 4.7). Several of the fish maintained fairly steady headings, indicating that they might be able to orient and swim in the direction of migration as well as drifting with the appropriate currents. De Veen (1970) had found evidence that soles used selective tidal transport at night and that they were oriented in the direction of migration (see Sect. 2.4.2.3).

It is difficult to tell how widespread the use of modulated drift is in fish migration. In part this is because it will be most effectively used by small or slow-moving fish such as larval fish or small non-commercial species and in part because the difference between accidental displacement and modulated drift as a form of migration will be hard to discern without painstaking observations and experiment. However, modulated drift has also been reported in cod, *Gadus morhua* (Harden Jones 1977), and in European and American elvers (Creutzberg 1961; McCleave and Kleckner 1982) so it may be fairly common.

Selective tidal-stream transport will be possible with cues derived from pressure, lunar illumination, chemical difference between ebb and flood and differences in light penetration to the bottom. Modulated drift in less predictable environments may depend on response to the velocity and direction of the current itself. One of the greatest problems may be recognizing the end of a period of suitable tidal flow when the fish is up in the water column away from sensory contact with the bottom (McCleave and Kleckner 1982). Turbulence may change with rate of current flow, but it is not clear whether chemical information would penetrate the moving column of water. McCleave and Kleckner (1982) suggest for their American elvers that circa-tidal rhythms or responses to electrical stimuli originating from the water currents might be involved in detected the changes in tidal currents. Even in the use of tidal streams an animal able to sense the direction and the velocity of the current might be in a better position to save energy than an animal using secondary indicators such as water chemistry. Plaice, for example, can apparently maintain constant headings, indicating a sense of direction, based on some unknown source of information. When they are on the bottom they should also have information available on the direction and velocity of water currents, through tactile or visual responses if they drift with the current or through tactile, visual or lateral line responses to the water flowing past if they remain in place on the bottom. Particles in the flowing water, for example, could indicate the direction of current to the fish visually if it remained on the bottom.

4.4.1.3 Rheotaxis

Fish may orient to the direction of water currents instead of merely drifting with them. The term rheotaxis has often been used to refer to directed responses to water currents. Arnold (1974) proposes that rheotaxis should be restricted to such oriented or directed responses and also recommends the use of the term rheotropism as a more general term covering both drift and oriented responses. Many of the oriented responses of fish to currents are in fact visual or tactile responses to relative movement of the substrate or other objects. Such responses have been termed pseudo-rheotropism (Gray 1937), but in modern usage they are referred to as rheotaxis or rheotropism, depending on the degree of orientation involved (Arnold 1974).

Fish may respond to the sight of the bottom or shore moving past as the current carries them along. They can orient downstream, negative rheotaxis, by maximizing the rate of movement, hold position by swimming so that the visual reference point remains stationary, or they can move upstream by swimming so that the reference point moves in the opposite direction to that produced by drift. Both of these latter cases are referred to as positive rheotaxis.

It was early established that these responses could be produced by objects moving in the visual field of the fish without the presence of moving water. Lyon (1904) found that "Silverside minnows" in a corked bottle swam to the upstream end of the bottle when it was allowed to drift downstream and to the downstream end when the bottle was towed upstream. There would be no current in the bottle, so the fish were responding visually to the apparent movement of the surroundings and swimming so as to maintain their position, positive rheotaxis. In earlier experiments Lyon found that a patterned background moved under a glass-bottomed tank or past the sides of a tank would also induce responses consistent with

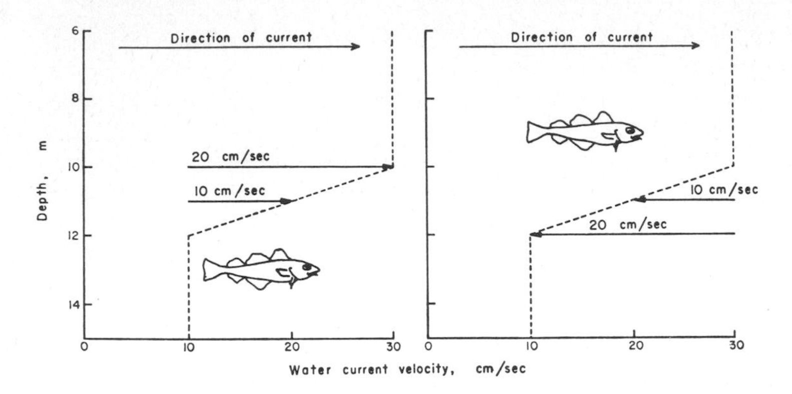

Fig. 4.8. Orientation at the rheocline. A fish lying below or above the rheocline could see particles moving with the relative velocities indicated *alongside the arrows* (Harden Jones 1968)

visually mediated positive rheotaxis. The fish would move along with the background, maintaining their position. Such "optomotor" responses have since been found in a number of fish species (Arnold 1974).

Fish are, however, able to hold their position in a current even in darkness. In some cases this may be accomplished through response to tactile stimuli. Lyon (1904) blinded *Fundulus* by enucleation and placed them in a wooden trough or in a flowing tideway. They drifted downstream until they touched some solid object – the bottom or vegetation – then turned upstream in response to the tactile stimulus. If they lost contact with the bottom they drifted downstream again until contact was re-established.

Lyon (1904) argued persuasively that fish in a uniform stream of water would not be able to perceive their movement over the ground unless they were in visual or tactile contact with some static reference such as aquatic vegetation or the bottom or sides of the stream. Harden Jones (1968) suggests a variant of this view. Fish in a uniform current but near a "rheocline," a boundary layer between currents of different velocities, might be able to see particles in the adjacent current and use their relative movement to orient to the flow (Fig. 4.8). This is still visual rheotaxis. Social interaction between the fish in a school could extend the distance at which fish are able to maintain reference to the bottom or other fixed objects. Hemmings (reported by Harden Jones 1968) has observed a school of *Chromis chromis* stretched out so that only those closest to shore could be in visual contact with shore but the whole school was holding position by the "visual chain" response to each other.

Since large uniform movements of water generate electrical currents as they move through the earth's magnetic field, fish might be able to perceive these electric currents and use them in orientation (see Sect. 6.4.1).

Small-scale water currents, in contrast to large uniform water currents, might provide stimuli to the lateral line or to touch receptors in the skin, allowing the fish to localize the current. Lyon (1904) found that blind stickleback (species not reported) were able to orient in midwater to a gushing stream of water entering

Fig. 4.9. Trout holding position downstream and to one side of a boulder in a stream (Sutterlin and Waddy 1975)

quiet water. In such a case there would be velocity gradients over the length of the fish, which might be detected by lateral line, touch, proprioception or by acceleration detectors in the inner ear, or some combination of these receptors. The observation of Deelder (1958) that elvers orient to areas of strong currents near sluice gates might be an example of this type of orientation.

Brook trout in a flowing stream take position just downstream and to one side of boulders or similar obstacles and hold position there with very little active swimming (Sutterlin and Waddy 1975) (Fig. 4.9). This entrainment leads to considerable energy saving over swimming in an unobstructed flow and allows the fish to hold position against a current twice as strong as that which will sweep them downstream without the obstruction. This response can be maintained in daylight even when the posterior lateral line nerves have been removed bilaterally. Unilateral removal of the nerve leads to asymmetrical position holding even in daylight. In the dark, fish with intact lateral lines are able to hold position, but those with the lateral line nerve removed are either unable to entrain on the object or do so by tactile cues, touching their nose to the object. These results of Sutterlin and Waddy (1975) indicate that there is an interaction of visual and lateral line stimuli in the control of object entrainment. Vision predominates under illuminated conditions, although the lateral line still has an effect, as indicated by the asymmetrical response when one line is not functioning. The lateral line predominates in the dark with tactile responses serving as a back-up system. In an experimental flume, cylinders as small as 3 mm in diameter were used by the trout.

The value to migrant fish of such responses to small-scale hydrographic conditions is illustrated by the calculations of Osborne (1961). Using engineering principles, he calculated the energy required to move a streamlined rigid body of the same length and surface area as a salmon along the migration routes taken by chinook and sockeye salmon moving up the Columbia River and by chum salmon migrating up the Amur River. He then compared this energy cost with the energy utilized by actual migrating fish, calculated by comparing weight and body composition of fish at the river entrance with those reaching the spawning areas. Pacific salmon do not eat during the freshwater migration. The energy used by migrating fish was so much less than the energy requirements calculated for rigid bodies that Osborne (1961) considered the relationship almost paradoxical. He suggested that the fish are able to reduce energy costs of migration by achieving lower drag than rigid bodies, and by selecting favourable velocities in the turbulent flow of the stream.

Small-scale turbulence may allow free-swimming animals to detect the direction of flow of the major current. Arnold (1974) discounts the possibility of turbulence as a directional cue for fish, but Nisbet (1955) contends that birds might use wind turbulence to detect wind direction in the absence of visual reference points. He states that wind gusts tend to be strongest in the direction of the mean flow and terms this effect gust anisotrophy. Gusts would be isotrophic if they showed equal velocity distributions in all directions. Gusts are apparently also asymmetric, tending to be elongated in the direction of the main flow. These conditions occur up to 1000 m above ground level and might be detected by birds through the effects of acceleration on the inner ear and the flight adjustments required to maintain flying speed in variable wind velocities. The same general principles should apply in water currents moving over variable bottom terrain, although the magnitude of the effect would be altered by the fluid characteristics of water.

In areas where surface currents are driven by prevailing winds, the movements of surface waves might provide an indication of current direction. Such waves could be detected visually either directly, or indirectly through their effects on underwater light patterns. They can also be detected by lateral line organs (Schwartz 1965; Schwartz and Hasler 1966; Bleckmann 1980).

Of course an animal with the ability to determine its position in space from celestial or other information, as homing pigeons may be able to do, could use the changes in position to determine drift caused by currents. No established case of such behaviour has been documented in fish.

The use of water current as a directional cue in river and stream migration has seemed so obvious that it has received little analysis. If a fish is capable of perceiving the direction of flow, by any of the several mechanisms outlined, then it can use stream direction as a reliable indicator of direction. In any anadromous fish an initial downstream movement will take it to the ocean and a subsequent upstream movement will return it to upstream spawning areas. The complexities of deciding which stream to re-enter have been discussed in Chap. 3. The situation is reversed for catadromous fishes such as the Atlantic eels; in animals moving in and out of lakes, there is also the added complexity of inlet streams and outlet streams (see Chaps. 2 and 3). In each case, however, positive or negative rheotaxis at the appropriate season or developmental stage will serve as a reliable mechanism for achieving movement in the right direction.

The use of the movement of ocean currents for directional information is more difficult to establish. Experimental work in the open ocean is difficult. Although migrating fish may move along ocean currents (Arnold 1974; Harden Jones 1968), it is not clear if any use the current direction as an orientation cue.

There is no reason to assume that orientation to water currents is limited to upstream and downstream alone. Animals are able to maintain angular orientations to light, and may also be able to use the current as a directional cue for cross-stream movements. Within a river or stream, the direction of current flow will be more constant than the direction of sunlight, requiring no time compensation. At least for short foraging trips, equivalent to the foraging of bees or pigeons, current flow might provide sufficient information. I know of no examples of such an orientation mechanism, but the most likely habitat for its occurrence would be in major rivers where cross-river distances might be sufficiently great to require more than pilotage.

Current Velocity as a Landmark. Velocity changes themselves may serve as indicators of location. Sockeye fry that have moved downstream from inlet incubation streams would be swept out of their nursery lakes if they continued to show negative rheotaxis in the region of the lake outlet. Brannon (1972), working with several different genetic strains of sockeye fry, found that inlet stocks showed a threshold response to velocity, switching from negative rheotaxis to positive rheotaxis as velocity declined. This system allowed downstream orientation in the swift waters of nursery streams, but positive rheotaxis in the low velocities found in the region of lake outlets. The transition threshold varied from stock to stock. The outlet spawning Chilko stock, in contrast, showed positive rheotaxis at all velocities tested, a result that could be interpreted as indicating a very high transition threshold. The appropriate velocity threshold could account for the changes in rheotaxis that occur on the migration route of some sockeye fry: downstream movement in swift tributaries, which empty into outlet rivers, then upstream movement in the slower-flowing rivers. Weaver Creek fry, for example, move downstream in Weaver Creek and Morris Slough, but turn upstream in the Harrison River, which has a lower velocity than either Weaver Creek or Morris Slough. It is not just a matter of inability to make headway against the faster currents as all the velocities tested are well within the swimming capabilities of the fry. Velocity is not the only factor affecting the behaviour of these fry, they are probably also responding to chemical (see Sect. 3.4.4.2), celestial (see Sect. 2.4.2.2.4), and magnetic (see Sect. 6.8) stimuli, and the many different stocks of fish seem to differ genetically in their response (see Sect. 7.4).

The response of salmonid fry and smolts to velocity changes may be one of the factors contributing to the success of vertical louvers in guiding downstream migrants. The changes in speed and direction of flow immediately upstream of the louvers may stimulate fish to show avoidance responses, which ultimately move them toward the bypass. That the response is not purely visual is indicated by high guiding efficiency at night with reduced efficiency during daylight and full moon (Ruggles and Ryan 1964). The velocity of water entering the bypass is also critical. If the flow slows too much, the migrant juvenile salmonids turn upstream (Ruggles and Ryan 1964), showing positive rheotaxis just as one would expect from the work of Brannon (1972). A gradually accelerating flow works best.

Orientation of Jumping Fish at Waterfalls. The orienting factors that affect the jumping of salmonids and minnows, at waterfalls, both natural and artificial, have been studied by Stuart (1962). When a nappe of water adhered to the face of an obstacle in an experimental flume, both minnows and salmonids ("trout" and "salmon" – presumably brown trout, *Salmo trutta,* and Atlantic salmon) attempted to swim up the side of the barrier in the falling water, with occasional success (Fig. 4.10 A–C). When the flow rate was sufficient for a jet of water to break free of the face of the obstacle, the fish tended to jump from the standing wave just downstream of the point of entry of the falling water (Fig. 4.10 F). In some cases this would position them in an appropriate location for clearing the

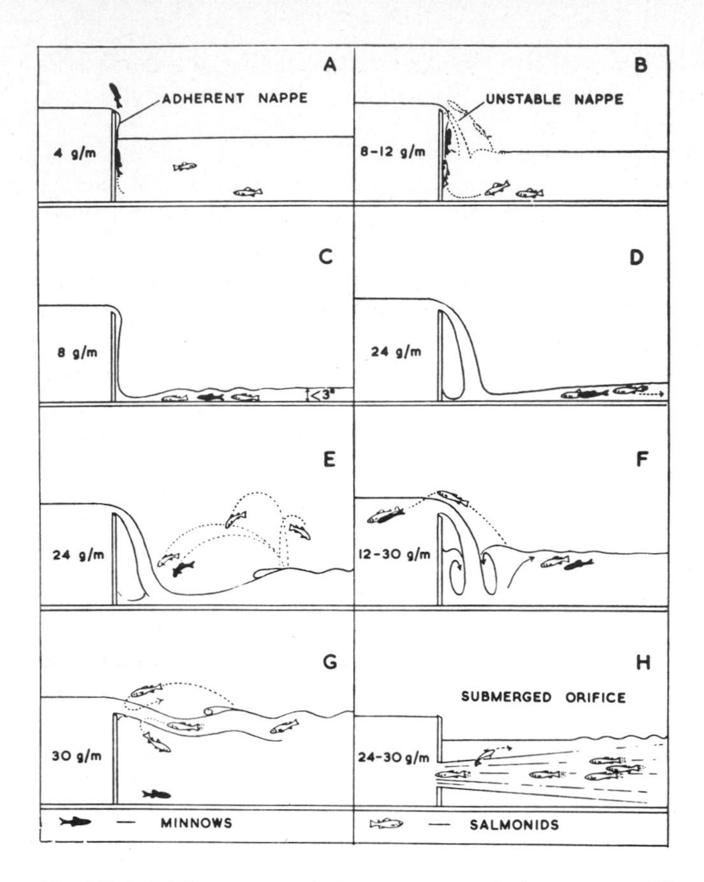

Fig. 4.10 A–H. The reactions of minnows, trout, and salmon parr to different types of overfall at various rates of discharge and pool depth (Stuart 1962)

obstacle, in other cases it would not, and the fish performed futile jumps too far downstream from the obstacle (Fig. 4.10 E, G).

The appropriate size of spillway for the flow rate of the stream can greatly assist the leaping fish by placing the standing wave in the best position for a successful leap. Pool depth was important in providing sufficient space for the development of a standing wave. However, salmon do not, as earlier supposed, require a deep pool for a "take-off run"; instead they start their leap from near the surface. Water containing a high proportion of entrained air bubbles was avoided, perhaps because it would reduce the propulsive force of swimming movements. Stuart (1962) speculates that sound reception may be one of the factors allowing the fish to locate falling water. Visual responses to the height of the barrier and lateral line stimuli may also be involved, although there is no experimental support for this suggestion.

The fish were also attracted by fast-flowing water, as through a submerged orifice (Fig. 4.10 H), although there was a curious relationship between attraction and ability to swim against the flow: fast-flowing water was most attractive although the fish were unable to penetrate the orifice against it, as the velocity was reduced to the point where the fish could make headway, the current also became less attractive to them.

4.4.2 Human Modification of the Physical Environment

The physical habitat of migrating fish is now often modified by the presence of man-made obstacles such as dams or by man-made diversions such as the intakes for cooling water for power plants. Entrainment in the cooling water intake of many power plants leads to mechanical damage from sudden pressure changes, cavitation, rapid acceleration, strong shearing forces, abrasion, and collision as the fish is swept through a series of pumps, pipes, and valves (Marcy et al. 1978). Mortality is usually high. Downstream migrants swept through hydroelectric plants will face similar risk of mechanical damage (Semple and McLeod 1975; Hanson et al. 1977).

Several methods have been attempted for diverting fish away from the entrances to power plants or turbines. Physical barriers to migration may be effective in situations where behavioural barriers are ineffective. Screens and similar physical barriers represent a compromise between interference with water flow and the blocking of fish entry. The more complete the barrier to fish the greater the loss of flow (Hanson et al. 1977).

Barriers are not equally effective against all species. The torrent sculpin, *Cottus rhotheus,* may bypass salmon screens by sub-gravel movement (Thomas 1973). Young salmonids migrating downstream tend to stay near the top of turbine intakes and can be diverted upward into gatewells, but lamprey larvae move near the bottom of the intake and are not diverted (Long 1968). The tendency is to concentrate on techniques that protect the valuable commercial or game species but not necessarily the other species, which may still be important components of the ecosystem. Often forage fish are smaller and therefore harder to block or divert, and their value is not readily understood by the public or by financial decision-makers. This effect was well illustrated by the public ridicule heaped on those who wished to protect the snail darter, *Percina tanasi,* from the effects of the construction of the Tellico Dam by the Tennessee River Authority.

Dams are, of course, barriers to upstream movement as well as hazards to downstream migrants. Radio-tracked adult chinook salmon were delayed as much as 100 h at Little Goose Dam on the Columbia River (Haynes and Gray 1980). Although upstream migrant salmon are larger than downstream migrants and thus easier to deal with, other species, whether stickleback, suckers or sturgeon are seldom taken into account in the construction of fishways. If they happen to be able to negotiate salmon ladders they may continue to migrate. As well as requiring physical exertion to negotiate, fishways may also be sources of infection. Fujihara and Hungate (1971) noted that *Columnaris,* a salmon pathogen, was abundant in fishways on the Columbia River.

Dams influence fish migration by altering the flow of river systems as well as by presenting physical obstacles. The impoundment of a stream coverts flowing water into water that is essentially still. This may slow and disorient downstream migrants that depend on the flow of the stream to carry them to their destination and/or on the direction of flow as an orienting cue to direct their movement downstream. One might expect that the reduction in water flow above a dam would facilitate upstream movement of fish through the impoundment by reducing the velocity that they have to counteract by swimming. If the stream flow itself is an important directional cue, however, migrants may become disoriented in the quiet water above a dam.

4.4.3 Hydrostatic Pressure

Water pressure varies in a rigidly predictable manner with water depth: approximately 1 atm of pressure for each 10 m of water. The range of pressure available to aquatic organisms is from approximately 0.5 atm in mountain streams to 1100 atm in the deepest ocean trenches. Fish are known to inhabit most of this range and are sensitive to pressure changes. It seems reasonable, therefore, to expect that pressure may serve as a directing stimulus for vertical migration.

Vertical migrants may use gravity receptors to direct their movements upward or downward, but baroreceptors would also be useful in recognizing the goals of the migration: the appropriate depths for spawning or feeding. In some circumstances light might also be a useful indicator of both direction and depth, but it cannot be used at night, at great depths, or in circumstances where bioluminescence is stronger than sunlight. There is evidence from a shallow-water species that fish can make appropriate behavioural adjustments to pressure change. Larval plaice swim upward or toward the light when they are subjected to increases in hydrostatic pressure (Rice 1964). Swimming up would bring them to regions of reduced pressure, behavioural baroregulation. This type of response to pressure has been suggested by Riley (1973) as one of the potential cues used in the shoreward migration of displaced O-group plaice. Similarly, Winn et al. (1964b) found that juvenile parrot fish, *Scarus guacamia* and *S. coelestinus,* and some adults headed directly for shallow water when displaced out to sea and released. Although fish can sense pressure changes and the pressure gradient would be a reliable indicator of depth, we do not as yet have direct evidence that it is used by vertical migrants.

4.4.4 Sound

Aquatic environments are noisy. The sound transmission qualities of water ensure that sounds produced by such natural forces as wind, surf, rain, water turbulence, and the movement of substrates such as gravel or boulders in flowing water will be carried substantial distances. Just as a surf line or waterfall can be heard in the air, it can also be heard under water. Sounds are also produced by organisms, as byproducts of their normal activities such as swimming and eating, and also as specific social signals analogous to the calls and songs of birds or mammals.

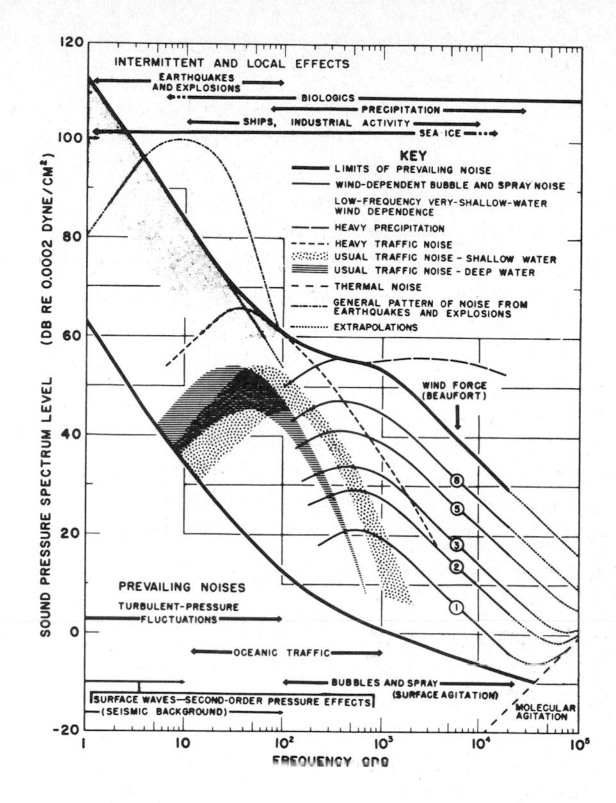

Fig. 4.11. A summary of sources of ambient noise in the ocean. *Horizontal arrows* show the approximate frequency band of influence of the various sources. With permission from Wenz GM (1964) Curious noises and the sonic environment in the ocean. In: Tavolga WN (ed) Marine bioacoustics. Pergamon, NY, pp 101–119. Copyright 1964, Pergamon Press Ltd.

Sounds of human activity are now a major component of the background noise in many aquatic environments, particularly the propellor noise of boats and ships but also such interesting variations as the sounds of trawling, dredging, pile driving, and oil drilling. In general, oceanic sounds are loudest in the relatively low frequencies (Fig. 4.11). Presumably freshwater sounds are similar, although the sounds associated with rapids, waterfalls and similar phenomena will be more important than in the ocean.

4.4.4.1 Sounds as Landmarks

The role of sounds in fish migration and orientation is little studied. Stober (1969) attempted a study of the use of auditory stimuli in locating stream entrances by cutthroat trout in Yellowstone Lake, a system in which the use of olfactory and visual stimuli was in doubt (Jahn 1966, 1969; McCleave 1967). Noise from water flow and bubbles near stream entrances was largely below 4 kHz, and thus perhaps within the hearing range of the trout, while surf sounds, which were also present, were largely above 5 kHz and probably out of their hearing range. Unfortunately the system noise in Stober's equipment masked most natural sounds. His attempts to condition trout to sounds using electric shock and light as reinforcers were not successful.

4.4.4.2 Social Signals

Sounds produced by the swimming movements of conspecifics may help maintain schooling behaviour in hogmouth fry, *Anchoviella choerostoma,* since blind fish are able to keep place in the school when it is moving but not when the school is at rest (Moulton 1960). Fish are capable of producing sounds by a variety of means (Tavolga 1976) and use these sounds in social communication such as courtship and territory defense. The role such social communication might play in migration and orientation is open to speculation. Maintenance of school integrity and social facilitation are the most likely functions.

4.4.4.3 Echolocation

Animals that hunt in the dark have often evolved echolocation as a means of locating prey and other objects in their environment. This ability is best known in bats and toothed whales, but echolocation has also been reported in birds, seals, shrews, and tenrecs. Fish would seem to be prime candidates for the use of echolocation: water is a good sound transmitter, visibility is often poor and fish are capable of both producing and receiving sounds. The apparent lack of echolocation in fish has been attributed to their supposed inability to localize sounds and to the low frequency of the sounds they produce.

We now know that many fish do possess the ability to localize sound direction, at least in the near field, and Tavolga (1971) showed that the low-frequency sounds produced by the sea catfish, *Arius felis,* produce reflections and reverberations, which should provide information about objects in the environment. The sea catfish produces a variety of sounds in its social interactions with conspecifics but one sound is most frequent, a short 5–10-ms pulse burst varying from below 100 Hz to over 1500 Hz with the low frequency portion of the sound being the most regular (Fig. 4.12) (Tavolga 1976). Lone catfish seldom produce the sound and tend to spend most of their time resting passively. Groups of catfish produce

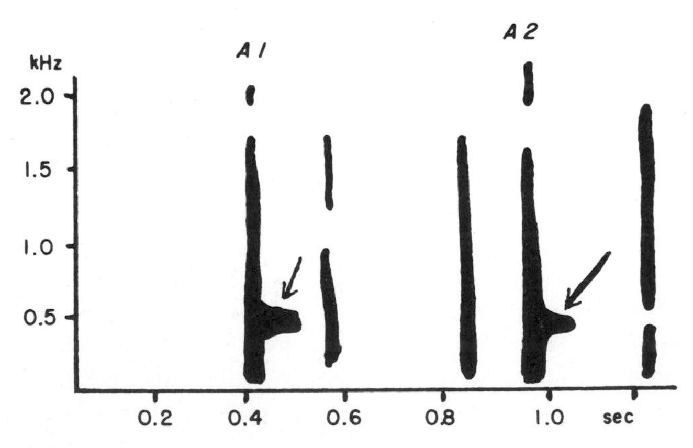

Fig. 4.12. Sound spectrogram of sounds produced by the sea catfish (*Arius felis*). The two sounds *A1* and *A2* were made by an individual in a concrete aquarium as it approached to within about 10 cm of a transparent plastic barrier, and within about 50 cm of the hydrophone. *Arrows* indicate a strong reverberation and resonance at about 500 Hz (Tavolga 1971)

sound almost continuously and spend most of their time moving slowly about the tank (Tavolga 1976). When groups become separated the rate of sound production increases, so it appears that these short pulses of sound function to maintain social contact in the school.

These short pulses also appear to allow a form of echolocation. Tavolga (1976) placed clear acrylic plastic barriers in a circular tank. Clear plastic is almost invisible in water. Single catfish did not produce the pulsed sounds, due to lack of social facilitation, and bumped into the plastic barriers in 182 of 200 approaches. Groups of 8–10 catfish produced sound constantly and very seldom bumped into the barriers during slow swimming. If they became separated from the group, they tended to become agitated and swim more rapidly and this increased the number of collisions with the barriers. Blinded catfish in groups produced sound and were also able to avoid the plastic barriers but fish that had been muted, by cutting the sound-producing muscles, tended to bump into the barriers frequently even though their eyes were intact (Fig. 4.13). These results provide evidence for the use of the low-frequency pulsed sounds in obstacle avoidance at close range. The avoidance reactions occur well within the near field of the sounds produced and the plastic sheets would be almost transparent to the far field pressure waves but would provide a strong reflection of particle displacement near field sound (Tavolga 1976). The catfish apparently use an interaction of ear, swim bladder and lateral line stimuli in their detection of obstacles, with the swim bladder acting as a directional sound source and directional receiver (Tavolga 1977).

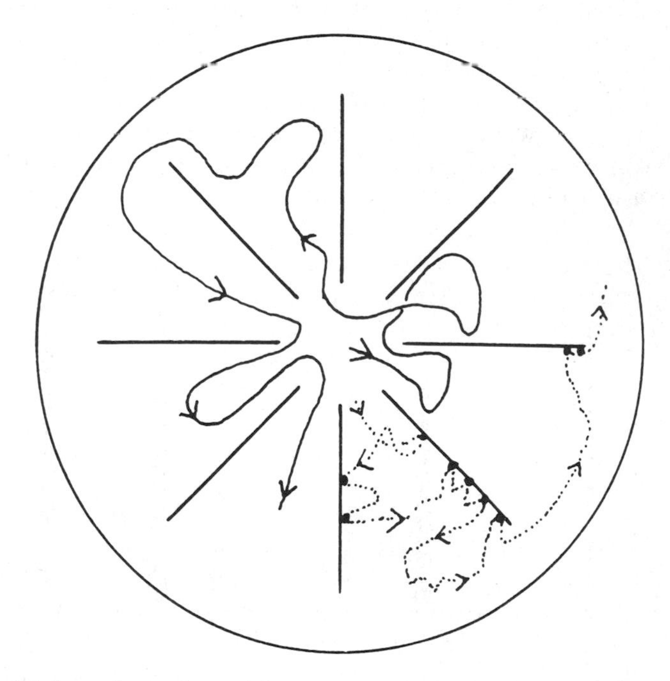

Fig. 4.13. Tracks of two sea catfish in an observation tank fitted with clear plastic barriers. *Solid line* track of a normal individual, producing sound as it swims. *Dotted line* track of a muted individual (Tavolga 1976)

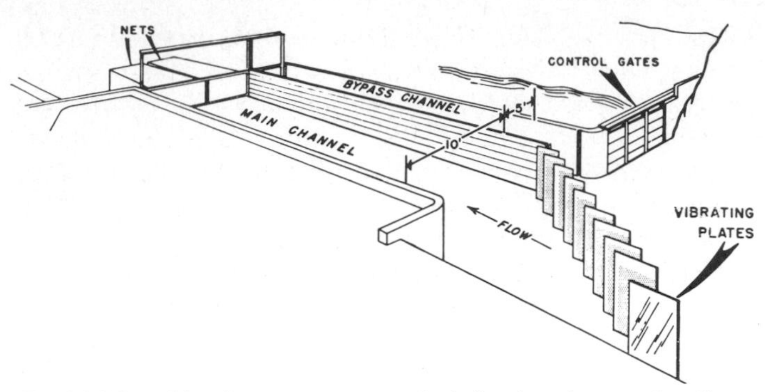

Fig. 4.14. Sound barrier system composed of vibrating plates used to divert migrant steelhead trout in an irrigation canal. Reprinted with permission from Vanderwalker JG (1967) Response of salmonids to low frequency sound. In: Tavolga WN (ed) Marine bioacoustics, Vol 2. Pergamon, NY, pp 45–58. Copyright 1967, Pergamon Press Ltd.

This catfish echolocation system is very short-range and crude compared with the echolocation abilities of the odontocete whales. It is probably not useful in detecting or identifying predators or prey or in long-range orientation. It might provide landmark information if different locations have typical echo characteristics. Its chief significance is in showing that fish are capable of sonic emitted-energy orientation, thus spurring the search for other, perhaps more impressive examples.

4.4.4.4 Sound and the Directing of Fish Migrations

It would seem reasonable to use loud frightening noises to scare migrant fish away from dangerous locations such as turbine intakes and to use attractive sounds to lure them into fishway entrances and other desirable locations. Attempts to accomplish these objectives have not proved very effective (Hanson et al. 1977).

Brett and Alderdice (1958) attempted to divert downstream migrant sockeye salmon smolts to one side of a trough by "beaming" a sound generator producing sounds of 1–10 kHz at 25 W across one side of the trough. The smolts were not diverted. Similarly, Moore and Newman (1956) using "young salmon" and Burner and Moore (1962) using small rainbow and brown trout found only startle responses and no evasion with a variety of loud noises including explosions. In contrast, Vanderwalker (1967) was able to divert downstream migrant steelhead, *Salmo gairdneri,* (size not given) in an Oregon irrigation canal by suspending vibrating metal plates at an angle across the flow (Fig. 4.14). When the plates were absent 11% of the fish entered the bypass. With the plates in place but not vibrating 33% entered and when the plates were vibrated at 270 Hz, 77% entered the bypass.

4.4.5 Inertial Guidance

Animals might be able to maintain a heading or even to navigate from one location to another by using information on their angular and linear acceleration (turns and changes in speed) derived from the vestibular organs (review: Barlow 1964). Mechanical accelerometers linked to computers allow man-made inertial guidance systems to accurately guide vehicles, such as missiles, over distances of thousands of kilometers without reference to outside information. Each change of speed and direction, whether caused by the propulsive system of the vehicle or by the actions of tide or wind, is received and integrated to calculate movement from the original starting point. An accurate timing mechanism is required to determine the duration of each acceleration. As well as such self-contained systems there are inertial guidance systems that include information from outside sources, such as the position of celestial objects or Doppler-radar information on the speed of movement over the ground, in their calculations. This can either increase the accuracy of the system or allow the same accuracy to be maintained with less precise accelerometers. Navigation by inertial guidance does not require a compass because the compass direction of a course could be determined by calculating the successive positions reached by a moving vehicle and then turning until the changes in position corresponded with the desired course. However, it is much more efficient to combine the guidance system with a compass system such as a magnetic or solar compass in order to set approximately the correct course without an initial period of trial and error. The force of gravity interferes with the accelerometers of inertial guidance systems. They may be accurate with respect to the horizontal position of the animal because gravity, at right angles to all horizontal directions can be cancelled out, but inertial guidance systems are inaccurate for vertical movement, changes in depth or altitude.

Animals possess accelerometers; the vertebrate ear, for example, seems capable of detecting angular acceleration of the order of $0.2° s^{-2}$ and linear accel eration of $6 cm s^{-2}$ (Barlow 1964). Barlow suggests that these thresholds are too high to allow accurate inertial navigation for more than a few minutes. Animals possess a time sense or biological clock, another necessary component of inertial guidance systems. The brain would presumably play the role of computer, integrating the various durations and intensities of acceleration. Independent compass systems, celestial or magnetic have been found (see Chaps. 2 and 3), and external information on speed of movement relative to the substrate and the position of celestial bodies or other indicators of position could reduce the accuracy demanded of the inertial guidance system. The poor vertical accuracy of inertial systems may be an important factor for the guidance of rockets, but it is probably not important for many animal migrations that are essentially in two dimensions, with a very minor vertical component. Pressure sensitivity should provide independent depth or altitude information to migratory animals in any case.

Despite the promising picture painted above, there is little direct evidence of the use of inertial information in animal orientation. For example, experiments in which pigeons were anesthetized, rotated on turntables or had their vestibular apparatus interfered with did not seem to indicate the use of inertial navigation, although in none of these cases was the use of other information totally excluded (Barlow 1964).

In fish, Kleerekoper et al. (1969, 1970) found that goldfish, tested in a photocell array, showed a tendency to balance their amount of turning so that the amount of turning to the right (cumulated angles in radians) was almost the same as the amount of turning to the left. Despite the near-balance of cumulative turning, most fish tended to turn a little more in one direction than in another. Some were "left-handed" others "right-handed." Despite the use of the term "inertial guidance" in the title of Kleerekoper et al.'s 1969 paper, they present no data on fish with impaired sense organs. So, it seems to me that the fish might be using visual or other non-inertial senses to maintain a roughly constant course.

Although inertial navigation is an interesting potential navigation method and it may be used by vertebrates over very short distances, there is no compelling evidence that it plays a role in fish migration.

4.5 Summary

Water currents, sounds, pressure, gravity, touch, and the perception of body movement and posture are accessible to fish mechanoreceptors. The use of these stimuli may allow fish to make efficient use of water currents, to communicate with conspecifics and to respond to "mechanical landmarks" and timing cues available in the environment.

The mechanical senses share the common attribute of responding to changes in the shape of the receptor, bending or compression of the sensory element, for example. Several different types of mechanoreceptors occur in fishes. These include touch, proprioception sensitivity to water pressure, and perhaps barometric pressure and the acoustico lateralis system, which includes the lateral line and ear.

Timing by mechanical stimuli may be largely restricted to situations in which local or unpredictable events are correlated with mechanical changes in the environment. Tidal conditions, for example, are often quite local in their amplitude and timing.

Barometric pressure changes often precede or accompany weather changes and rainfall and associated flooding can provide suitable conditions for migration in streams. Chinook salmon, for example, tend to enter rivers when the barometric pressure is falling, indirect evidence that barometric pressure changes may serve as timing cues to allow anticipation of rainfall and suitable conditions for migration. Small earthquakes, microseisms, are associated with barometric depressions, and usually with rainy weather, off the Dutch coast. Eels tend to migrate downstream on nights characterized by microseismic activity, even if there is no rain associated with the depression.

Floods and freshets are also characterized by changes in water depth and flow rate and increasing turbulence. There are numerous reports of salmonids initiating upstream migration in response to freshets and of eels migrating downstream under flood conditions. The exact stimuli releasing these responses have not been determined.

Mechanical stimuli may affect the direction of fish migration through a number of mechanisms. High velocities of water flow may act as limiting factors excluding fish from some portions of the habitat. Within tolerable ranges of velocity, water currents can carry fish along in the appropriate direction for their migration. Use of this mechanism depends on behavioural selection of appropriate water currents. In some young fish this selection is handled by the parents when they place the fertilized eggs in an area with consistent current direction.

Tidal currents change direction with each tide. In order to use tidal currents effectively in migration, fish must "modulate" their behaviour to synchronize movement into midwater with the phase of the tide that is flowing in the appropriate direction for their migration. During the opposing tide their behaviour should minimize the ground lost, for example, by resting on the bottom where flow rates are low. North Sea plaice use this type of selective tidal stream transport in their spawning migrations.

Rheotaxis, the orientation to water currents, is largely a visual response in fish. They detect the direction of flow by their movement relative to fixed objects such as a stream bottom. Small-scale water movements, however, are detectable by the lateral line, allowing fish to adjust their position to take advantage of a turbulent flow. Rheotactic responses can be reliable mechanisms for orienting a fish along a migration route.

Hydrostatic pressure will limit fish distribution in depth through the effects of pressure on physiology. Within physiological limits, pressure sensitivity will provide accurate depth information. This may be useful in orienting and terminating vertical migrations. Pressure changes could serve as landmarks to bottom-swimming fishes.

Sound could provide landmark information to migratory fishes. Surf, rapids, waterfalls, and biological sources such as shrimp choruses all produce underwater sounds that would predictably indicate particular types of habitat. Sound production by fish themselves could be involved in social interactions related to migration, maintenance of schools or dispersion onto breeding territories at the end of a migration, for example.

Inertial guidance based on the vestibular senses may be possible over short distances but there is no evidence that it plays a role in the long-distance orientation of migratory fishes.

Chapter 5

Temperature

5.1 Introduction

Just as bodies of water vary in their chemical characteristics, providing a chemical landscape for orientation, so they differ in their temperatures, providing a thermal "landscape" which might serve as an orienting cue in fish migration.

Freshwater and marine environments are characterized by predictable horizontal and vertical temperature gradients as well as daily and seasonal temperature changes. Rivers and streams change temperature as they flow from cool high altitudes to warmer low altitudes. Ocean currents show similar changes with latitude, cooling as they move toward the poles and warming toward the equator.

Water reaches its greatest density at about 4 °C. Denser water tends to sink beneath lighter, usually warmer, water, thus forming layers or strata of different temperatures. The most common situation is a surface layer of warmer water separated from a deeper cold layer by a thermocline, a region in which the temperature changes markedly over a short distance.

Rivers and streams do not normally stratify because of the mixing effects of turbulence, but lakes and impoundments can alter the thermal characteristics of a stream by allowing stratification so that warmer surface waters flow on downstream from the outlet, while cooler water fills the deep portions of the lake. Conversely, impoundments may cool downstream rivers through release of deeper, cool water, e.g., through deep-inlet turbines.

Thermal stratification is a permanent feature of deep tropical lakes where water temperatures never fall to 4 °C. In such lakes the cooler layer below the thermocline often becomes deoxygenated because it is never in contact with the surface and is too deep for much photosynthesis.

In temperate lakes, stratification occurs seasonally. Spring and fall turnovers occur as the temperature passes through 4 °C and the water is uniform in density. In the summer, temperate lakes have a warm surface layer and a cooler deep region, that sometimes becomes deoxygentated. In winter the situation is reversed with the water below 4 °C being lighter.

These predictable thermal gradients might be used by fish as orienting stimuli in their migrations. Additionally, the daily and seasonal warming and cooling of the water could serve as timing stimuli. This may be particularly important for the synchronization of migratory behaviour with events such as spring thaw, which might vary from year to year in their dates of occurrence.

5.2 Temperature Reception

There is ample evidence that fish are capable of showing behavioural responses to changes or differences in temperature (Brett 1970; Coutant 1977). The actual receptors involved in these behavioural responses are not known. Unlike vision, smell, taste or hearing, there are no morphologically prominent sense organs associated with temperature perception. The situation appears to resemble that of touch, with a diffuse array of rather unspecialized receptors.

Bull (1936) was able to condition 18 species of fish to respond to changes of 0.03° to 0.07 °C, using food as a reward, and several other conditioning experiments confirmed the general result in a variety of fishes. Since, in general, the deep body temperature of fishes gradually follows the temperature of their environment, it is not absolutely essential that temperature receptors be on the outer surface of the body. In fact, as in other vertebrates, portions of the central nervous system will respond to temperature changes. Researchers have implanted "thermodes" (the thermal equivalent of eletrodes) in the forebrain of Arctic sculpins, *Myoxocephalus* sp. (Hammel et al. 1969), and in the forebrain of the brown bullhead, *Ictalurus nebulosus*. Warming of the rostral forebrain with these thermodes led to selection of lower body temperatures by the fish in a thermal gradient. Heating the spinal cord of two cyprinid species, *Cyprinus carpio* and *C. carassius,* led to an increase in heart rate (Iriki et al. 1976).

There is also good evidence for thermoreception on the surface of the body (Murray 1971). Dijkgraaf (1940, 1943) trained minnows, *Phoxinus laevis,*to respond to streams of warm or cold water applied to the flank, and Bardach (1956) trained goldfish to move if touched by a rod 2 °C warmer than the surrounding water and not to move if touched by a rod at the same temperature as the tank water. The goldfish were responsive to thermal stimuli over the whole body surface. Additionally, behavioural responses such as avoidance of extreme temperatures occur too rapidly in laboratory studies to be based only on deep receptors.

Although the whole body surface of fish may be capable of detecting changes in water temperature, some portions may be more sensitive than others. In the skipjack tuna, for example, thermal stimuli were more effective in conditioning heart rate when they were applied to the oral-branchial cavity than when delivered to the nostrils or the dorsal-anterior quadrant of the body (Dizon et al. 1974). It is not clear whether this is typical of fish in general or peculiar to the tuna.

The actual receptors on the body surface or in the oral cavity may be the same nerve endings responsible for touch (Murray 1971). Cooling decreases the sensitivity of the touch receptors, while warming increases their sensitivity over a range of temperatures close to the acclimation temperature of the test animal. Temperatures well above the acclimation temperature sometimes cause a decline in sensitivity. Such touch-temperature units are found in other vertebrates and the CNS must be able to separate the two forms of information. Perhaps, in fish, the swimming movements provide a standard mechanical stimulus allowing the animal to detect temperature-induced changes in the response to that stimulus.

Portions of the lateral line have also been proposed as thermoreceptors. In teleosts, the experiments reported above by Dijkgraaf (1940, 1943) and Bardach (1956) showed that the lateral line was not required for temperature perception. In the elasmobranchs, touch-temperature receptors are present but it is not clear if there are pure temperature receptors as well. The Ampullae of Lorenzini, modified lateral line organs found in elasmobranchs, show a variable response with temperature but Murray (1971) argues that they are anatomically unsuitable as thermoreceptors. They do have an established function in electroreception.

Fish, like other cold-blooded vertebrates, regulate their body temperature by behavioural responses. The CNS thermoreceptors provide the basic thermostat but the peripheral thermoreceptors, possibly dual touch-temperature receptors, are used in the behavioural selection of preferred temperatures. The CNS receptors are also likely to control thermal timing of endocrine and behavioural responses to seasonal or diel changes in the environment, by analogy to the extra-optic photoreceptors in the CNS which can control similar timing by photoperiod (see Sect. 2.2.2.1.5).

5.3 Timing by Thermal Stimuli

5.3.1 Diel Timing by Thermal Stimuli

Shallow waters often show a marked diel temperature rhythm, as they warm up during the day and cool down at night. In theory this thermal cycle might serve as a timing mechanism for behavioural rhythms, just as the daily cycle of light and dark does. Similarly, the seasonal change in the daily length of the warm and cool portions of the thermal cycle might serve as a seasonal timing mechanism, comparable to photoperiod. In fact, there is little evidence for either of these roles for temperature in the biology of fishes. It may be that in most cases the light cycle is more reliable than the temperature cycle and has been favoured by natural selection as an environmental timing cue.

5.3.1.1 Migration and Daily Temperature Peaks

There are several reported cases of diel periods of upstream migration correlating with the maximum temperature in daily thermal cycles. European elvers, migrating upstream, seem better able to move against fast water at higher temperatures (Sorensen 1951). This may indicate a general relationship. Higher temperatures up to the thermal optimum for swimming performance will increase the swimming speed of fish and hence their ability to move upstream against strong currents. Adult alewives migrating upstream in the Annaquatucket River, Rhode Island, move upstream during the afternoon temperature maximum of 8 °C during the early part of the migratory season. Later in the summer they move only during the morning temperature minimum of 18 °C, shifting their activity period so that it occurs within their preferred temperature range (Richkus 1974). Similarly, rainbow trout fry in the outlet of Loon Lake, B.C. move upstream in the greatest numbers near the daily temperature maximum (Northcote 1962). Trout migration increased on warm days, while fewer migrated during a series of cooler days. In an experimental stream tank, increasing temperature served as a stimulus for upstream movements by rainbow fingerlings (Northcote 1962), probably another case of temperature-timing behaviour through the relationship between temperature and swimming performance.

Pink salmon fry are better able to swim up to the surface and escape out of the overflow of incubation tanks at warm temperatures than at cooler temperatures (Coburn and McCart 1967), but this may be an emergency response to high temperature since even relatively immature fish left the tanks at the highest temperatures. A 2.2 °C increase in temperature-induced increased emigration of chinook salmon fry from simulated incubation channels (Thomas 1975), so this temperature effect is probably widespread in Pacific salmon species. Small streams may overheat in warm years and emigrating to larger bodies of water would be adaptive.

In contrast to the examples cited above, sockeye salmon in the outlet of Babine Lake move upstream over a wide variety of temperatures, whether steady or fluctuating (McCart 1967).

5.3.1.2 Temperature Preference and Diel Movements

Daily fluctuations in the distribution of water temperature will lead to shifts in the distribution of fish with strong thermal preferences. The desert pupfish (*Cyprinodon macularius*), for example, moves out of the shallow portions of pools as they heat up during the day (Barlow 1958).

Some fish undertake daily movements in order to take advantage of different temperature zones for different activities. For example, sockeye salmon fry make daily excursions between the warm surface water where they feed and deep cold water where they use less energy during the non-feeding portion of the day (Brett 1971). Diurnal cycles of temperature preference have also been reported in a number of other fish, although in most cases their adaptive significance is not clear (e.g., Reynolds and Casterlin 1976a, b, 1978a, b). These are cases of fish selecting different temperatures at different times of day rather than examples of temperature serving as a proximate timing stimulus. We might speculate that temperature is one of factors involved in the ultimate causation of the timing of these diel movements.

5.3.1.3 Summary of Diel Timing by Temperature

The general picture which emerges is one of a limited role for temperature in diel timing. Daily temperature cycles may time periods of intense exertion such as upstream migration by restricting these activities to the time of day at which the best temperature for activity occurs. Daily temperature extremes may exclude fish from some habitats during part of the day, as in the case of the temperature-preference behaviour of the desert pupfish. In general though, demonstrated effects of daily temperature cycles as timing cues for migration seem to be much less common than the effects of the daily light cycle.

5.3.2 Seasonal Timing by Thermal Stimuli

There are at least two ways in which temperature changes might time seasonal events: in shallow waters, the relative length of the daily periods of warming and cooling might serve as seasonal cues just as do the relative length of the periods of light and dark in the photoperiod cycle, acting on the CNS and endocrine system to control behaviour and physiology. Or, the seasonal change in average water temperature could serve as a timing mechanism either by acting through the CNS/endocrine route or by limiting the distribution of the temperature zones preferred by the fish.

5.3.2.1 Diel Thermal Cycles and Seasonal Timing

Despite the theoretical possibility, there is little to indicate that the thermoperiod is actually used as a timing cue by fish. There are several studies of the effect of thermal cycles on temperature tolerance and acclimation (Heath 1963; Shrode and Gerking 1977), and a study indicating that growth is improved in rainbow trout by diel temperature fluctuations within the normally selected temperature range (Hokanson et al. 1977) but few studies of thermal rhythms as timing mechanisms. Gonadal regression in the estuarine gobiid fish *Gillichthys mirabilis* does seem to be determined by the amount of time that the diel temperature cycle spends above 24 °C, a possible case of endocrine events being timed by a thermal cycle (de Vlaming 1972). Similarly, the gonadal growth of goldfish is stimulated by heat treatment in the last 4 h of darkness but may be inhibited by heat at other times of the day (Spieler et al. 1977). It is not clear whether the lack of evidence for or against diel thermal cycles as timing mechanisms for seasonal events is due to lack of study or to the rarity of the phenomenon itself.

5.3.2.2 Seasonal Temperature Trends as Timing Cues

5.3.2.2.1 Endocrine Cycles

There is ample evidence that fishes use seasonal temperature changes as environmental cues to regulate their endocrine cycles (e.g., Peter and Hontela 1978; Hontela and Peter 1978; Peter and Crim 1979). Goldfish, for example, show ovulation more readily at warm temperatures (21 °C) than at cool temperatures (13 °C) when stimulated by long photoperiod and aquatic vegetation or by HCG injection (Stacey et al. 1979), and a sharp rise in temperature from 10 °C to 20°–30 °C leads to an increase in circulating gonadotropin levels (Gillet and Billard 1977). The endocrine system in turn can time migratory behaviour. Most experimental work on this subject has used constant temperatures or a few stepwise changes rather than thermal cycles or rhythms, although the natural situation is a gradual shift in mean monthly temperatures with considerable day-to-day fluctuation. It seems that temperature and photoperiod are both used by fish, with the balance between the two shifting from one species to another.

5.3.2.2.2 Temperature Preference and Seasonal Timing

There are many reports of seasonal changes in fish habitat preference or migratory behaviour that seem to be correlated with seasonal temperature changes. Migration coincides with a specific temperature which occurs at different times in different years. The photoperiod present at the time of, say ice break-up, will be shorter in a year with an early spring than in a year with a late spring. Thus if migration begins at break-up, it appears that temperature is the final trigger or releasing stimulus, although these observations do not exclude priming roles for photoperiod or other timing mechanisms such as endogenous circannual cycles. Unfortunately in the studies discussed below these possibilities can only occasionally be speculatively separated by the apparent time scale of the change; in fact, the implication of temperature as a controlling stimulus is often only by speculation or inference.

Often fish tend to appear in a region at about the same time as the temperature reaches a particular level. Elvers, for example arrive on the Dutch coast when the spring water temperature reaches about 4.5 °C (Deelder 1952) while bluefish, *Pomatomus saltatrix,* arrive off New England shores when the rising spring temperature reaches 12°–15 °C, and are described as "following the warmer water" (Lund and Maltezos 1970). They retreat to the South again when the water temperature drops below 15 °C. Similarly, black sea bass, *Centropristis striata,* winter in the southern portion of the Mid-Atlantic Bight and move North in the spring, apparently following warmer water, with the greatest catches being reported in water of 8 °C or higher (Musik and Mercer 1977). Yellowfin and skipjack tuna seem to follow the 21 °C isotherm in their seasonal movements up and down the Californian coast (Fig. 5.1) (Blackburn 1970). A coldwater species, the winter flounder, *Pseudopleuronectes americanus,* seems to avoid temperatures above 15 °C. In summer the flounders leave the inshore areas of the Canadian Atlantic

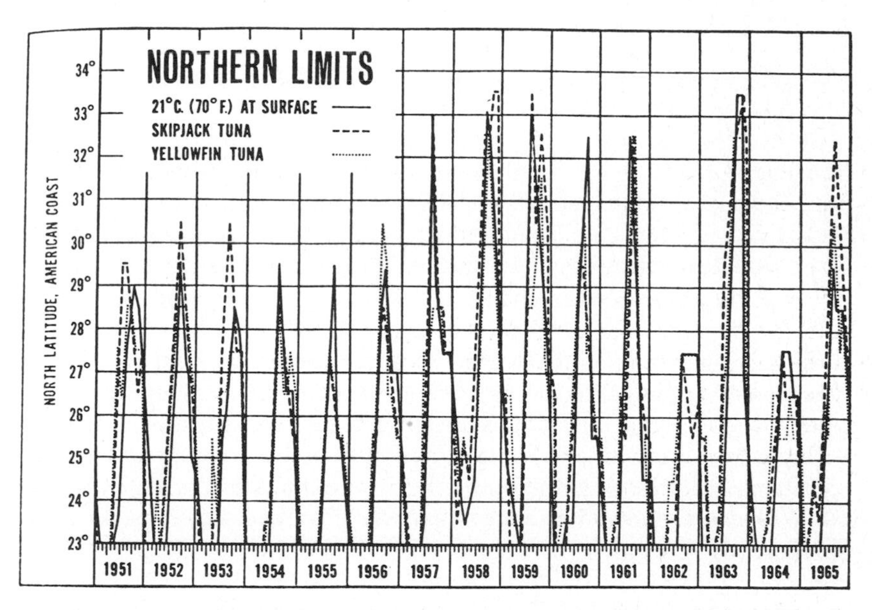

Fig. 5.1. Changes in the latitudinal position of the 21 °C surface isotherm and the nothern limits of commercially caught yellowfin and skipjack tunas of the coast of Baja California and California 1951–1965 (Blackburn 1970)

coast where the bottom temperature reaches 15 °C, but not the regions where the temperature remains below 15°, and return to shore when the temperature drops below 15° in the fall (McCracken 1963). The preferred 12°–15° range is well below the upper lethal range of 19°–26.5 °C, which is not reached in Canadian waters. Light and salinity do not seem to affect the flounders' seasonal distribution, although this has not been examined experimentally, so we can only say temperature predominates within the normal seasonal ranges of photoperiod and salinity.

These are cases in which the appearance or disappearance of fish in a region coincides in time with the occurrence of particular temperature conditions. In a sense, therefore, temperature may be both timing the movement and controlling the fishes' location if the fish follow or stay in particular thermal conditions which move seasonally. To a stationary coastal observer it appears to be a matter of timing, to an observer who follows the fish and the isotherms it appears a matter of control of location.

5.3.2.2.3 Thermal Timing of Changes in Response to Non-Thermal Stimuli

Temperature sometimes serves to time changes in the response of migratory fish to stimuli such as current, salinity or habitat. Adult common shiners, *Notropis cornutus,* tested in an optomotor apparatus during their upstream migration phase, showed the strongest positive rheotaxis at spring (12L:12D) photoperiods and warm (20 °C) temperatures, although substantial positive rheotaxis also occurred under summer photoperiods (16L:8D) and cooler temperatures (Dodson and Young 1977). On the short term, sudden increases of 5 °C were followed immediately by stronger positive rheotaxis. Another case occurs in the salinity preference in three-spined stickleback, *Gasterosteus aculeatus:* higher temperature increased the preference of adult fish for freshwater, corresponding to the prespawning migration from the sea to freshwater streams (Baggerman 1957). The shift in salinity preference is probably affected by the thyroid (Baggerman 1957) (see Sect. 3.4.2.2). In steelhead trout, a delayed annual temperature cycle lengthened the migratory period for downstream migrant smolts and an advanced temperature cycle shortened the migratory period, perhaps through an inhibitory effect of low temperatures on the smoltification process (Wagner 1974).

Habitat preference can change in response to temperature. European yellow eels, for example, hide in the mud at temperatures below 8 °C, emerge and move to hiding places among rocks at 8°–9° and above 9° spend more time with their heads sticking out of the hiding place (Nyman 1972). Feeding begins as temperatures reach 14° and territorial aggression at 17°. These responses are apparently independent of photoperiod at least over the narrow photoperiod range at which such temperatures would naturally occur. Redbreast sunfish, *Lepomis auritus,* come together to form winter aggregations of passive, tightly packed fish at temperatures of 5 °C or less (Breder and Nigrelli 1935) while longear sunfish, *Lepomis megalotis,* disappear from their summer home ranges at low temperatures (Berra and Gunning 1970). Instead of coming out to forage at night, fingerling channel catfish remain in their shelters all day if the temperature dips below 4 °C (Brown et al. 1970) and adults ignore feeding machines until the spring temperature reaches 12° (Randolph and Clemens 1976). These changes in behaviour probably reflect the reduced activity and food requirements of these fish at low temperatures. They tend to wait out cold periods in relatively safe locations.

The winter ecology of young salmonids has been the subject of intensive study. The species that overwinter in freshwater streams as fry typically change from active territorial defence to hiding in cover or in deep water as the temperature drops. Hartman (1965) found that underyearling coho formed aggregations in the deeper portions of his stream tank under winter conditions, while underyearling steelhead hid in the substrate. This corresponds with the findings of Chapman and Bjornn (1969) and Bustard and Narver (1975 a) that steelhead move into hiding places in rubble or similar cover at lower temperatures of the order of 5 °C, regardless of the photoperiod at the time the temperatures naturally occurred. Juvenile coho (Bustard and Narver 1975 a) and chinooks (Chapman and Bjornn 1969) also seek cover at low temperature. Artificial side bays built to simulate the natural side pools used by coho fry in Pacific coast streams for winter refuge were attractive to fry at temperatures in the 2°–5 °C range, but at higher temperatures the fry moved from the artificial pools out into the main stream (Bustard and Narver 1975 b). These changes in habitat preference are associated with other changes in behaviour, reduced feeding, loss of territoriality and passive hiding and with cryptic colouration replacing the more conspicuous colouration associated with territorial behaviour (Hartman 1965; Chapman and Bjornn 1969; Bustard and Narver 1975 a).

5.3.2.2.4 Temperature and Spring Migration of Salmon Smolts

The break-up of ice cover in the spring is probably an appropriate timing cue for the initiation of some migrations. Ice may block migration prior to break-up and moving to feeding areas as soon as possible in the spring will allow full use of the summer season. For example, the initiation of lakeward migration by sockeye fry seems to be related to the break-up of lake ice in the Wood River Lakes of Alaska (Fig. 5.2, Burgner 1962 b), while the upstream migration of Chilko River fry coincides with temperatures warming to 3.3° to 4.4 °C, which also corresponds roughly to ice break-up and spring turn-over (Brannon 1972). The Chilko sockeye fry also show peaks of upstream migration corresponding with sharp increases in water temperature. It is conceivable that upstream migration is facilitated by the effect of warmer water on swimming speed.

The termination of the spring lakeward migration has been less studied. Hartman et al. (1962) found no short-term correlation between water temperature and the seasonal decline of fry migration into Brooks Lake, Alaska.

The seaward migration of sockeye smolts also seems to be timed by temperature, at least in part. The dates at which different stocks migrate coincide with general climatic conditions over the North–South range of the species, with more southern, and hence warmer, populations migrating earlier than more northern ones (Fig. 5.3) (Hartman et al. 1967). The actual date of the start of migration coincides with lake surface temperatures in the range of about 4°–5 °C in Cultus Lake (Foerster 1937 b), Wood River Lakes (Burgner 1962 a) and other Southwestern Alaska Lakes (Hartman et al. 1967) and with ice break-up, which will

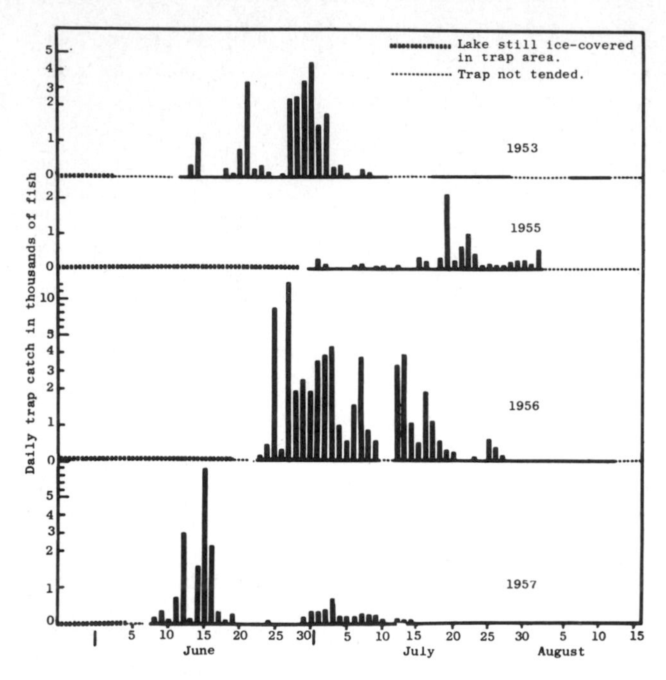

Fig. 5.2. Daily catches of sockeye fry in Anvil Bay Lake trap, 1953, 1955, 1956, and 1957 seasons (Burgner 1962 b). Migration begins shortly after ice break-up each year and hence may be controlled by temperature

give approximately the same temperatures, at Babine Lake (Groot 1965). Short-term elevations of temperature stimulate pulses of migration and cool temperatures decrease migration (Hartman et al. 1967). However, temperature is obviously not the only factor timing the migration, since the Cultus Lake migration in 1931 and 1934 started before the temperature had begun to warm appreciably (Foerster 1937 b) and the 1962 Babine migration began before the outlet temperature began to rise and before it reached the critical 4°–5 °C range (Groot 1965). Each of these years was characterized by a late spring.

The onset of sockeye smolt migration before the critical temperature is reached suggests the presence of an alternate timing mechanism which will initiate migration if the temperature stimulus has not occurred by a certain time or stage of development. Coho smolt migration seems to be timed by lunar control of thyroxine levels (Grau et al. 1981) (see Sect. 2.3.3.2.2). Similar mechanisms may occur as back-up timing systems in sockeye.

The termination of the seaward smolt migration from the nursery lakes sometimes coincides with surface water of the lake reaching about 10 °C (Ward 1932; Foerster 1937 b). Ward (1932) orginally suggested that the formation of a "temperature blanket" of warm water on the lake surface prevented migration by confining the sockeye smolts to the lower depths of the lake. However, it is now known that smolts move into the warm surface waters at night to feed. Burgner (1962 a) was able to capture large numbers of sockeye fry in surface waters when

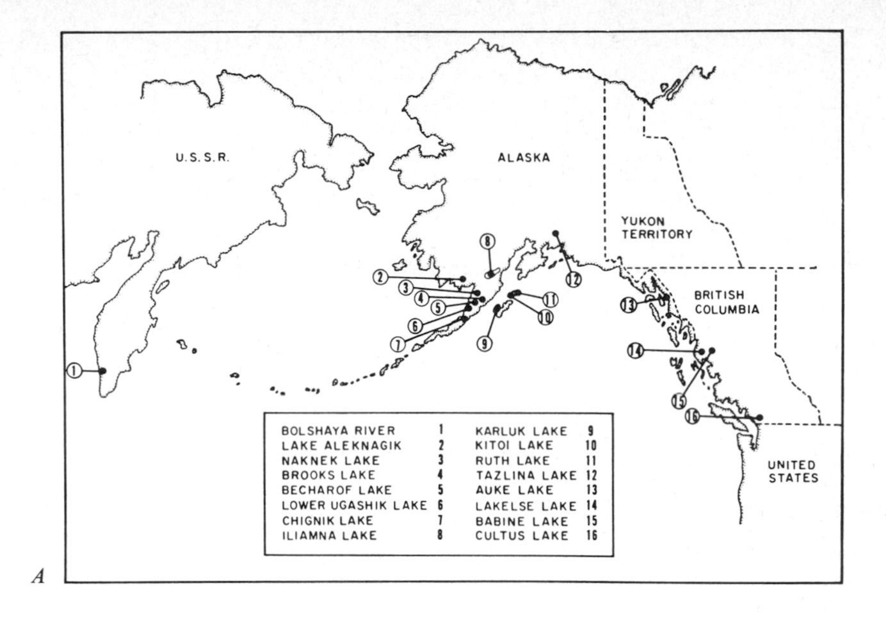

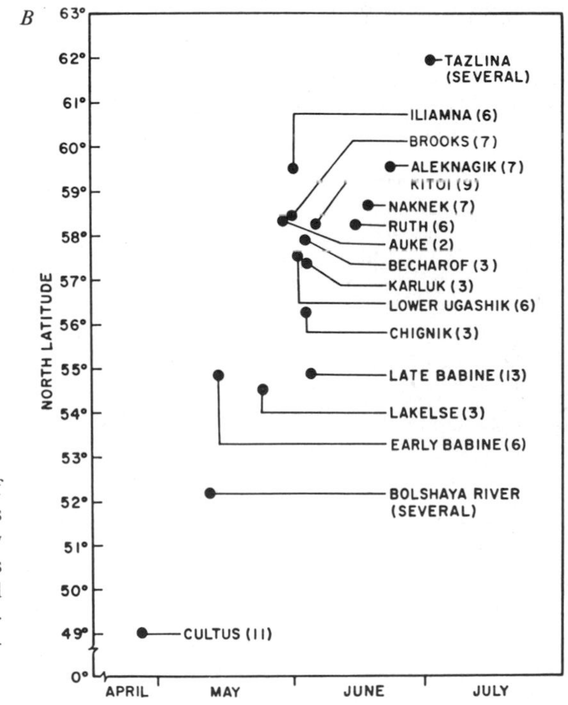

Fig. 5.3. A The geographic location of sockeye salmon-producing systems treated in Fig. 5.3 B. *B* Average date by which 50% of the spring sockeye smolts had migrated from lake systems located at different latitudes. *Numbers in parentheses* show number of years of observation (Hartman et al. 1967)

the temperature was above 10 °C, but it is not clear whether they were feeding or migrating. The rising temperature may initiate a resident phase in the behaviour of those smolts that have not yet departed from the lake. There is, however, no direct evidence of this.

Summer steelhead smolts in the Columbia River face delays in their downstream migration because the increased impoundment of the river slows the current flow (Raymond 1968, 1969). The effect of these delays is to keep the fish in the river long enough for high temperature and advancing season to hasten their reversion to a non-migratory state (Zaugg et al. 1972). The relationship between temperature and reversion to a non-migratory state was tested by subjecting smolts to an increase from 6° to 13 °C at the end of May. This treatment reduced behavioural downstream migration in a raceway and caused a decline in gill ATPase, a physiological indicator of smoltification (Zaugg 1981). This temperature-induced reversion was more effective in fish which had been subjected to advanced photoperiod, suggesting that advancement of their seasonal condition by photoperiodic manipulation increased their susceptibility to the temperature stimulus. Under natural conditions, the high temperatures typical of summer rivers would serve as a reliable indicator that the fish has failed to reach the sea and that reversion to a physiological state facilitating freshwater survival is appropriate.

Chum fry, coho fry, and smolts and sockeye smolts all switch from positive rheotaxis to negative rheotaxis in response to sudden increases in temperature when tested in small circular tanks (Hoar 1951; Keenleyside and Hoar 1955). On a longer term, temperatures in the range of 6° to 9 °C were associated with predominantly positive rheotaxis while 14° to 21 °C led to predominantly downstream swimming. Since the natural downstream migration of the populations tested occurs at temperatures below 10 °C, the function of this relationship between temperature and rheotaxis is not clear. It may again be an emergency response leading fish downstream out of rapidly warming streams into cooler lakes or oceans.

Seaward migration of Atlantic salmon smolts in the spring often starts when the temperature warms to 5° to 10 °C (e.g., Fried et al. 1978; McCleave 1978). McCleave (1978) suggests that photoperiod acts as a primer, preparing the fish for migration, but that temperature may be the releasing stimulus, since the migration seems to occur within the same temperature range each year regardless of the actual date and hence to some extent independently of photoperiod.

In contrast to the studies reported above, Bjornn (1971) concluded that the downstream movement of juvenile chinooks and rainbow or steelhead trout out of two Idaho streams in the fall was not related to temperature. He felt that photoperiod or maturation were probable initiators of migration, having found that migration did not consistently correspond to changes in temperature, food abundance, stream flow, cover or population density.

5.3.2.2.5 Thermal Timing of Migration by Non-Salmonid Fishes

Non-salmonid migrants also move predictably at certain temperatures. A cyprinid, the redside shiner, *Richardsonius balteatus,* first migrates into its spawning streams when the temperature increases to 10 °C and peaks of upstream migration follow temperature maxima by 1 day (Lindsey and Northcote 1963). Brook stickleback, *Culaea inconstans,* move upstream when water in the Roseau River, Manitoba, is warmer than lake water, but migration stops when the stream temperature reaches 19° to 20 °C, near the upper limit for this species (MacClean and Gee 1971). Similarly, walleye, *Stizostedion vitreum,* in Lac La Ronge, Saskatchewan, move into tributaries for spring spawning when the temperature rises above 3.3 °C, with the highest levels of migration occurring between 4.4° and 5.6 °C and spawning at 7.7° to 10 °C (Rawson 1957). The upstream migration of the clupeid twaite shad, *Alosa fallax,* in the Severn Estuary, begins in the spring when the river temperature warms to 7 °C and the autumn downstream migration peaks soon after the temperature falls to 19° and stops at 9 °C (Claridge and Gardner 1978). Again, each of these examples shows an apparent association between temperature and migration time but there is no convincing experimental test of the causal nature of the relationship.

American shad, *Alosa sapidissima,* migrate upstream before spawning. The timing of these upstream migrations appears to be based on a combination of seasonal river temperatures and the location of suitable temperatures in the adjacent coastal waters (Leggett and Whitney 1972). River entry corresponds with temperatures between 10° and 20 °C and peaks at 13 °C, on the average. This means that there are considerable differences in timing between rivers at different latitudes (Fig. 5.4). The St. James River in Florida only reaches suitably low temperatures in mid-winter and the fish begin to enter on the dropping fall temperatures and spawn near the annual temperature minimum. In the more northern York River, Va. and Connecticut River, Ct. they enter with rising spring temperatures. The timing of the river migration coincides with the period during which sea-surface temperatures of 13° to 18 °C occur off the stream mouths (Fig. 5.5). The seasonal excursions of the 13° and 18 °C isotherms correspond to the seasonal changes in peak shad catches and the extreme latitudes reached by these isotherms correspond to the extremes of the shad's breeding range. However, Neves and Depres (1979), using catch data over 14 years, feel that the deep-water temperatures of 3° to 15 °C provide a better indication of shad distribution than do Leggett and Whitney's 13° to 18 °C surface temperatures. Shad introduced to the Pacific Coast of North America in 1871 seem to be bound by similar temperature limits, although in both oceans some shad are captured outside these limits, suggesting that these are probably preference limits rather than survival limits.

Adult tautog, *Tautoga onitis,* migrate offshore from their summer territories along the coast of Long Island, New York, when the fall temperature drops to between 13° and 9 °C (Olla et al. 1980). In aquaria, migratory restlessness and an associated change from diurnal to predominantly nocturnal activity occurred under a combination of natural fall-winter photoperiod (12.5L:11.5D declining to 10L:14D) and temperature (17.9° declining to 2.6 °C) when the temperature reached about 10 °C. These experiments were carried out during the normal October to December fall migration period and the behavioural changes of the laboratory fish coincided with the migratory period of fish in the wild. The migratory activity continued until the temperatures reached about 5 °C, at which point activity began to decline and the fish gradually became torpid (Olla et al. 1980). In

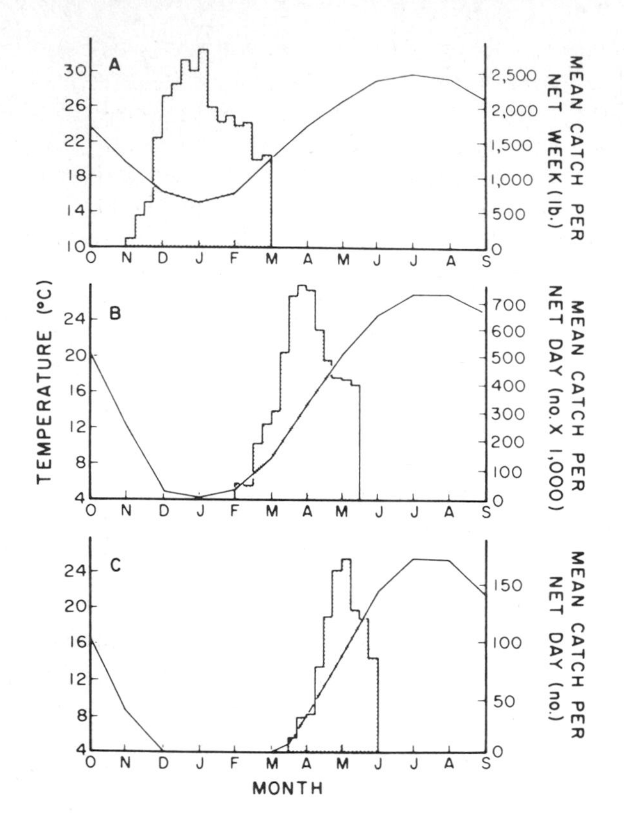

Fig. 5.4. Shad catches in relation to season and temperature in three Atlantic coast rivers. *A* St. Johns River, Florida, temperatures are averages of 1960 to 1967, shad catches 1962 to 1967. *B* York River, Virginia, temperatures are averages of 1953 to 1962, shad catches 1953 to 1956. *C* Connecticut River, Connecticut, temperatures are averages of 1958 to 1969, shad catches 1944 to 1964 (Leggett and Whitney 1972)

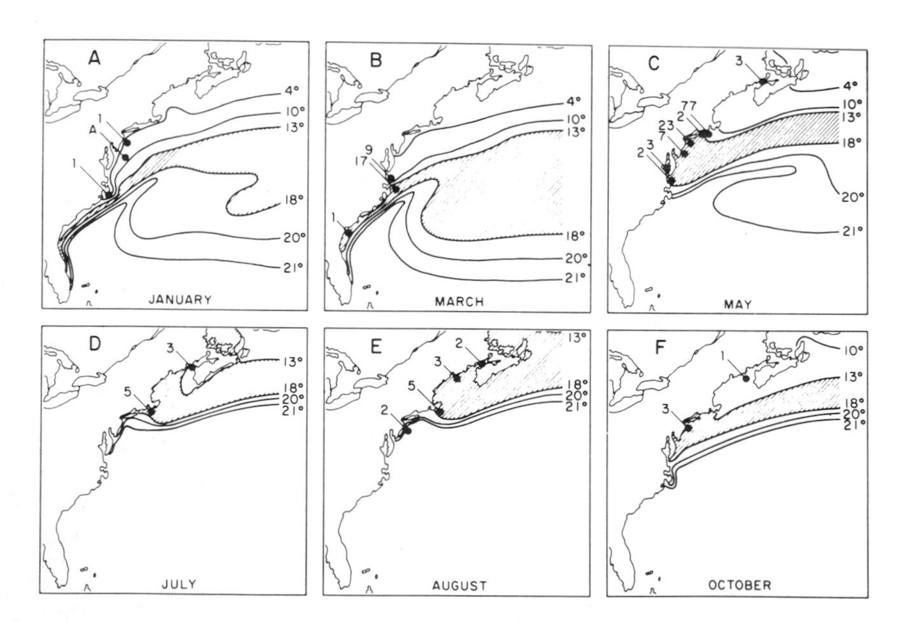

Fig. 5.5. Atlantic ocean average sea surface temperatures and associated locations of recaptures of marked shad for selected months. Number of recaptures is shown (Leggett and Whitney 1972)

August and September, a combination of summer photoperiod (15.7L:8.3D to 14.2L:9.8D) and fall–winter temperatures (15.8 °C declining to 2.3 °C) did not induce migratory restlessness, nor did a combination of fall–winter photoperiod (12.5L:11.5D to 10L:14D) and summer temperatures (19.5 °C rising to 20.1 °C) administered in December. Olla et al. (1980) interpreted this to mean that temperature was an important triggering stimulus for the fall migration but that an appropriate shortening photoperiod was also required. The experiments did not exclude an annual cycle in responsiveness to temperature.

Temperature changes coincide with periods of onshore wind which are important to the survival of newly emerged capelin, *Mallotus villosus,* and thus serve as an appropriate timing stimulus for emergence (Frank and Leggett 1981, 1982). The capelin eggs are laid in beach gravel and the young hatch and remain in the gravel until the occurrence of strong onshore winds when a synchronized emergence from the substrate occurs. Offshore winds prevail during the emergence season in eastern Newfoundland where the study was done. Onshore winds bring warmer, less saline surface water inshore to replace the cold, saline bottom water which is normally drawn to the surface along the shore by the prevailing winds (Frank and Leggett 1982; Fig. 5.6). The warm surface water contains fewer predators than the cooler water, and thus constitutes a "safe site" for the newly emerged capelin to begin their period of larval drifting. Frank and Leggett (1982) cite unpublished experimental data as showing that the increase in water temperature associated with onshore winds is a significant stimulus to larval emergence, although decreased salinity and increased wave action also correlate with onshore

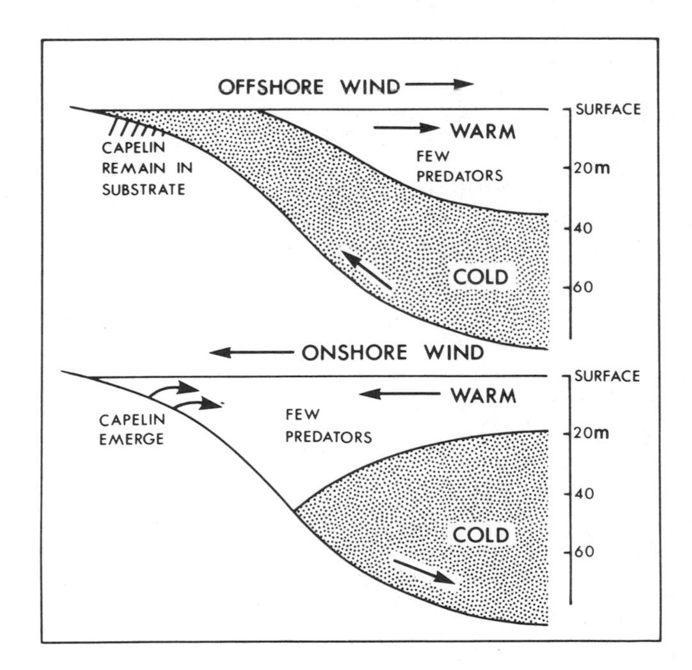

Fig. 5.6. The effect of offshore and onshore winds on the summer temperatures and water layers in the nearshore zone along Newfoundland's east coast (Frank and Leggett 1982)

winds and might serve as timing stimuli. As with such events as ice break-up, these temperature changes correlate with a biologically important event that cannot be predicted precisely from such regular environmental changes as photoperiod or tides.

5.4 Thermal Stimuli Controlling Direction and Distance

5.4.1 Temperature and Fish Distribution

Temperature can affect the distribution of fish in the environment in a number of ways. Fish are killed by water that is too hot or too cold: the upper and lower lethal temperatures. This will set outer boundaries on a species' distribution in the environment but, in fact, the actual limits of distribution seem to be well within the extremes set by the adult lethal levels (Brett 1956). This is probably because growth and reproduction require a narrower temperature zone than survival, and juvenile stages are often less tolerant than adults. This limiting effect of temperature may account for some seasonal movements as fish follow suitable temperatures. The seasonal excursions of the 21 °C surface isotherm and the northern limits of commercially caught yellowfin and skipjack tuna (Fig. 5.1) referred to earlier provide a good example. However, in such cases it has not been demonstrated experimentally that temperature is the only, or even the main causal factor responsible for the changes in distribution pattern shown by these fish. If temperature is a primary factor in some cases, then the means by which temperature alters or controls the distribution is unknown. This state of affairs is not surprising. Temperature responses are difficult to study. One cannot, for example, experimentally block the receptor system in the way that one can destroy vision or olfaction.

5.4.2 Temperature Preference

As well as providing limits to distribution, temperature may affect the movements of fish through individual fish seeking or remaining in their preferred temperature. There is ample evidence that fish are capable of selecting a preferred temperature when given a choice. Coutant (1977) lists over 100 species in his table of fish temperature-preference studies and recent interest in the effects of thermal pollution has spurred increased study of fishes' ability to select or avoid temperature zones, and caused a dramatic increase in publications on the subject (Beitinger and Fitzpatrick 1979). Temperature preference is worthy of a brief examination here because the mechanisms (such as the kineses or predictive responses discussed below) by which fish locate preferred temperatures in temperature gradients may be similar to the mechanisms used in thermal orientation during migration.

5.4.2.1 The Measurement of Thermal Preference

Temperature preference is usually studied with one or the other of two different types of apparatus: spatial gradients and temporal gradients. In a spatial gradient the fish are placed in a tank that contains a range of temperatures distributed horizontally, vertically or radially (McCauley 1977). The fish can then swim from one portion of the tank to another and locate and stay in a restricted portion of the available temperature range – the preferred temperature zone.

In the temporal gradient apparatus the fish changes the temperature of its container through a switching mechanism. These mechanisms include levers that the fish push and various shuttle box arrangements in which the fish break a light beam as they swim from one chamber to another (McCauley 1977). The most commonly used variation of this type of apparatus is a two-chamber tank based on a design by Neill et al. (1972). The two chambers with a passage between them are maintained at different temperatures. When a fish moves through the passage, photocells record the direction of movement and activate a control mechanism to change the temperature of the whole apparatus, while still maintaining the temperature difference between the two chambers. When the fish swims from the warmer chamber to the cooler one the apparatus is cooled slowly. When it swims to the warmer side the apparatus is warmed. Since the control system depends on the fish making biologically appropriate responses to the temperature difference between the chambers, moving to cooler water to lower the temperature of the system and to warmer water to heat it, the fish learn the task quickly. Neill et al. (1972) tested a naive bluegill in their shuttle system without a temperature difference between the chambers and the fish was still able to learn to regulate the temperature. Obviously it was able to relate moving through the passage or location in the chambers with future changes in temperature. Since one of the chambers would always be associated with warming and the other with cooling, this may have been an example of spatial learning.

Both the spatial and temporal gradients give essentially similar results, with fish selecting temperatures in one apparatus that are within a few degrees of those selected in the other (McCauley 1977). The small differences reported are not surprising when comparing studies carried out at different locations and times of year and with different stocks of fish. What is clear, regardless of the details of the method used, is that fish are capable of responding to water temperature and of either seeking and remaining in preferred temperatures or avoiding areas of aversive temperatures. It is often difficult to distinguish between these two processes.

5.4.2.2 Mechanisms for Selection of Preferred Temperature

Despite the overwhelming evidence that fish are able to select a preferred temperature in laboratory gradients and in the wild, there is relatively little information on the mechanisms used in the process. Neill (1979) distinguishes between two major types of mechanism, which might be used in thermal preference or thermoregulation: predictive and reactive responses. Since these two types of mechanism could also apply to temperature-guided migration we will review their characteristics.

In a predictive response the fish moves to a place where the preferred thermal conditions have been found in the past, either in the past experience of the individual fish or in the evolutionary past of the species, according to whether the response is based on individual learning or natural selection. An individual might learn that the inlet stream of a lake was cool and move there when the lake became too warm. The ability of a bluegill sunfish to associate location in a tank with warming or cooling is good evidence that fish can learn to associate temperature and location (Neill et al. 1972). Natural selection would probably favour fish that moved to deeper water when temperatures were too high since, in general, cool water will lie below warmer water. Such responses might evolve even without temperature receptors: if, for example, fish that move to deep water during the summer had higher survival rates than those that remained in shallow water. In fact, these responses are probably based on the known temperature sensitivity of fish. A simple set of "rules of thumb" would probably be adequate in many situations in nature: Too cool? Move upward or toward the equator. Too warm? Move downward or toward the poles (Neill 1979).

Predictive mechanisms are vulnerable to the unpredictability of the environment. Neill (1979) infers from a study by Gallaway and Strawn (1974) that Gulf menhaden, *Brevoortia patronus,* living in the discharge area of a power station are "fooled" when cooler water comes to lie over the warm discharge, leading to mortality.

In many cases in nature the distribution of temperatures will be too inconsistent for predictive thermoregulation, or the fish will not have sufficient information for accurate prediction. In these cases the fish must use reactive methods of thermoregulation in which temperature differences lead to differences in behaviour.

Orthokinesis, a simple relationship between swimming speed and temperature, with the lowest speed occurring at the preferred temperature and progressively higher speeds as the fish encounter temperatures above or below the preferred temperature, would theoretically lead to congregation of fish at the preferred temperature (Fraenkel and Gunn 1940, 1961). It is doubtful, however, that this mechanism is sufficient to account for the actual performance of fish in temperature gradients. Sullivan (1954) found that fish in changing thermal conditions showed less movement at the preferred temperature, but that equilibrated fish were more active at the preferred temperature. Reynolds and Casterlin (1979) found reduced activity by goldfish and bluegills near the final preferendum, but they considered this to be a minor anomaly in the upward trend of activity with temperature. These results might indicate a basis for orthokinesis in fish, but in an unbounded temperature gradient simple orthokinesis will eventually lead to the exposure of every fish to lethal temperatures (Fraenkel and Gunn 1961; Neill 1979). At the very least one must add a turning component to the response to avoid mortality. Neill (1979) proposed a klinokinetic avoidance mechanism in addition to orthokinesis. If fish increased their rate of turning as they entered areas of unsuitable temperature and showed a decreasing turning rate as conditions improved they would tend to avoid the temperature extremes.

Fish might be able to respond to sharp temperature gradients by sensing the temperature differences between different parts of their bodies. The thermoclines of lakes, for example, are often so sharp that a human diver can feel the temperature difference over the length of a finger; fish should be at least equally sensitive. More gradual gradients will be more difficult to detect. Neill (1979) suggests that memory of recent thermal events could be one mechanism for orientation in gradual gradients. If the fish knew its direction of swimming (through some compass mechanism) and remembered recent temperatures, it might reverse its course if the immediate temperature was less suitable than the temperature experienced earlier on the same course. If the temperature was improving it could maintain its course. Memory, in the normal sense, might not be necessary. For the short term, the thermal inertia of the fish might allow it to detect its recent thermal history by comparing surface receptors with deep body receptors such as those found in the central nervous system (Neill 1979). If, on a straight course, the surface receptors were cooler than the deep receptors then the fish would be moving into cooler water. If the surface receptors were warmer than the deep receptors it would be moving toward warmer water.

Fish, then, are known to have the basic requirements necessary for several different types of temperature orientation. They have a temperature sense, memory, sense of direction, and ability to relate spatial location to temperature. These abilities will allow them to use reactive mechanisms like orthokinesis, klinokinesis, and comparison of temperature changes with swimming direction to position themselves in temperature gradients and to make predictive responses to the seasonal and diel temperature changes that occur in nature.

5.4.3 Temperature as a Guidance Mechanism in Migration

In spite of the abundance of evidence that fish can sense and respond to environmental temperatures, there is very little evidence that they use temperature as a directional cue in migrations. This may be a reflection of the way in which thermal differences are distributed in the environment. Locations will not have a unique thermal signature in the way that they may have a unique chemical signature, so temperature sensitivity is probably less useful than the chemical senses for identification of a home site. Thermal gradients are often disrupted by turbulence so that they are unreliable indicators of general direction. Celestial or magnetic compass are probably more reliable. We are left with relatively few cases where temperature will be the most appropriate cue for migratory direction. During summer, for example, water draining from the surface of a lake may be reliably warmer than water running into the lake from mountain tributaries. Similarly, in summer, inshore water may be consistently warmer during the day than offshore water. We might therefore expect that fish migrating inshore or toward lakes would use thermal cues in their orientation.

5.4.3.1 Lakeward Orientation by Temperature

As already noted several times, salmonid fishes, that spend part of their life cycle in lakes but spawn in streams, often face the "problem" of different rheotactic requirements for inlet and outlet fry. Rainbow trout, for example, spawn in both the inlet and outlet streams of Loon Lake, British Columbia (map in Chap. 2, p. 24) and the young move to the lake during the first or second summer of their lives. The young fish move downstream at night in Hihium Creek, which drains into Outlet Creek, and in Inlet Creek. In Outlet Creek the fish move upstream during daylight. Northcote (1962) found that Inlet and Hihium Creeks were generally cooler (less than 15 °C) than the Outlet Creek which drained the warm surface waters from the lake and was usually over 15 °C during the migration period. He argued that the trout in Loon Lake were of one genetic stock and that the two different rheotactic responses were controlled by a temperature-dependent switching response: upstream in water over 15 °C, downstream in water less than 15 °C. Transfer experiemnts seemed to confirm this hypothesis. Inlet fry transferred into warmer outlet water did not move downstream at night as they would in cool inlet water, but instead held their position or moved upstream. Additionally, fry in cool water remained off the bottom at night, which would facilitate downstream drift, while fry in warmer water often maintained tactile contact with the bottom in the dark which would help them hold position in the current. The experiments performed using lake and stream water to provide the warm and cool temperatures would confuse thermal and chemical stimuli (Brannon 1972) (see Chap. 3), but this objection is countered by Northcote's observation that downstream movement occurred in Outlet Creek when cold periods in June led to outlet temperatures below 15 °C.

Kelso, Northcote, and Wehrhahn (1981) raised the Loon Lake and Hihium Creek rainbow stocks and inlet and outlet stocks from Pennask Lake, B.C. in captivity and tested their response to current at high (11.5°–13.5 °C) and low (4.2°–5.2 °C) temperatures. Low temperature did not significantly increase downstream migration, in fact there was a slight tendency for more upstream movement at the lower temperatures. Kelso et al. (1981) concluded that genetic differences between inlet and outlet stocks were more important than temperature in determining lakeward migration (see Chap. 7).

Northcote's (1962) findings cannot be applied without modification to all rainbow trout populations. Large rainbow trout from Kootenay Lake move upstream in the Lardeau River to spawn just below the outlet of Trout Lake (the Gerrard Stock). Some of the emerging fry moved upstream at low temperatures, to enter Trout Lake, the opposite of the response shown by Loon Lake rainbow fry. However, most fish went downstream early in the season when the water was cold but as the season progressed and the water warmed, more and more moved upstream. In an experimental flume more fish moved upstream in warm water from the Lardeau River (12°–15 °C) than in cold water from nearby Gerrard Creek (7.5°–11.7 °C) (Northcote 1969a). Unfortunately, this flume experiment confuses thermal and chemical stimuli, since it could also be interpreted as showing more upstream migration in lakewater than in streamwater.

Temperature is probably not the general controlling mechanism for lakeward movement of salmonids. Brannon (1972) concluded that temperature was not a major controlling factor in lakeward migration of his sockeye fry, since upstream migrating genetic stocks from outlet streams retained their positive rheotaxis even at lowered temperatures (see Chap. 7). Raleigh and Chapman (1971) found that Yellowstone Lake cutthroat trout fry from inlet streams were negatively rheotactic, while those from outlet streams were positive when both stocks were tested at the same temperature. In fact, decreasing temperature increased the positive rheotaxis of outlet fry, while temperature had no significant affect on the rheotactic responses of inlet fry, suggesting a genetic difference between the two stocks (see Chap. 7).

Temperature may play a guiding role in the migration of another group of lake-seeking fish; alewives, *Alosa pseudoharengus,* and glut herring, *A. aestevalis,* spawn in freshwater pools and lakes after a period of growth at sea or in large bodies of freshwater. Streams with suitable spawning lakes may tend to be warmer than streams without lakes or pools. Collins (1952) tested these fish in an experimental flume in which the river water could be warmed by immersed heaters. He found that they consistently preferred the warmer water in a choice situation.

5.4.3.2 Onshore Migration

Onshore migrations from offshore spawning areas might also be guided by thermal stimuli. Norris (1963) suggested that prejuveniles of the California opaleye, *Girella nigricans,* move inshore in response to a thermal gradient. This suggestion is based on his observation of the behaviour of prejuvenile opaleye in long, thin tide pools which formed natural temperature gradients along the length of the pool. The fish moved along the pool, following their preferred temperature as it moved during the day. Similar behaviour in the ocean would bring them near shore, where their preferred temperature occurred and where the young opaleye tended to congregate.

5.4.3.3 Summary of Thermal Guidance Mechanisms

Considering the apparent potential of temperature as a guiding stimulus, the foregoing is a paltry list of examples. It is worth remembering that the lack of evidence for the use of thermal information in migration may also reflect the difficulty of experimental work with thermal receptors. The importance of olfaction, for example, was first indicated by studies on fish with blocked nostrils. Ablation of the thermal receptors is not feasible at the moment.

5.4.4 Effects of Thermal Effluents

The increasing use of electrical power has led to the construction of more large thermal generating plants. These plants require cooling water, whether fueled by nuclear or conventional fossil fuels. Once this water has been heated by passage through the generating plant, it is often discharged back into the body of water from which it was taken, several degrees warmer and often with algicides such as chlorine added (Stober et al. 1980). We might expect that such warming would have profound effects on migratory fish moving through the area of the thermal discharge.

There seem to be remarkably few direct studies of the effects of thermal effluents on migratory fish. One which stands out is the work of Leggett (1976) on

American shad migrating up the Connecticut River past the effluent (about 8 °C above ambient) of the Connecticut Yankee Atomic Power Plant. Fish fitted with ultrasonic tags moved past the outfall area along the opposite bank of the river with only a minority showing brief meandering near the outfall area before continuing upstream. In other studies on the effects of thermal effluents, young shad avoided rapid increases of about 4.5 °C (Moss 1970) and juvenile chinook salmon avoided temperatures of 9° to 11 °C above ambient (Gray et al. 1977). It appears that migratory fish may be capable of avoiding plumes of thermal effluent when there are suitable alternative routes available past the heated water.

Thermal pollution has been the subject of intensive research. Raney and Menzel (1969) listed 1870 references in their bibliography of the effects of heated effluents on fish. There seems, however, little reason to change the conclusions of Alabaster and Downing (1966), that most fish are able to detect and avoid lethal temperatures. In fact, resident fish in a heated outfall area may use the predictable warmth of the effluent in their thermoregulatory behaviour, moving in and out of the heated water in order to maintain their body temperature as close as possible to optimum levels. Of course, this relatively benign situation will only persist as long as the thermal load on the water body is low enough that there are areas near the outfall to which the fish can retreat in warm weather. The thermal characteristics of a water body are a natural resource which must be protected from excessive alteration (Magnuson et al. 1979).

5.5 Summary

There are predictable relationships between thermal phenomena and other biologically important features of the environment. These relationships include such things as the seasonal cycles of temperature, the tendency for warmer water to lie above cooler water and the association of sudden temperature increases with ice breakup in rivers and streams.

Fish have the sensory capability necessary for the detection of these thermal stimuli. However, this sensory ability is not associated with prominent morphological sense organs but rather with small structures widely scattered on the body surface and in the central nervous system. The diffuse nature of this sensory system precludes experiments based on the blocking or the destruction of the thermosensory system. This makes it more difficult to establish the importance of thermal stimuli in migration.

Temperature is, along with photoperiod, an important component in the timing of seasonal endocrine cycles, and hence in timing seasonal changes in migratory behaviour. Additionally, certain events that cannot be predicted precisely from seasonal light and temperature cycles are often associated with temperature changes, which can serve as releasing stimuli for migration. Such events include freshets and offshore winds; freshets often provide favourable conditions for migration by increasing stream depth, while offshore winds provide suitable environmental conditions for the emergence of capelin fry. Extreme temperatures, unseasonable hot spells for example, may require emergency avoidance responses such as early emergence from spawning gravel or evacuation of shallow streams.

Temperature-preference behaviour limits fish to areas bounded by zones of acceptable temperature. These regions of preferred temperature are usually well within the limits of tolerance or survival. Thus extreme temperatures can block entry into a river system and seasonal movements of coastal isotherms correspond with the seasonal appearance of migratory fish in a region. In some cases temperature relationships could be appropriate guiding stimuli. For example, predictable differences between the temperature of the inlet and outlet of mountain lakes may be one among several factors directing lakeward movement of juvenile salmonids.

The thermal characteristics of natural waters are vulnerable to modification by damming, which allows thermal stratification, and by the addition of heated waste water from power plants and industries.

Chapter 6

Electrical and Magnetic Stimuli

6.1 Electricity in Water

Natural waters conduct electricity. Natural processes generate electrical fields and many species of fish have receptors that respond to electric currents (Bennett 1971 b; Fessard 1974). There is, therefore, the potential for fish to use electrical information in their migratory behaviour. Electric orientation and communication may serve as effective alternatives to vision in situations where light is limited by turbidity or depth.

6.2 Electroreception

Any animal receptor will respond to electrical stimulation if the stimulus is strong enough. What makes fish unusual is the presence in many groups of specialized sense organs with very low thresholds for response to electrical stimuli. These electroreceptors are linked with areas of the central nervous system that are adapted for analysis of electrical information.

The study of fish electroreception has a long but interrupted history. Parker and van Heusen (1917) demonstrated that brown bullheads, *Ictalurus nebulosus,* would respond differently to metal rods and to rods of insulating material placed in the water. The freshwater bullheads were indifferent to the insulators, but were repelled when the metal rods were deeply immersed in the water. When just the tip of the conducting rod was immersed, however, the fish were attracted to it. Parker and van Heusen attributed these responses to the fish sensing the weak galvanic currents around the metal rod. Despite this promising beginning, the study of fish electroreception languished until the discovery by Lissmann (1958) that fish with weak electric organs used electroreceptors to detect the electric field generated by their own electric organs. Changes in the electric field caused by objects in the nearby environment could then be perceived and used in short-range orientation and electrical discharges could be used in social communication.

6.2.1 Passive Electroreception

Electroreceptive fishes are sometimes divided into two categories, active and passive. Active electroreceptive fishes generate electric currents by means of modified muscles or motor nerve end-plates. The passive electric fishes do not have specialized organs for producing electricity, but do have electroreceptive organs. This will allow them to respond to natural electric fields, for example those generated by prey or predators or fields generated by the movement of water currents or the fish itself through the earth's magnetic field. This ability could be useful in allowing fish to orient the direction of water flow or to use a magnetic compass system to orient along a migration route.

The ranks of the passive electric fishes include the brown bullhead (Roth 1968, 1969) and other silurid catfishes including *Clarias batrachus* and *Heteropneustes fossilis* (Srivastava et al. 1978), the transparent catfish *Kryptopterus bicirrhus* (Wachtel and Szamier 1969) and three marine catfishes, *Bagre marinus, Galeichthyes felis* (Szamier 1974), and one species of *Plotosus* (Obara 1976). The distribution of passive electroreception in catfish has not been fully determined, but it would not be surprising to find it almost universal in this group, which possesses other specializations for sensory reception at night or in murky water. Many, perhaps all, elasmobranchs, including both sharks and rays, are also electroreceptive, through the ampullae of Lorenzini (Bennett and Clusin 1978; Kalmijn 1978 a, b) and the related ratfishes, Chimaeridae, are also equipped with electroreceptive ampullar organs which are structurally and functionally homologous with the ampullae of Lorenzini (Fields and Lange 1980). Electroreceptive ampullary organs, probably homologous with ampullae of Lorenzini, also occur in the chondrosteans, including the sturgeon, *Scaphirhynchus platorynchus,* (Teeter et al. 1980) and the paddlefish, *Polyodon spathula* (Jorgensen et al. 1972) and in the lungfishes (Dipnoi) and reedfishes (Brachiopterygii) (Pfeiffer 1968; Roth 1973).

There are a number of migratory species within this large group of electroreceptive fishes. Dogfish sharks (*Squalus acanthias*), for example, have been tagged in British Columbia and recaptured in Japan, 7600 km away (e.g., Holland 1957), transatlantic movements of dogfish have also been recorded (Templeman 1976) and tag recoveries indicate that the species contains both migratory and non-migratory populations (Holland 1957). Similarly, tagging studies indicate migratory behaviour in several other sharks including blue sharks (*Prionace glauca*) and mako sharks (*Isurus oxyrinchus*) (Beckett 1970). Sturgeons undertake migrations associated with spawning (Scott and Crossman 1973), as do paddlefishes (*Polyodon spathula*) (Pfleiger 1975). Unfortunately, beyond establishing that migrations occur, little or no experimental work on orientation mechanism has been done on these groups.

The discovery by Bodznick and Northcutt (1981) that the anadromous lamprey, *Lampetra tridentata,* has a threshold of response to electric stimuli as low as that of freshwater electric fishes indicates that electroreception is a primitive condition in the vertebrates. Bodznick and Northcutt suggest that electroreception in the lamprey is homologous with electroreception in primitive jawed fishes such as the sharks and rays, but that teleost electroreception probably evolved independently.

The electroreceptors of the passive electric fish are probably all tonic receptors, responding to prevailing direct current (d.c.) electrical stimuli (Bennett 1971 b). Tonic receptors are spontaneously active, even when apparently unstimulated (Hoar 1975). The rate of activity changes in response to external stimuli. Phasic receptors, in contrast, are inactive unless stimulated. Tonic receptors found in weakly electric freshwater fish are insensitive to the high frequency discharge of the electric organs, and presumably carry on passive electroreception independently of the phasic receptors used in active electroreception.

Among these passive electric fishes there is a general relationship between the sensitivity of the receptors and the conductivity of the water. The saltwater species are more sensitive, responding to stimuli in the μV range, while freshwater fish are sensitive only in the mV range (Szamier and Bennett 1980). This relationship appears to hold even across taxonomic groups, with the freshwater ray having the low sensitivity more typical of freshwater teleosts, while the marine catfish, *Plotosus*, a teleost, is as sensitive as the marine elasmobranchs. Background electrical noise is higher in freshwater (Hopkins 1973; Kalmijn 1974), eliminating the possibility of using very weak signals and hence any need for highly sensitive receptors (Szamier and Bennett 1980).

6.2.2 Active Electroreception

The active electric fishes, which produce their own electric currents, include several skates and rays, mormyrids, gymnotids, a silurid catfish and marine stargazers (Bennett 1971 a). The electric organs of these fish serve at least three different functions: (1) Weak electric organs, producing discharges of fractions of a volt, are used in social communication such as threat and courtship (Black-Cleworth 1970; Bell et al. 1974; Westby 1974). (2) Weak electric organs produce electrical fields used in locating objects and orienting to them. (3) Strong electric organs, producing discharges of 10's or 100's of volts, are used in defense and in predatory behaviour (Bray and Hixon 1978). Electric fish often have a variety of electric organs and receptors specialized for these different functions. In terms of migration, electric social signals could serve as timing cues in the same way as other social signals. The weak electrolocation system could be used in local orientation or perhaps in landmark orientation.

Active electric fishes usually have both tonic and phasic receptors. The phasic receptors are sensitive to changes in the stimulating voltage and respond to relatively high frequency changes in stimulation. (The electric elasmobranchs have only tonic receptors.) These two categories of receptor differ in their morphology. The tonic receptors have an obvious canal leading to the exterior of the animal, while the phasic receptors have no obvious canal (Bennett 1971 b). Tonic receptors are further divided into two major types. One type is used in electrolocation, using an electric field to detect objects in the environment. The other type is used in electric communication, sending electric signals to conspecifics.

Since only tonic receptors are found in the weakly electric rays, it seems probable that tonic receptors can also function in electrolocation, although Bennett (1971 b) points out that the ray discharges its electric organ infrequently and that the tonic receptor is probably functioning passively most of the time. It is also possible that the ray's electric organs function in communication rather than electrolocation (Kalmijn 1974).

Electrolocation systems are subject to interference from neighbouring fish, "jamming," and the phasic receptors used in these systems are often modified so as to reduce this jamming effect. The problem of jamming and its avoidance has attracted a great deal of attention (e.g., Heiligenberg 1977). Electric fish commonly shift to a different frequency when they encounter another fish using a frequency close to theirs. This is termed a jamming avoidance response. They may also physically attack other individuals which use a frequency close to their own. *Gymnotus carapo*, for example, attacked "model" electric fish most vigorously when their frequency was close to that of the attacker (Westby 1974). In addition to these mechanisms, jamming can be reduced by tuning the receptors to the frequency of the individual's own electric organs. In *Eigenmannia virescens* the phasic receptors are tuned to the individual "signature" discharge rate of each individual, which may vary from 250 to 500 Hz, and which can only be varied about 10% for short periods in a jamming avoidance response (Viancour 1979). The tremendous variation in electrical discharge patterns both within and between species may be another mechanism for reducing interference (or perhaps for insuring privacy) (Hopkins and Heiligenberg 1978). This situation is analogous to the use of radio and television frequencies by humans.

6.2.3 Electrical Sensitivity in "Non-Electric" Fish

Despite the literature dealing with the specialized electric fish, one of the most interesting questions to the student of fish migration is the sensitivity to electrical stimuli of fish without specialized electroreceptors. This interest arises from the fact that the most intensively studied and most economically important migratory fish fall into this category. The threshold sensitivity of non-migratory goldfish to d.c. current was found to be as low as 5 μamps cm^{-2} while codlings, *Gadus callarias*, reponded to 15 μamps cm^{-2} d.c. and 2 μamps cm^{-2} a.c. (Regnart 1931). Cutting of the lateral line nerves in the codling greatly increased the threshold,

Table 6.1. Some response thresholds of eels to electrical stimuli

Species	Voltage threshold	Current threshold	Method	Ref.
Anguilla rostrata	2.5 μV cm^{-1} (f.w.) 2.5×10^{-2} μV/cm (s.w.)	10^{-2} μA cm^{-2} 10^{-2} μA cm^{-2}	Turning Turning	McCleave and Power (1978)
A. rostrata	25 μV cm^{-1} (f.w.)	10^{-1} μA cm^{-2}	Turning	Zimmerman and McCleave (1975)
A. rostrata	6.7×10^{-2} μV/cm	1.7×10^{-4} μA cm^{-2}	Cardiac cond.	Rommel and McCleave (1973b)
A. anguilla	0.97 mV cm^{-1} (50 Hz., a.c.)	1×10^{-2} μA cm^{-2}	Cardiac cond.	Berge (1979)
A. anguilla	0.4–0.6 mV cm^{-1}	4×10^{-2} μA cm^{-2}	Cardiac cond.	Enger et al. (1976)

suggesting that the lateral line receptors may have been responsible for the lowest thresholds observed. Rommel and McCleave (1973b) conditioned Atlantic salmon parr to show cardiac responses to and 0.167×10^{-2} μamps cm^{-2}. The results of experiments on eels have been contradictory (Table 6.1). McCleave and his co-workers found response thresholds in the μV range for American eels, in both cardiac conditioning tests and tests on turning direction. In contrast, Enger et al. (1976) and Berge (1979), working with European eel, found thresholds for cardiac conditioning in the mV range. The difference is important, since the lower thresholds would allow these animals to detect the electric fields generated in oceanic currents (Rommel and McCleave 1973a) but the higher thresholds would not (see Sect. 6.4.1). These studies used different species, ages, and methods. It remains a moot point whether or not fish without specialized electric receptors can use electrical information for orientation in natural conditions.

6.3 Electrical Timing

Since electric signals can be used in social communication in the electric fishes, we could surmise that social aspects of timing could be mediated by electrical stimuli. These would include synchronization of the activity of a social unit, as in schooling (Moller 1976), so that fish started or stopped migratory activity at the same time.

I am not aware of any cases of electrical events in nature being used in diurnal or seasonal timing of migration although such use might be possible. The basic requirements for timing are present, predictable changes in electrical stimuli and sensitivity to these stimuli. For example, Peters and Bretschneider (1972) found that electrical phenomena in natural catfish pools were more stable at night than during the day. Such differences could conceivably provide timing cues. Timing of behaviour with regard to storms might be possible, since lightning provides stimuli which would be detectable by electroreceptive fishes (Hopkins 1973). The common association of heavy rainfall and changes in river level with thunderstorms might provide an adaptive advantage to storm prediction or detection. However, since lightning-generated signals propagate for great distances, thousands of kilometers (Hopkins 1973), it may be difficult to distinguish between nearby and distant storms.

6.4 Electrical Stimuli Affecting Direction and Distance of Migration

We shall deal first with electrical stimuli which might be used in the migratory behaviour of passive electric fish or fish without specialized electroreceptors, keeping in mind that active electric fish are also capable of passive electroreception.

6.4.1 Ocean Water Currents

The direction of ocean currents should be a useful piece of information to a migrating fish. An individual that could identify an appropriate water current, perhaps by chemical characteristics, then align itself so as to swim along the flow, could use this system to guide its oceanic migrations (Royce et al. 1968; Harden Jones 1968).

Any conducting material moving through a magnetic field will generate an electrical field. This includes oceanic water currents moving through the magnetic field of the earth. The magnitude of the electrical field induced by an ocean current will depend on the velocity of flow, the strength of the magnetic field, the angle between the direction of flow and the magnetic field and the influence of nearby moving water, the proximity of the bottom and the conductivity of the bottom (Fig. 6.1) (Kalmijn 1974). Measured oceanic voltage gradients are of the order of 0.05–0.5 $\mu V\ cm^{-1}$. In the English channel, for example, the voltage gradient was 0.25 $\mu V\ cm^{-1}$ when tidal flow was at maximum velocity (Kalmijn 1974). These values are well within the range of detection of the ampullae of Lorenzini of marine elasmobranchs and might allow these fish to align themselves along the direction of water flow, and to select regions of maximum or minimum velocity according to whether they were moving upstream or downstream. Apparently the presence of the water surface reduces the development of voltage gradients by movement of the water through the horizontal component of the geomagnetic field. This means that surface currents primarily induce voltage gradients by movement through the vertical component and that electrical information will be less available near the magnetic equator where the magnetic field is largely horizontal (Kalmijn 1974). Perhaps, however, deep water currents away from the surface would still generate perceivable voltage gradients as they moved through the horizontal component of the earth's magnetic field.

Rommel and McCleave (1973a) used a mathematical model to predict the electric field in six cross-sections of the Gulf Stream. The velocity of movement of the Gulf Stream through the geomagnetic field was predicted to generate elec-

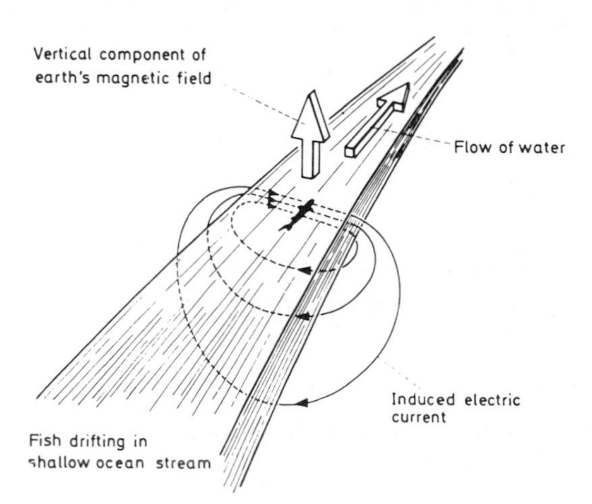

Fig. 6.1. Ocean water currents flowing through the earth's magnetic field yield electrical voltage gradients that may inform drifting animals of upstream and downstream directions (Kalmijn 1974)

trical currents near the top and bottom of the flowing water. If the estimates of the electrical sensitivity of the American eel and Atlantic salmon by Rommel and McCleave (1973a) and by McCleave and Power (1978) are correct, then oceanic electrical fields could be detected by these migratory fish. As mentioned before, however, Enger et al. (1976) and Berge (1979) have been unable to confirm these high sensitivities (see Sect. 6.2.3).

In freshwater, other electrical fields from ionic differences and similar phenomena seem to predominate (see Sect. 6.4.2) and would probably obscure electrical fields arising from motion through the magnetic field (Kalmijn 1974).

6.4.2 Local Electric Fields

There are several other processes that can lead to electrical fields that are within the perception levels of electroreceptive fishes. Any time that two dissimilar chemical media come into contact a potential difference arises across their interface (Kalmijn 1974). In aquatic habitats differences in chemical composition, salinity, oxygen content, pH, and temperature are widely present and give rise to complex electrical current and ohmic potential fields. There are also streaming potentials arising from the relative movement of water running through river channels or one layer of water moving over another. Local electrical conditions might therefore serve as landmarks allowing recognition of a particular location.

Peters and Bretschneider (1972) found such local hydroelectric fields of unknown source in freshwater catfish pools in the Netherlands. The fields, of the order of 10 to 15 $mV\ m^{-1}$, were within the range of sensitivity of the brown bullhead, a passive electric catfish with a response threshold of 600 $\mu V\ m^{-1}$. The pools were located well away from man-made electrical interference and the electric fields were probably of natural origin, perhaps the chemical or physical interaction of regions of the pool which differed in pH, ion concentration or temperature. In a subsequent experiment, Peters and van Wijland (1974) were able to train these catfish, using a food reward, to orient in electric fields of the same strength as those found in nature. The fish were capable of both an electrotaxis response, seeking food at the cathode, and an electro-compass response, moving at an angle to the electric field in order to find food. They also showed an electrocompass response in seeking shelter, even when visual cues contradicted the electrical information. These results indicate that catfish can use local electric fields for orientation over short distances within their home range. The potential therefore exists for migratory orientation by other passive electroreceptive fishes using such electrocompass and electrotaxis responses.

There is evidence that at least one other passive electroreceptive fish is influenced by local electric fields. Sturgeon, *Acipenser guldenstadti,* tracked by means of ultrasonic transmitters, all slowed down and in several cases made course changes as they passed under a high voltage electrotransmission line (Fig. 6.2) (Poddubnyi 1969). Of course it is not clear in such a situation that the fish are responding to the electrical or magnetic effects of the transmission line. A visual response to the wire or supporting structures is not impossible.

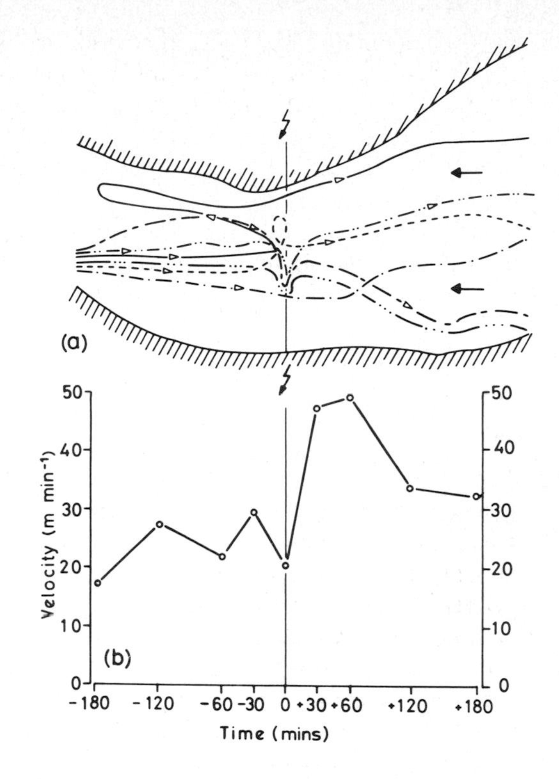

Fig. 6.2. Changes in swimming speed and direction by sturgeon, *Acipenser guldenstadti*, as they passed under a high-voltage electrotransmission line. *a* Traces of six fish as they passed under the line while being tracked with ultrasonic transmitting tags. *b* Swimming speed (m min^{-1}) recorded over a period of 6 h. Modified by Tesch FW (1975) Ch 8.2 Orientation, fishes. In: Kinne O (ed) Marine ecology. Copyright 1975. Reprinted by permission of John Wiley and Sons Ltd. Originally from Poddubnyi AG (1969) F.A.O. Fishery Reports 62:793–801. Food and Agriculture Organization of the United Nations.

Poddubnyi's (1969) observations, and Peters and Bretschneider's (1972) finding of strong 50 Hz noise in water near the laboratory and strong electrical currents which they felt were associated with electric railways, raise the possibility of electrical "pollution" that might interfere with orientation mechanisms of electroreceptive fishes.

6.4.3 Bioelectric Effects

Living plants and animals generate bipolar electric fields in water. Peters and Bretschneider (1972) were able to detect the electric field of organisms as small as chironomid larvae (100 μV cm^{-1}) and presumably electroreceptive fish would be able to detect them as well (e.g., bullheads with a threshhold of 600 μV m^{-1} or 6.0 μV cm^{-1}). This ability has been demonstrated in a shark, *Scyliorhinus canicula*,, and a ray, *Raja clavata*, (Kalmijn 1971, 1979). These animals could detect live prey (plaice) when the prey was covered with electrically transparent agar and buried, but not when the prey was dead or when the agar was covered by a thin electrical insulating layer of plastic film. Electrodes duplicating the electric field of the plaice elicited the same feeding responses as the live prey and attracted the sharks more strongly than a piece of dead fish.

The bioelectric potentials from plants or animals could conceivably serve as local landmarks allowing recognition of a location.

6.4.4 Active Electroreception

The subject of active electroreception is too specialized to discuss in detail particularly since it is not known to play a role in migratory behaviour. In summary, the fish with weak electric organs, including those which also have strong electric organs, are able to generate electric fields around their bodies and then to detect the changes in these fields which result from objects in the nearby environment (Bennett 1971 a, b; Heiligenberg 1977; Scheich and Bullock 1974). This is essentially a short-range sense. The sternarchid *Apteronotus albifrons,* for example, can detect a rod 3 mm in diameter at a distance of 40 mm (Bastian 1976). It could be used in local orientation or recognition of location, as well as in predator–prey and social interactions.

6.5 Magnetism

Recent research now indicates that some animals use magnetism in orientation. Magnetism is an obvious orientation stimulus to people familiar with magnetic compasses, but early attempts to test for magnetic-compass responses in animals foundered because of two problems: lack of known magnetic receptors and the inconsistent response of animals to magnetic stimuli.

We now seem to be on the point of resolving both of these problems. Plausible receptor mechanisms have been postulated and magnetic perception has been demonstrated in animals, including fishes (see Sect. 6.6). The apparent inconsistency of response has been resolved by our more sophisticated understanding of animal orientation. In the past, researchers tended to think in terms of a single orientation mechanism for the migration of a species, but it has become obvious that animals often possess several orientation mechanisms and use them in a hierarchical fashion. Experienced pigeons, for example, use a time-compensated sun compass on clear days but switch to magnetic cues on overcast days, and may have several other types of orientation information available as well (Schmidt-Koenig and Keeton 1977). This means that disruptive effects of magnets could be masked if the animal used alternative orientation mechanisms.

The earth's magnetic field has several properties which would be useful to migrating animals. The well-known orientation of the polarity of the earth's field could be used, as humans use a magnetic compass, to maintain a constant direction. Local anomalies caused by ore bodies, for example, could be used as landmarks for identification of a location. The inclination of the field relative to the surface of the earth is used by European robins and other birds for maintaining a compass heading (Wiltschko and Wiltschko 1972; Viehman 1979) and the inclination or other parameters might also provide positional information for navigation (Gould 1982). The magnetic field undergoes regular diurnal, lunar, and annual cycles (Dubrov 1978). These cycles could serve as timing mechanisms, Zeitgebers, for diurnal, tidal or seasonal rhythms of behaviour. In addition to these regular cyclical changes there are also irregular changes, magnetic storms, associated with sunspot activity. These may cause considerable disruption of the magnetic field with unpredictable timing and duration. There is a 27-day cycle in magnetic storm activity related to the rotation of the sun and an 11-year cycle related to sunspot activity, but the occurrence of any given storm is unpredictable.

There is then a potential source of useful information to migratory animals, if they are in a position to make use of it. This would require the ability to perceive the magnetic stimuli through a sensory system.

6.6 Magnetic Senses

Over the last two decades there has been a gradual accumulation of evidence that animals can perceive and react to magnetism. The list of magnetic organisms now includes, among others, mud snails (Brown et al. 1960), honey bees (Lindauer and Martin 1972; Gold et al. 1978), cave salamanders (Phillips 1977), European robins, pigeons, blackcaps (Viehman 1979) and perhaps even human beings (Baker 1980, but see Gould and Able 1981). Fish with electroreceptive organs are also responsive to magnetism, through the effects of magnetism on local electrical fields, and there is evidence of magnetic reception in fish which lack specialized electroreceptors (e.g., Quinn 1980).

6.6.1 The Leask Hypothesis of Magnetoreception

With the exception of the electroreceptive fishes, the site and mechanism of magnetic reception in animals has been a mystery. Leask (1977) proposed a hypothesis for magnetic reception based on the perception of changes in polarization patterns in the eye, but his mechanism required visible light for magnetoreception. In the cave salamander, at least, magnetoreception occurs in total darkness (Phillips 1977) eliminating the Leask hypothesis for that organism. Sockeye salmon fry are able to maintain an appropriate compass heading under black plastic which may exclude the Leask hypothesis for that group as well (Quinn et al. 1981).

6.6.2 Magnetite-Based Magnetoreception

A more promising line began with the discovery of particles of the magnetic compound magnetite in magnetoreceptive bacteria (Blakmore 1975) and later in bees (Gould et al. 1978), pigeons (Walcott et al. 1979; Presti and Pettigrew 1980), white-crowned sparrows (Presti and Pettigrew 1980), and yellowfin tuna (Walker et al. 1982). Magnetic material, possibly magnetite, has also been found in guitar fish (*Rhinobatos rhinobatos* and *R. productus*) (O'Leary et al. 1981).

Although the exact mechanism of action is not known, it seems reasonable that deposits of magnetic material might be responsible for the magnetic sense of these animals. Several potential mechanisms have been suggested (Kirschvink and Gould 1981), including the coupling of the ferromagnetic material to muscle receptors (Presti and Pettigrew 1980).

One interesting feature is the variable anatomical location of the magnetite. Walcott et al. (1979) found magnetite deposits on the inside of the skull of their pigeons, while Presti and Pettigrew (1980) found it in the necks of their pigeons and white-crowned sparrows. In the elasmobranch guitarfish the magnetic material is found in the utricular and saccular otoliths interspersed with the calcium carbonate crystals (O'Leary et al. 1981). Magnetic forces on these concentrations of magnetic material might conceivably be detected by the sensory hair cells of the vestibular system. The magnetic material of the guitarfish may be of exogenous origin. The endolymphatic duct is open to the outside and sand particles, for example, sometimes enter the system. The magnetite deposits in the tuna are in the ethmoid complex of bones near the nares (Walker et al. 1982), so the sensory mechanism there would not appear to be homologous with the elasmobranch situation.

6.6.3 Magnetic Sensitivity in Non-Electric Fish

A magnetite-based sensory system might account for reports of magnetic responses in fish without known electroreceptors. Kholodov (1966), for example, found that the activity of three-spined stickleback increased in response to increasing magnetic field strength and a similar increase in activity ocurred in freshwater European eels subjected to increased magnetic fields (Branover et al. 1971; Vasil'yev and Gleiser 1973). Varanelli and McCleave (1974) found increased activity in 3 of 12 Atlantic salmon parr subjected to increased magnetic field.

The two North Atlantic eels were the subject of most the early work. Gleizer and Khodorovsky (1971), using a hexagonal maze (similar to that illustrated in Fig. 2.26), found that young European eels took up a preferred direction in the maze under uniform lighting, and that cancellation of the geomagnetic field with Helmholtz coils eliminated that directional preferrence. Exposing the eels to a very strong magnetic field, 5.4×10^3 times normal, eliminated their ability to respond. Five-year-old Atlantic eels were tested in a similar maze by Branover et al. (1971). They also took up preferred directions in the maze and orientation became random when the geomagnetic field was neutralized with Helmholtz coils. European eels have also been reported to increase their activity in response to increases in the strength of the magnetic field (Branover et al. 1971; Vasil'yev and Gleiser 1973). Tesch (1974) tested adult European and American eels in a circular tank in darkness, except for photographic flashes. The eels took up a preferred direction in the normal geomagnetic field and the direction changed when the field was cancelled with Helmholtz coils.

In contrast to these positive results, Rommel and McCleave (1973a) were unable to condition adult American eels to show changes in heart rate in response to magnetic stimuli. When American elvers were tested for turning direction upon release, they seemed unaffected by orientation of release relative to the magnetic-compass direction. Negating the horizontal component of the geomagnetic field had no effect on turning direction (Zimmerman and McCleave 1975). Similarly, no change in turning direction occurred when the horizontal component was cancelled (McCleave and Power 1978). Zimmerman and McCleave (1975) also tried to duplicate the maze-orientation studies of Gleizer and Khodorovsky (1971), using American rather than European elvers, but were unable to obtain a preferred direction in the normal magnetic field or to demonstrate any change when the horizontal vector was cancelled using Helmholtz coils.

We are left with a confusing and inconsistent picture. Perhaps the use of two different species, and differences in method account for the apparent disparity. We have at least three studies by different researchers in which *A. anguilla* took up non-random orientation which changed when the geomagnetic field was neutralized with Helmholtz coils (Branover et al. 1971; Gleizer and Khodorkovsky 1971; Tesch 1974). Cardiac conditioning and turning direction might not have the same thresholds of response as orientation in a maze or circular tank.

Rommel and McCleave (1973a) attempted cardiac conditioning of Atlantic salmon parr as well as eels, with no success. Varanelli and McCleave (1974) did find changes in the activity pattern of salmon parr when the magnetic field was increased to four times the normal strength.

Sockeye fry change orientation in response to changes in the magnetic field (see Sects. 6.8 and 2.4.2.2.4) (Quinn 1980), but examination of 20 specimens found no detectible natural remanent magnetization and isothermal remanent magnetization could be induced by a 2000 gauss magnetic field in only four of the specimens (Quinn et al. 1981). These authors concluded that magnetite was not present, and hence was not necessary for the magnetoreception system of sockeye fry. However, subsequent studies with more sensitive equipment and different techniques have revealed weak but detectable isothermal remanant magnetism in newly emerged fry (Quinn pers. commun.).

Yellowfin tuna have been conditioned to discriminate between two earth-strength magnetic fields (Walker et al. 1982). Four fish were first trained to swim through a frame lowered into the tank, using a food reward. The frame was then presented to the fish in association with a 10 to 50 μT increase of the normal 35 μT Hawaiian magnetic field and the fish were rewarded if they swam through. If they swam through when the magnetic stimulus was not presented they were not rewarded. Discrimination occurred after two 30-trial sessions. Walker et al. (1982) suggest that the "ease" with which these wild-caught fish were conditioned to magnetic stimuli indicates a well-developed magnetic sense. The ingredients, detection of magnetic fields, potential magnetoreceptors based on magnetite, and a migratory nature combine to suggest that the tuna may use magnetic orientation or navigation techniques.

6.6.4 Magnetoreception in Fish with Specialized Electroreceptors

Unlike the fish above, the mechanism of magnetoreception in the electroreceptive fish can be clearly linked with a known sense receptor, the electroreceptor. It had been known since the early work of Lissmann that the active electroreceptive fishes respond behaviourally to the proximity of magnets, presumably through

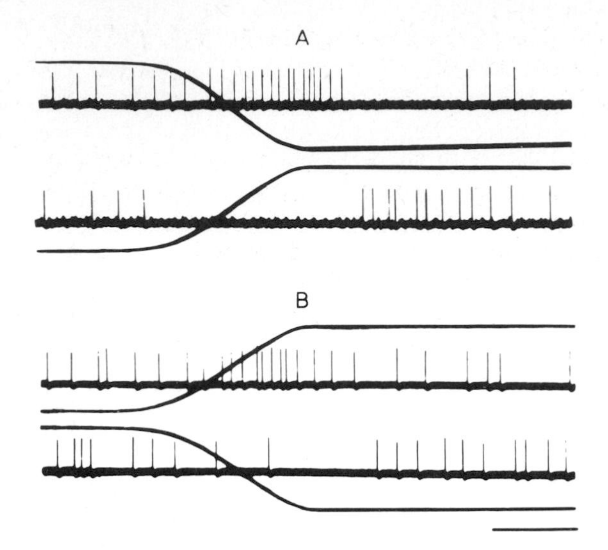

Fig. 6.3 A, B. Responses of dorsal ampullae of Lorenzini in the left half of the skate, *Trigon pastinaca,* to magnetic stimuli. *A* ampullar canal goes in the rostral direction, *B* in the caudal direction. The upward deflection of the beam (*solid trace*) represents the northern direction of the magnetic field. Downward deflection represents the southern direction of the field. Rate of change of the magnetic field: 93 Gs s^{-1}. Time bars: 500 ms (Akoev et al. 1976)

the effect of the magnet on their electric field, but Lissmann and Machim (1963) also mention in passing that several passive electric catfish of the family Siluridae, including *Clarias, Saccobranchus, Ictalurus,* and *Silurus,* respond to moving magnets.

If the passively electroreceptive silurids are sensitive to magnetism, one might expect that the elasmobranchs with their more sensitive electroreceptors would also be able to detect magnetism. A group of Russian workers have shown that Black Sea skates, *Trigon pastinaca* and *Raja clavata,* respond to changing levels of magnetic flux with changes in the activity of nerves leading from the ampullae of Lorenzini to the central nervous system (Brown et al. 1974; Adrianov et al. 1974; Akoev et al. 1976; Brown and Ilyinsky 1978). In a stationary ray in still water, these receptors only respond to changes in the magnetic field, not to the steady field. Some units responded to increasing magnetic stimulation with a decrease in spontaneous firing, while others showed an increase in activity. This effect was reversed by a reversal in the polarity of the field (Fig. 6.3).

The response of the receptors depended on the position and orientation of the ampullae. Those with canals oriented laterally or caudally showed an opposite response to those oriented rostrally or medially, while ampullae on opposite wings of the skate also showed opposing responses to magnetic polarity (Akoev et al. 1976; Brown and Ilyinsky 1978). The longer the ampullary canal, the greater was the magnetic sensitivity of the receptor and sensitivity was reduced by surgical shortening of the canal. When the skate moved, or when water moved over a stationary skate, the receptors responded even in a constant magnetic field (Brown and Ilyinsky 1978). This indicates that the receptors were responding to electromotive forces in the ampullary canals induced by movement through the magnetic field. Neutralizing the vertical component of the earth's magnetic field with Helmholtz coils largely eliminated the response (Brown and Ilyinsky 1978). The demonstrated sensitivity of these skates and rays would be sufficient to allow them to detect the earth's magnetic field and orient in it and, as Akoev et al. (1976)

point out, behavioural thresholds are usually even lower than the thresholds measured neurophysiologically. Therefore, behavioural responses to the geomagnetic field should be well within the grasp of these elasmobranches and other electroreceptive fishes.

The interaction on oceanic currents with magnetic fields causes predictable "telluric" electric currents which might be perceptible to electroreceptive fishes. Brown et al. (1979) tested this possibility by keeping rays, *Raja radiata,* in tanks which were in electrical contact with the Baerents Sea and recording the activity of nerve fibers from the ampullae of Lorenzini (Fig. 6.4). A magnetic substorm on 26 January, 1978 was clearly associated with responses by the ampullae (Fig. 6.5) indicating that variations in the geomagnetic field, such as those caused by magnetic storms, were detectable.

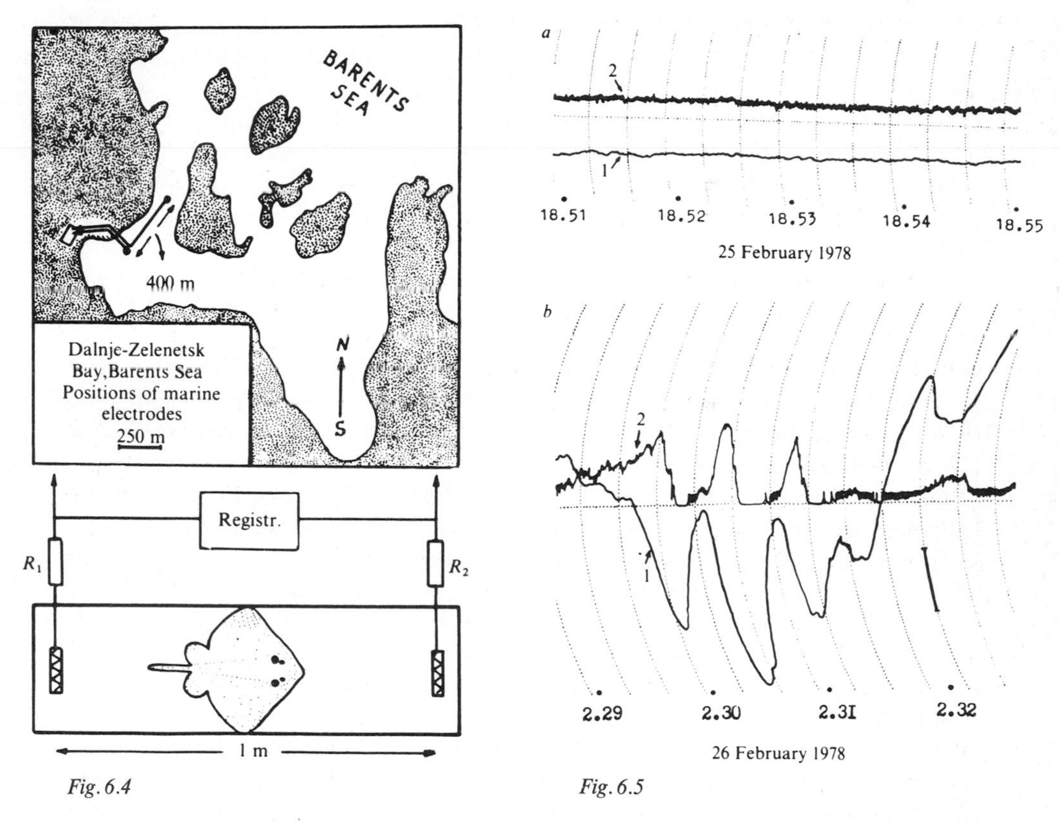

Fig. 6.4

Fig. 6.5

Fig. 6.4. A skate in an experimental tank linked electrically to the Baerents Sea so that natural electrical currents in the sea are mimicked in the tank (Brown et al. 1979)

Fig. 6.5 a, b. Synchronous recording of the induced electric field in the sea (*1*) and the time course of spike frequency of a single electroreceptor (*2*) before (**a**) and during (**b**) a magnetic substorm. *Vertical bar* 0.2 μV cm^{-1}; 15 imp s^{-1} (GMT) (Brown et al. 1979)

Figs. 6.4 and 6.5: Reprinted by permission from Nature, Vol 277, No 5698, pp 648–649. Copyright © 1979, Macmillan Journals Limited

We have, then, a situation in which fish with electroreceptors, an imposing group which includes sharks, rays, ratfish, sturgeons, paddlefish, reedfish, silurid catfish, gymnotids, and mormyrids, are probably sensitive to the earth's magnetic field and to its variation. There is, moreover, evidence of magnetoreception in such non-electroreceptive fishes as eels and salmon.

6.7 Magnetic Timing

The basic requirements are present for geomagnetic phenomena to influence the timing of animal migration: periodic variation in the magnetic field and receptors capable of perceiving the variations. What is lacking is any direct evidence that magnetic information is used in this way by fishes. This may be an accident of our recent discovery of the magnetic sense and the difficulty of separating magnetic effects from those of illumination. The general success of rhythm experiments in which the light cycle but not the geomagnetic cycle is altered would suggest that light is the dominant timing mechanism in many fish.

6.8 Magnetic Stimuli and Direction and Distance

When we think of compass orientation we automatically think of magnetism, and so it seems reasonable to expect that fish with the ability to perceive a magnetic field would use this ability to orient themselves during migration. So far there is remarkably little evidence for this, although sockeye fry do seem to orient appropriately to magnetic stimuli. As with magnetic timing, the paucity of evidence may not mean very much. We have only recently begun to examine seriously the possibility of magnetic orientation. The finding of magnetic-compass orientation in the six species of birds which have been seriously examined (Viehman 1979) would tend to support the idea that magnetic orientation is widespread in the animal kingdom, but was being overlooked until very recently.

The findings of Tesch (1974) and others on the orientational responses of eels to magnetic fields were summarized above in the section on receptor mechanisms. In addition, round stingrays, *Urolophus halleri,* have been trained to enter a shelter facing magnetic East rather than one facing magnetic West (Kalmijn 1978a). The best fish made correct choices about two thirds of the time in double blind tests with frequent random magnetic field reversals to control for position learning.

One study of fish magnetic orientation stands out from the others because of the way it fits in with other known aspects of migration. Quinn (1980) (see Sect. 2.4.2.2.4) examined the orientation of fry of two races of sockeye. Cedar River fry migrate downstream at night into Lake Washington and then move uplake in a north-northwesterly direction, while Chilko fish migrate upstream during the day into Chilko Lake and spread through the lake in a south-southeasterly direction. The fry were tested in a tank allowing a choice of four directions. When the tank was uncovered and there was no change in magnetic conditions, they chose the appropriate compass directions for their race. When the tanks were covered, the fry still oriented appropriately but the addition of Rubens coils, even when turned off, altered the orientation. When the coils were turned on, altering the magnetic field by 90°, the orientation of the fish, in covered tanks or at night, changed by 79° to 122°. Fish tested in daylight without the translucent cover did not change direction when the magnetic field was altered. A subsequent series of tests on Chilko and Weaver Creek fry indicated that each stock took up appropriate orientation when celestial information was available even if the magnetic field was changed by 90° (Quinn 1982a). When celestial cues were cut off by complete overcast or by translucent covers, the fish altered their orientation by about 90° in response to a 90° change in the magnetic field. These results indicate the use of both a magnetic compass and a sun compass, with the sun compass taking precedence when both sources of information are present and in conflict.

This finding is similar to the situation in pigeons. Experienced birds use the sun compass in preference to the magnetic compass when the two are in conflict (Keeton 1979). This similarity in the hierarchical relationship between sun and magnetic compasses in two very different animals may indicate that the sun compass is inherently more accurate than vertebrate magnetic compasses. This view is strengthened by the increase in the scatter of headings observed in both pigeons and sockeye when they are forced to shift from sun compass to magnetic compass.

Quinn (1980) suggests that if sockeye use a magnetic compass at one stage of their life cycle they are also likely to use it at other stages such as the migration of smolts out of the nursery lakes and the oceanic migrations. The "X-orientation" mechanism (see Sect. 2.4.2.2.4) found in sockeye smolts by Groot (1965), for example, has turned out to be based on magnetic orientation (Quinn and Brannon 1982). Sockeye smolts caught at the Babine Lake outlet oriented toward the outlet regardless of the magnetic field as long as they had a view of the sky, cloudy or not. When the tanks were covered, the smolts showed orientation along the axis of migration in a normal magnetic field and a 56° change in direction when the magnetic field was rotated 90° (Quinn and Brannon 1982).

The results of a study on the orientation of chum salmon (*Oncorhynchus keta*) fry from the Conuma River B.C. in a magnetic field were not so clear-cut. The fry oriented in an appropriate westerly direction, whether the tanks were covered or not, under unaltered magnetic conditions (Quinn and Groot 1983). However, when the magnetic field was rotated 90° clockwise the fry changed orientation by 25° counterclockwise, regardless of their view of the sky. This change was significant with a test based on treating individual fish as independant samples, but not with a more conservative test based on differences in groups of fish. One interesting conclusion is that the chum fry can orient in approximately the correct direction even in covered tanks with an altered magnetic field. This ability may be based on learning of local cues or landmarks in the tank. The apparatus consisted of 12 tanks with a central inlet and 8 equidistant outlets. Two hundred fry were placed in each tank 30 min after sunset (chum fry normally migrate downstream at night), water then ran through the tank all night and the fry were trapped if they left through an outlet port. Covers were placed on some of the tanks immediately after the fish were introduced, but the tanks were apparently not turned, so transfer of directional orientation to local cues in the tank might allow the fish to maintain an appropriate heading and act on it later in the night.

Small metal tags inserted into the flesh of the snout are often used to mark juvenile Pacific salmon, including chum fry. Quinn and Groot (1983) also examined the effect of magnetized and unmagnetized tags on the orientation of their Conuma River chum fry. A magnetized tag would produce a magnetic field of approximately earth strength in the head region of the fish. The tags did not alter the orientation of the fish under normal magnetic conditions and the minor changes that occurred under altered magnetic conditions were not clearly significant. Thus it appears that this tagging procedure does not interfere with chum fry orientation.

Animals that are capable of navigation, such as the homing pigeon, need information on their position (latitude and longitude) to select an appropriate course for returning home in the case of a pigeon or for orienting toward the goal of their migration in the case of a migratory species. The source of this positional information has been one of the great mysteries of animal migration. To determine its location on the earth's surface an animal requires at least two axes. A recurring suggestion is that the characteristics of the earth's magnetic field might provide one or both of these axes (Gould 1982; Quinn 1982 b).

In pigeons, there is some evidence that magnetism is involved in navigation. Experienced pigeons orient by means of a sun compass on clear days, but if they are released near an anomaly in the earth's magnetic field, their orientation is deflected away from their home direction, even on clear days. Similarly, magnetic storms alter the direction of orientation on clear days and to a greater degree than would be expected on the basis of effects on magnetic-compass orientation.

In fish there is relatively little evidence of true navigation and no standard navigating test subject comparable to the homing pigeon. There are many reports of fish which home successfully, but it has not been established that they are able to determine their geographic position relative to a goal and then set a direct course toward their goal. An exception may be the radiated shanny, *Ulvaria subbifurcata,* which is capable of taking up an orientation toward its home range when blinded and tested under ice cover at night (Green and Fisher 1977; Goff and Green 1978). The shanny should be considered a candidate for use of magnetic map information in fishes, although odours or other cues need to be examined as well.

The migration of adult Pacific salmon from their ocean feeding grounds to the coastal region near their spawning streams suggests the use of navigational information (Quinn 1982 b). The fish of a single stock are widely scattered through the feeding area, yet they converge at the river mouth within a relatively short, predictable period of the year. Quinn (1982 b) proposes that position information might be available to the fish from integration of the inclination and the declination of the earth's magnetic field. The inclination is the angle of the field relative to horizontal. Inclination increases as one approaches the magnetic poles so that it gives an indication of latitude (Fig. 6.6). Declination is the difference between

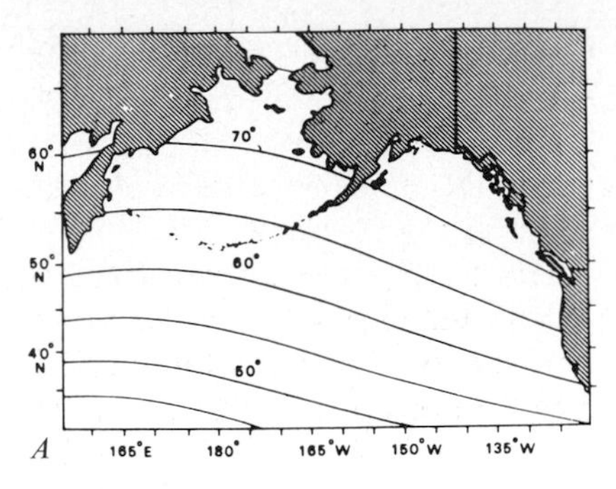

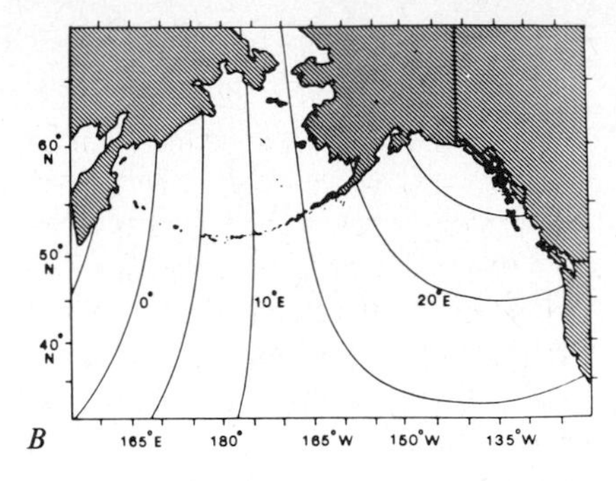

Fig. 6.6. A Inclination of the earth's magnetic field, in 5° increments, in the North Pacific Ocean. *B* Declination of the earth's magnetic field, in 5° increments, in the North Pacific Ocean (Quinn 1982 b)

the horizontal direction of the magnetic pole and the direction of the geographic pole. The two north poles are about 1500 km apart and the angle between them changes predictably with longitude in some areas, such as near the Aleutian Islands. In other areas there are sometimes substantial irregularities. The two coordinates, latitude and longitude, would allow the fish to fix its position.

If thc fish also possessed accurate seasonal information, based perhaps on photoperiod and circannual cycles, it could select the appropriate date to begin its homeward journey so as to arrive at its home stream near the optimal time for breeding. Seasonal control of the salmon's reproductive physiology will require the same type of information and the two systems could be linked. Maintaining the correct course, once it was selected on the basis of navigational information, could be based on a sun compass or magnetic compass. Both are found in the early life-history stages of Pacific salmon.

Several of the elements required for the navigational system proposed by Quinn (1982 b) have already been found in salmonids at various stages in their life cycle. Magnetic sensitivity is present in sockeye fry, although detection of inclination has not been established. Seasonal timing is indicated by the temporal patterning of returns to river-mouth fisheries. Compass orientation is established in juvenile salmonids and is also reported in adult pink salmon (Churmasov and Stepanov 1977; Stepanov et al. 1979). Experimental testing of the hypothesis is the next step.

6.9 Summary

Electrical currents are produced in natural waters by several sources, including the movement of fish and water currents through the earth's magnetic field and the physiological activity of living organisms. Fish that are able to detect electrical stimuli could receive useful information about such things as the direction of water flow, the orientation of the earth's magnetic field or the presence of other organisms.

Specialized electroreceptors occur in a wide range of fishes ranging from lampreys to teleosts. Most of the elasmobranchs have electroreceptors termed ampullae of Lorenzini, and the chondrosteans, such as paddlefish and sturgeons, have electroreceptors that are probably homologous to the ampullae. Some teleosts, including many catfish, have specialized electroreceptors. Other teleosts, such as the salmonids and eels, lack specialized electroreceptors and there are conflicting reports on their ability to respond to electrical currents at levels that may occur in nature.

Some electroreceptive fishes also possess electric organs that enhance the electrical field around their body. The so-called weak electric fish use their low-voltage discharges to communicate with conspecifics and to detect objects in the environment. The distortions of the field that occur when an insulating or conducting object enters the field can be analyzed by the electroreceptors and the CNS. The strong electric fish produce fields of 10's or 100's of volts that are used to stun prey and deter predators.

There is no evidence that electrical stimuli time the occurrence of migratory behaviour. However, the potential exists for social timing by means of electrical communication. Such a mechanism could synchronize the migratory activity of electric fish, for example. There are also cycles of electrical activity in natural waters that could possibly serve as timing cues.

Directional information may be available from the electrical currents induced when objects move through the earth's magnetic field. Large masses of moving water, ocean water currents like the Gulf Stream, for example, will generate electrical currents that lie within the range of sensitivity of fish with specialized electroreceptors and possibly of fish without specialized electroreceptors. Knowledge about the direction in which the ocean current was flowing could be extracted from the pattern of electrical activity. This information would allow fish to select currents flowing in a appropriate direction and to align their swimming activity along the direction of flow. The movement of a fish's body through the earth's magnetic field can also generate detectable electrical currents, providing a mechanism for magnetic-compass orientation in fish with sensitive electroreceptors, including skates and rays.

In natural freshwaters local electrical currents occur that could act as landmarks for home recognition or local orientation. Catfish, that have electroreceptors, can be trained to orient in weak electrical fields and are capable of showing an electro-compass response, moving at an angle across an electrical field to reach a feeding site.

Magnetic information could allow fish to select a compass heading even when the sky is obscured by cloud or turbidity. Fish with electroreceptive organs can detect the magnetic field by means of the electric currents induced when they move through the geomagnetic field. Several animals detect magnetic fields through some other mechanism, possibly based on magnetite or similar magnetic compounds. Some fish have been found to possess magnetite, and it may play a role in the magnetoreception of fish that lack specialized electroreceptors.

Several species of fish, including sticklebacks, eels, and salmon, have been found to change their activity patterns when subjected to changes in magnetic field, suggesting some magnetic effects on behaviour. Sockeye fry and smolts change direction of orientation when the earth's magnetic field is altered by magnetic coils and yellowfin tuna can be trained to respond to weak magnetic stimuli.

Periodic variations occur in the earth's magnetic field, in response to solar events, but there is no evidence that these are used in the timing of fish behaviour.

Electrical and magnetic stimuli are available in the environment and could provide useful information to migrant fish. Fish possess receptors that are capable of detecting these stimuli. Evidence is beginning to accumulate indicating that animals including some fish use electrical and magnetic information in their migration.

Chapter 7

The Past: Learning and Genetics

The events of the past have affected the migratory behaviour of fishes through two major pathways. During the life of individual fish, behaviour will have been altered by learning. On the time-scale of generations, natural selection will alter the behaviour of the population. These two mechanisms interact with each other. Natural selection will favour the survival and reproduction of those individuals with the ability to learn and remember specific important information such as the odour of a home stream. Life history traits such as life span, reproductive rate and development rate are also under some degree of genetic control and therefore are subject to natural selection (Cole 1954; Allendorf and Utter 1979; Schaffer 1979; Ryman 1981).

If local breeding populations of fish are reproductively isolated to a sufficient degree to accumulate genetic adaptations to the local situation, then genetic control over some of the smaller details of migration may evolve. These adaptations might include the timing of downstream migration to conform to local climatic conditions (climate as opposed to weather) or the matching of rheotaxis in the fry to the spawning location of adults above or below a nursery lake. On the other hand, we might expect learning to govern response to variables that would change within an individual life span. These would include the shoreward direction for fish dwelling in the shallows of small lakes and pools.

7.1 Learning

Learning is undoubtedly important in migration, for example in the learning of home characteristics, allowing subsequent home recognition or perhaps in the learning of a migration route that is frequently traveled. The best-studied example of learning in fish migration, home-stream odour imprinting, has already been discussed at length (see Chap. 3). Other comparable learned responses probably occur that have not yet been studied. For example, the return of migrant salmon to the area near the home stream seems to be independent of home stream odour, yet the fish reliably return to the area near the point of release (Cooper and Hirsch 1982). Since these fish are not necessarily from genetic stocks adapted to the release location some learning is probably involved in this open-water phase of homing. Another situation in which learning may be important is the response of some salmonids to conspecific odours. When Tinnsjø and Hammerfest stocks of char, *Salvelinus alpinus,* were kept together, the Hammerfest smolts were attracted by water that had contained the Tinnsjø fish (Selset and Døving 1980). This suggests learning of a population odour from tankmates.

Animals may learn migration routes by gradually extending their area of experience (Baker 1978). This mechanism may be important in fish, but its importance has not yet been demonstrated for long-range migration. Conditions do exist that are suitable for this type of learning. Hogchokers, *Trinectes maculatus,* for example, visit their spawning grounds each year even as juveniles not yet ready to spawn (Dovel et al. 1969). Such "dry runs" might reinforce memory of the spawning location and migration routes.

There is more evidence for learning in short-distance orientation. For example, tide-pool gobies learn the location of nearby pools during their movements at high tide (see Sect. 2.4.1.1) (Aronson 1951, 1971) and Y-axis orientation to the home shoreline is learned as part of the predator avoidance response of several species, e.g., *Gambusia* (see Sect. 2.4.2.2.2) (Goodyear 1973). These are appropriate uses for learning, since local topography, such as the position of adjoining tidepools and the Y-axis of the home shoreline, may vary within the lifetime of an individual and will be highly variable over the range occupied by a population.

In summary, learning is probably important in many aspects of fish migration but it has been little studied as a mechanism, perhaps because of a tendency to consider fish migration at the population level rather than the individual level. There are a number of cases such as the open-water migration of returning salmon and the acquisition of pheromone responses that are ripe for examination.

7.2 Genetic Differences Between Fish Stocks

The fishes are the most diverse group of vertebrates. This evolutionary plasticity should lead us to expect a great deal of genetic variation between fish stocks. The problem of the degree of genetic differentiation between fish stocks has long been important to fishery managers, who must decide whether to manage different breeding populations of fish separately or as one stock. It would certainly be easier to be able to treat all members of a species, in a region, as one management unit rather than to deal separately with several different populations. Scientific opinions have differed on the subject of the degree of genetic differentiation within fish populations. Williams (1975), for example, has argued that most of the morphological differences between adult populations of marine fish are due to environmental selection on the numerous and variable larvae of a single genetic stock rather than to the presence of genetically different stocks. On the other hand, biochemical evidence sometimes indicates that even morphologically similar fish may be genetically distinct. The two sibling species of bonefish, *Albula vulpes,* show a high degree of genetic divergence at the biochemical level but are sufficiently similar morphologically to have been considered a single species for many years (Shaklee and Tamaru 1977).

Evidence for the genetic separation of fish stocks should be most obvious in freshwater habitats with their high degree of geographic isolation. The anadromous salmonids have been the subject of intense study. They are found in separate breeding populations scattered in many different freshwater systems, this combined with home-stream fidelity by returning spawners, should encourage the accumulation of genetic differences between populations. Yet, despite these factors, for the first 30 years of the 20th century fisheries managers regarded all populations of the various salmon species to be more or less identical genetically (Ricker 1972). They acted on this assumption and transferred fish from stream to stream and river system to river system without regard to genetic differences. Very few of these transfers were successful in augmenting existing runs or establishing new runs (Ricker 1972; Blackett 1979). The assumption of genetic similarity between stocks allowed the same managers to ignore the loss of local stocks due to dam construction or similar causes, since they felt that such stocks could be replaced by fish from other locations.

Ricker (1972) has reviewed the evidence for genetic differences between stocks of Pacific salmon and trout. Even fish within the same river system often differ in such morphological traits as vertebral counts, fin rays, gill rackers, parr marks scales, pyloric caecae and colour at maturity. In many cases there were environmentally induced differences in these traits as well, but some differences persisted even when the environmental conditions were controlled.

Some of these morphological differences occurred between stocks that were closely associated geographically. Capilano River winter steelheads, for example, have more parr marks and vertebrate than Capilano summer steelheads, and these and several other differences persist when the two stocks are reared together under hatchery conditions (Smith 1969). This indicates that there are two genetically separate steelhead stocks in the same small river. The mechanism of genetic isolation between the two stocks is not clear. The winter fish arrive from December until March in a sexually mature state, while the summer fish usually enter freshwater between early May and late August with immature gonads. However, both races spawn at about the same time, the summer race having spent 7–10 months in freshwater. Smith (1969) suggested that differences in spawning colours between the silvery winter fish and the redder summer fish, or differences in spawning region within the river might serve to keep the stocks separate.

The example above illustrates the degree of subdivision that may occur between salmonid stocks. Ricker (1972) documents several other examples of genetic differences between stocks. The sockeye, for example, is divided in several river systems into two behaviourally distinct populations, the anadromous or normal sockeye population and the "kokanee" population that remains and matures in freshwater. The two populations often differ in merisitic characteristics and these differences persist when different stocks are reared under identical conditions in a hatchery.

Even the kokanee populations of a single lake may be genetically distinct. Kootenay Lake contains three demes that differ in vertebral counts and several other attributes (Vernon 1957). When reared under the same conditions in a hatchery, each of the three stocks retains the vertebral count typical of its deme (Table 7.1). This indicates that the differences are determined by genetics rather than by environmental differences between the regions of the lake.

Similar anadromous and non-anadromous stocks of the same species occur in other salmonids, such as the rainbow (non-anadromous) and steelhead (anad-

Table 7.1. Mean vertebral counts of kokanee from three different arms of Kootenay Lake. The parental fish were captured in the natural spawning streams. The progeny were reared in a hatchery under identical conditions (Vernon 1957)

Area		Mean vertebral count
North End	Parental	65.04
	Progeny	65.00
West End	Parental	64.10
	Progeny	64.29
South End	Parental	62.94
	Progeny	63.81

romous) stock of *Salmo gairdneri* that often spawn in the same river system. Ricker (1972) concludes that there may be of the order of 10,000 different salmonid stocks on the Pacific coast of North America, each genetically adapted to its particular location. The high fidelity of homing is probably one of the major factors responsible for this separation into stocks.

Homing may actually be more accurate than tag recoveries suggest. Many of the fish caught in the "wrong" river may have been overshooting or proving (Ricker 1972). An overshooting individual would be one that initially passed the appropriate turn in its upriver migration but would then turn back to the correct choice point. A proving fish would be one that entered the wrong stream, stayed there a while, then left to carry on to the correct home stream.

The Pacific coast is not the only region where several fish stocks have been reported in close association. Two different populations of brown trout inhabit the same small lake in Sweden. The upstream, inlet-spawning trout of Lake Bunnersjoarna are smaller and slower-growing than the outlet-spawning population and differ in the frequency of two alleles, LDH-1 and CPK-1, indicating reproductive isolation (Allendorf et al. 1976; Ryman et al. 1979).

Atlantic salmon are also good candidates for the formation of genetically isolated stocks. Child et al. (1976) tested sera from about 10,000 salmon from the British Isles and found polymorphism at the transferrin locus that indicated that at least two separate races of salmon inhabit the isles, a "celtic" race in southwestern England, Wales, and southeastern Ireland and a "boreal" race in the rest of the British Isles. They did not attribute adaptive significance to the racial differences.

North American stocks of Atlantic salmon differ from one another in ways that seem adaptive to their particular location. One would expect on theoretical grounds that the more difficult a river is to ascend, the older fish would be at their first spawning, since larger, stronger fish would have a better chance of spawning successfully. This prediction is borne out by the differences in age at first spawning of salmon on the east coast of North America. Fish from short, easy rivers breed early and those from long, difficult rivers breed when they are older (Schaffer and Elson 1975; Schaffer 1979).

Similarly, heavy commercial fishing tends to eliminate larger fish from the population and fish that breed early are not exposed to the fishery for as long. One would thus expect heavily fished rivers to have smaller, early breeding populations and again this is confirmed, in general, by the findings of Schaffer and Elson (1975) and Schaffer (1979). In a more specific example from the same region, two tributaries of the Southwest Miramichi River in New Brunswick differ in their physical characteristics. Rocky Brook is fast and cold, Sabies River is slower and warmer and has more food available over winter. The salmon in these two streams also differ. The Rocky Brook juveniles are more streamlined and have larger fins than the Sabies River fish and they migrate downstream in October as large parr, while the Sabies fish migrate sooner, in May, but as smolts (Riddell and Leggett 1981). Rearing the two stocks together in a hatchery indicated that the morphological differences were inherited (Riddell et al. 1981). The behavioural differences in migration time were not tested. The streamlined shape and larger pectoral fins of the Rocky Brook fish would adapt them to swift water. Atlantic salmon hold position on the stream bottom by using the pectoral fins as hydroplanes to hold themselves against the bottom. This single river system harbours different stocks adapted to their particular regions. Each would benefit from being managed separately and from being protected from contamination by less adapted stocks.

This type of adaptive specialization of fish stocks also occurs in non-salmonid species that show fidelity to a home stream. The shad, *Alosa sapidissima,* varies in morphology and life history over its geographic range on the east coast of North America (Carscadden and Leggett 1975). These variations seem appropriate to the differences between the various shad rivers. More northern rivers tend to have steeper gradients and the northern races of shad tend to have a larger body size (Leggett and Trump 1977), possibly as an adaptation to ascending faster rivers. Similarly the harsher conditions in the northern rivers could be expected to favour a higher frequency of repeat spawning, iteroparity, to counteract the greater variability of northern conditions. The shad in rivers south of 32° N are in fact semelparous – they spawn once and then die. North of 32° the percentage of repeat spawners increases with latitude (Leggett and Carscadden 1978).

The preceding examples demonstrate that, although there are undoubtedly many environmentally induced differences between fish stocks (Ricker 1972), there are also important genetic differences, even between stocks that are very close geographically. These genetic differences are often specifically adaptive to the conditions in the region. The specificity of homing allows this fine tuning of the genotype to the location by maintaining genetic isolation.

7.3 Genetic Control of Timing of Migration

Ricker (1972) summarizes a number of cases of apparent genetic control of the timing of migration. The age at the time of seaward migration, of sockeye in Cultus Lake B.C., tended to follow the age of parental migration. Fish that migrated

Table 7.2. Years of lake residence of fish of the Little Kitoi stock in Little Kitoi Lake and when transplanted to Ruth Lake, where a higher growth rate was achieved (Ricker 1972)

Years of lake res.	Ruth Lake		Little Kitoi Lake	
	% Leaving lake	Weight	% Leaving lake	Weight
1	66.6%	10.9 g	74.6%	2.5 g
2	30.2%	19.7 g	25.3%	3.6 g
3	3.2%	28.0 g	0.1%	6.0 g

Table 7.3. The relationship between the age at which parental fish return to spawn and the age of return for their progeny (Ricker 1972; from the work of L. R. Donaldson)

Year class	Parental ages		Mean progeny ages (completed years)	
	Male	Female	Male	Female
1955	2+	2+	1.63	2.39
1962	1+	2+, 3+	1.46	2.67
1963	2+	2+	1.83	2.63
1964	3+	3+	2.05	2.82

as smolts after 2 years of lake residence were more likely to have offspring that migrated after 2 years than were parents that migrated after 1 year of lake residence; the time of migration may be independent of size. For example, Ruth Lake on Afognak Island, Alaska was killed out with rotenone and stocked with sockeye from nearby Little Kitoi Lake (Meehan 1966). The age of seaward migration remained essentially the same in the donor and transplanted populations even though the Ruth Lake fish grew 4 to 5 times as fast as the Little Kitoi fish (Table 7.2). This suggests that age of migration is under genetic control in this population.

The age at which adult chinook salmon return also follows parental characteristics. Male chinooks mature as early as at 1 year of age and as late as 5. The female range is from 2 to 5 years. The offspring of young parents tend to mature and return to spawn at a younger age than the offspring of older parents (Table 7.3). Ricker (1972) attributes the decline in the age of returning chinook salmon observed in recent years to the selection and removal of larger, and hence later-maturing, fish by the commercial fishery, a form of genetic selection favouring early return by migrants.

The season of return for spawning Columbia River chinooks remained the same as that of the parental stocks even when they were transferred to new sites that supported resident populations which return at a different time. Fall chinooks transplanted to a spring-chinook river returned to the lower Columbia at the appropriate time for fall-chinook migration and spring chinooks transplanted to fall-chinook rivers returned to the lower river at the time of their normal spring migration (Rich and Holmes 1929). The native runs in each of these streams returned at their normal time. Thus the genetic stock rather than the stream environment seems to have the major effect in determining the season of return of chinook salmon. Ricker (1972) also reports a similar effect in pink salmon with the early running Lakelse River stock returning at the same time of year even after transplant to streams harbouring later-running stocks.

7.4 Genetic Control of Migratory Direction

Studies of the genetic control of the direction of migration have concentrated on the early migration of salmonids and the problems facing inlet and outlet spawning stocks that share the same nursery lake. Two alternate hypotheses have been proposed. The "environmental" hypothesis suggests that direction is controlled by environmental stimuli such as thermal or chemical differences between inlet and outlet water. The "genetic" hypothesis proposes that inlet and outlet spawners are different genetic populations with different behavioural responses to similar stimuli such as water current. Of course these two possibilities are not mutually exclusive.

Raleigh (1967) obtained eggs from inlet-spawning and outlet-spawning sockeye at Karluk Lake and reared the offspring under identical conditions from fertilization to testing. He then released the fry in mixed groups in the centre of a 12-m simulated stream with traps at either end. The fry were distinguished in the apparatus by staining one stock with Bismarck Brown. The staining was alternated between the two stocks and seemed to have no effect. Inlet fry moved primarily downstream (98%) and usually waited until after dark before entering the traps, while about 30% of the outlet fry moved upstream, about half during the day and half at night. There was, therefore, a significant difference between the two strains in the expected direction even when reared under the same conditions and tested together. The observation that 70% of the outlet fry went downstream is hard to interpret. It may be an artifact of the testing situation or may indicate that stimulus differences between natural inlet and outlet conditions play an important role in releasing appropriate rheotactic responses. In many systems, outlet fry show an initial short downstream displacement before starting active upstream migration. In Raleigh's apparatus this response could account for some of the downstream movement by outlet fry.

Both Raleigh (1971) and Brannon (1967, 1972) performed similar experiments on sockeye from the outlet-spawning Chilko River population and the inlet-spawning Stellako stock (see Sect. 4.4.1.3). Raleigh tested his fish in Idaho after rearing them both under the same conditions, and found that Stellako fry tended to move downstream regardless of temperature while a maximum of 60% of the Chilko fry moved upstream at 10 °C. At higher and lower temperatures fewer Chilko fish entered the upstream trap in the experimental stream tank (Fig. 7.1).

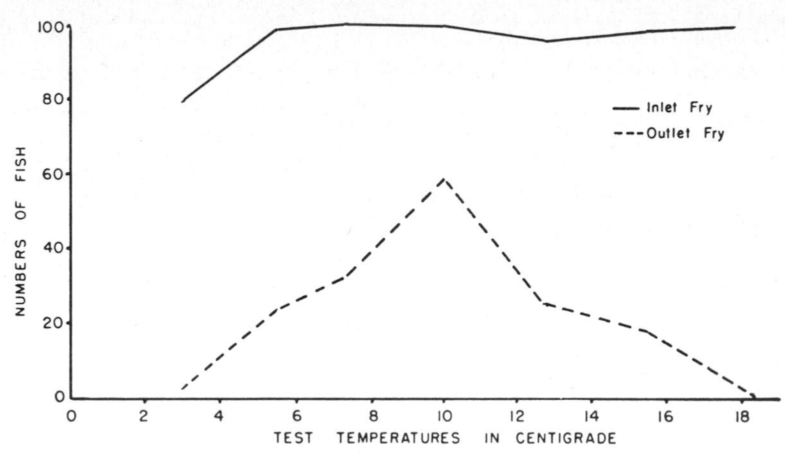

Fig. 7.1. Average effect of temperature on upstream migrations of outlet fry (*dashed line*) and downstream migrations of inlet fry (*solid line*) acclimated to temperatures of 3.4–4.5 °C. From "Innate control of migrations off salmon and trout fry from natal gravels to rearing areas" by R.F. Raleigh. *Ecology*, 1971, *52*, 291–297. Copyright © 1971 by the Ecological Society of America. Reprinted by permission

Table 7.4. Total number and mean response (per cent) of outlet Chilko and inlet Stellako genetic stocks of sockeye salmon fry responding to water currents in the laboratory (Brannon 1972). U = upstream. D = downstream. NT = not trapped. SD = standard deviation

Stock	No. of tests	Number of fry			Mean response (%)			
		U	D	NT	U	(SD)	D	NT
1967								
Chilko	20	1646	278	76	82.3	11.2	13.9	3.8
Stellako	20	309	1599	92	15.4	13.18	80.0	4.6
Stellako ♀ × Chilko ♂	20	969	955	76	48.4	31.29	47.8	3.8
1968								
Chilko	30	2750	173	77	91.7	6.86	5.8	2.6
Chilko ♀ × Stellako ♂	20	952	994	54	47.6	23.68	49.7	2.7
Stellako	30	302	2681	17	10.1	8.69	89.3	0.6
Stellako ♀ × Chilko ♂	20	928	1008	64	46.4	19.68	50.4	3.2

Brannon (1972) first tested each stock by capturing migrating fish at each location and testing them in their local water, using an experimental stream tank. Chilko fish moved upstream while Stellako fish moved downstream (Table 7.4). Two other stocks, which initially move downstream in their natural migration, Weaver Creek and seven-mile Creek, also showed downstream movement in the experimental apparatus. When these same stocks were reared under hatchery conditions in Cultus Lake water they still retained their appropriate directional preferences.

Crosses between the two stocks showed an intermediate response with about half of the hybrids going upstream and half downstream. Control tests indicated

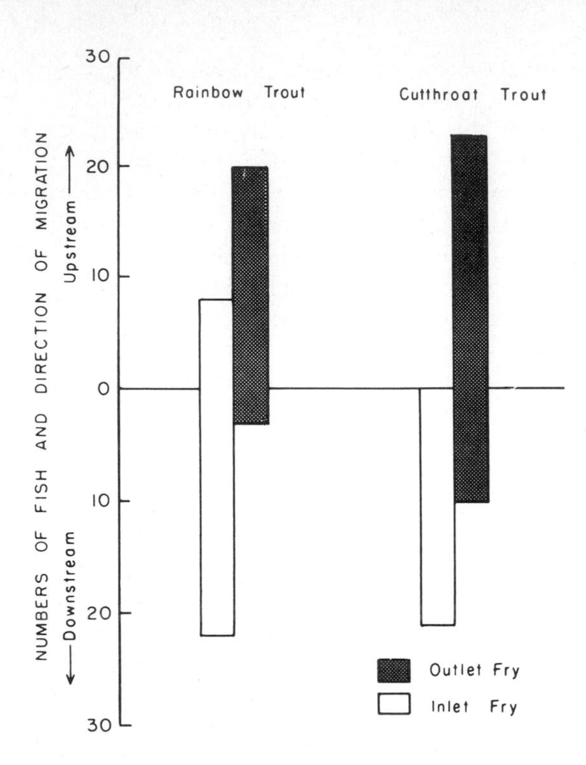

Fig. 7.2. Directional response to current of rainbow and cutthroat trout fry from inlet and outlet populations. Numbers of fish out of 50, not percentages. From "Innate control of migrations of salmon and trout fry from natal gravels to rearing areas" by R.F. Raleigh. *Ecology,* 1971, *52,* 291–297. Copyright © 1971 by the Ecological Society of America. Reprinted by permission

that size at emergence and age of the fry did not appreciably affect the results. These two studies by two different workers confirm that at least some inlet and outlet stocks of sockeye differ genetically in their rheotactic responses in a manner that would take them in the appropriate direction in their native streams.

The inlet/outlet problem has been investigated from the genetic point of view in cutthroat and rainbow trout as well as in sockeye salmon. Cutthroat hatching in Little Thumb Creek move downstream to Yellowstone Lake for a period of growth in the lake, while cutthroat from the Yellowstone River population move upstream into the same lake. The two populations differ in their serum antigens (Liebelt 1969), indicating some genetic separation. Raleigh (1971) and Raleigh and Chapman (1971) used a steam tank, or in one case a ditch, with traps at either end, and released mixed groups of the two stocks in the middle. Inlet fry tended to move downstream and outlet fry upstream (Fig. 7.2). This occurred even when both stocks were reared and tested in the same water.

A later study of Yellowstone cutthroat trout by Bowlen (1975) gave similar results with the exception that more inlet fry moved upstream when they were held in running water after emergence, than when they were held in still water. Perhaps there was a training effect due to the inlet fry being forced to hold position by swimming against the current in the holding troughs. Crosses between the inlet- and outlet-spawning stocks led to intermediate rheotactic responses.

Raleigh (1971) tested rainbow trout fry from inlet and outlet populations at Loon Lake, B.C. along with his sockeye and cutthroat. Again inlet fish showed more downstream movement than outlet fish even when reared under the same conditions (Fig. 7.2). Loon Lake rainbows and inlet and outlet stocks from Pennask Lake, also on the Fraser River system, were reared in captivity and tested in a chambered stream channel by Kelso et al. (1981). In crosses between stocks, fish with paternal parents from outlet stocks showed a greater tendency to move upstream than did fish fathered by fish from inlet stocks. The role of the male parent was emphasized by Kelso et al. (1981) because maternal influences might alter rheotaxis through non-genetic routes. For example, egg size might alter the rheotaxis of the offspring. Such influences could be controlled by comparing fish fathered by different stocks but with mothers from the same stock. Kelso et al. concluded that their results refuted Northcote's (1962) hypothesis that lakeward migration in inlet and outlet streams was controlled mainly by environmental factors.

Impassable falls pose problems similar to the inlet/outlet problem, with populations above the falls losing fish that are displaced downstream. In British Columbia, headwater rainbow trout populations tend to have a higher proportion of the liver lactate dehydrogenase phenotype H alpha A H alpha A while downstream, below the falls, H alpha B H alpha B is more common (Northcote and Kelso 1981). The homozygous H alpha A phenotype has about 2.3 times the swimming endurance of the H alpha B phenotype during the first 2 years of life. After that the difference becomes less. Fish of the two phenotypes captured from the same location at the Loon Lake inlet, above that nearest waterfall, differed in their rheotactic response when tested in a 25-compartment stream tank with traps at either end. The H alpha A homozygotes showed a greater tendancy to move upstream than did the H alpha B homozygotes and the difference was retained whether they were tested in the light or in the dark, although both phenotypes showed less upstream movement in the dark. Different phenotypes even within the same population may then have different behavioural responses to current. The difference in the distribution of the phenotypes probably originated through the continual loss of H alpha B individuals downstream over the falls.

Rainbow trout from stocks that occur above and below an impassable waterfall on Kokanee Creek, British Columbia, were raised in captivity and tested in an experimental stream tank (Northcote 1981). Under most conditions, neither stock showed much downstream movement. However, in Autumn under dark conditions the below-falls stock showed significant downstream migration while the above-falls stock did not. This indicates that the difference in migratory behaviour is retained even when the two stocks are reared under the same conditions in captivity.

The stock-specific differences in preferred direction of orientation found by Groot (1965) in sockeye smolts migrating out of Babine Lake (see Sect. 2.4.2.2.4) may be another case of genetic control of orientation. Morrison Lake smolts showed a seasonal change in orientation direction even after being raised under artificial conditions (Simpson 1979). This lends support to the genetic hypothesis. Brannon (1972) and Quinn (1980) found directional preferences by sockeye fry in still water (see Sect. 2.4.2.2.4), which might also be under genetic control, although Quinn points out that his fish could also have learned appropriate differences in orientation direction during their initial migration prior to capture.

An example of the influence of genotype on migration success occurs in pink salmon. The Headwaters Creek Hatchery on the Tsolum River in British Columbia reared and released fish whose parents came from the Kakweiken River in Knight inlet and the progeny of a cross between Kakweiken females and native Tsolum males (Bams 1976).

Both the pure Kakweiken genotype and the crossed genotype contributed equally to the fishery near the river mouth, indicating that both returned in equal numbers to the coastal region. However, at the junction of the Tsolum and Puntledge Rivers, downstream from the hatchery, returning hybrid fish outnumbered pure Kakweiken fish by 10:1 when the expected ratio of return was 1:1. Within the river system as a whole the Kakweiken fish scattered much more widely than the hybrids, which showed a stronger tendency to return to the hatchery area. Homozygous Tsolum fish appear to home even more accurately, although they were not specifically tested in this experiment. None of the Kakweiken fish or hybrids from the Tsolum release turned up in the Kakweiken River.

Home stream imprinting and open-water orientation mechanisms were sufficient to bring the foreign stock back to the stream of release rather than to the donor stream but the presence of some of the home stream genotype from the Tsolum parent significantly improved homing to and within the Tsolum River (Bams 1976).

Bams (1976) suggested that if stocks must be augmented by addition of foreign fish, then an admixture of the native genotype may improve homing. The use of native males and donor females would allow maximum use of a depleted native stock. He also emphasizes that these results indicate a significant degree of genetic adaptation to the local situation, and that introduction of any foreign genetic material may be disadvantageous.

The general conclusion that arises from this handful of experimental tests is that genetic differences in migratory behaviour between local populations have been found wherever they have been sought. This should serve as a warning that a great deal of valuable genetic material is present in fish populations and that it should be protected from loss through extinction and from contamination by genetic material from other populations. The use of hatchery strains of fish may be a particularly severe problem (Smith and Chesser 1981). Often these fish are of uncertain ancestry and have become genetically adapted to the specialized conditions of the hatchery. The encouragement and protection of the locally adapted native stocks seems a much better policy than the planting of stock from other sources.

The time required for the evolution of regional strains of salmonids may be quite short in a geological sense. Much of the northwest coast of North America emerged from glaciation less than 12,000 years before the present and the present shape of particular lakes is more recent (Prest 1969). The Babine Lake area, for example, was glaciated until 10,000 B.P. and the present outlet passes through a region that melted about 8500 B.P. so the modern arrangement of outlet and arms of the lake cannot be more than about 2000 sockeye generations old. There is a good deal of disagreement on how much salmon evolution has occurred since the last ice age. For example, Miller and Brannon (1982) suggest that separation of the genus *Oncorhynchus* from *Salmo* has occurred since the last glaciation, while Thorpe (1982) argues that the various species of *Oncorhynchus* evolved prior to the last glaciation. Obviously if the species separated since deglaciation, then there has been even less time for the evolution of local stocks. The evolution of specialized stocks is probably linked to precise homing ability and to the presence of a semelparous life cycle (Miller and Brannon 1982). Both these atributes would reduce gene exchange between stocks.

7.5 Summary

Past events can affect migratory behaviour through two fundamentally different routes. Within the life span of an individual, learning may alter behavioural responses to stimuli, in response to positive or negative reinforcement. On the scale of generations, genetic selection will favour genotypes controlling behaviour patterns that improve the survival of their carriers. These two processes, learning and genetic selection, interact with one another, for example the timing of critical learning periods may be genetically programmed and the determination of which stimuli are reinforcing is probably under genetic control.

Learning is most appropriate for situations that will change in the short term, within the life span of a single fish, or situations that may be unique to a single generation. The learning of a home location is a good example. The best-studied case of learning in fish migration is the home-stream odour imprinting of juvenile salmonids, discussed in Chap. 3. Learning the characteristics of a home area also occurs in the sun-compass orientation of inshore fish to their home shoreline. Learning may be involved in recognition of population-specific pheromones.

Fish stocks differ genetically, even within the same river or lake in some cases. These genetic stocks are kept separate by the homing fidelity of the species. This genetic isolation allows local stocks to become genetically adapted to local conditions, including conditions related to migration. The most advantageous time for migration, for example, may vary between locations. Selection will favour individuals that migrate at the optimal time for the location of their stock. Genetic differences in migratory season have been demonstrated. Atlantic salmon in two different tributaries of the same river migrate in different months and these differences persist when they are reared under identical conditions.

Different stocks may also require different directional orientation for successful migration. Inlet-spawning and outlet-spawning stocks of lake-dwelling fish are such a case. Juveniles hatching in inlets will reach a lake by moving downstream, while juveniles hatching in outlets must move upstream. In sockeye salmon and rainbow and cutthroat trout, inlet and outlet populations retain appropriate rheotactic responses even when raised under identical conditions. Crosses between inlet and outlet fish are intermediate in their rheotactic responses. Similar genetic differences occur between populations above and below waterfalls. The seasonal change in orientation shown by some migrating sockeye smolts is also retained when the fish are reared under hatchery conditions. Pink salmon with one parent from the resident stock of the imprinted stream home more accurately than fish with neither parent from the imprinted stream, indicating a genetic effect on homing success.

The genetic adaptation of fish stocks to local conditions means that the loss of local stocks or their contamination with foreign genetic material could decrease the productivity of a region. Where local stocks have already been destroyed, replacement should be made with fish from a stock matched to the conditions – outlet fish stocked in outlets, for example. Where a native stock is still present it should probably be enhanced rather than replaced or supplemented, since other stocks will seldom be exactly matched to local conditions.

Chapter 8

Conclusion

The study of fish migration is characterized by both strengths and weaknesses. The weaknesses are largely the products of limited study and resources. Few of the many species of fish have been studied. The studies that have been carried out have concentrated on freshwater and inshore aspects of migration. The major offshore oceanic migrations are still virtually untouched in terms of experimental research into controlling mechanisms. Within the areas which have been intensively studied there has been some tendency for research to advance unevenly. Over 2000 papers have been published on fish temperature preference, while other areas, such as genetic variation in migratory behaviour, have received little study until very recently.

Perhaps the greatest area of strength in this subject is the comprehensive picture which is emerging from the sustained research on salmon migration, particularly studies on the sockeye, coho, and other Pacific salmon. The combination of economic value and accessible freshwater phases in the life cycle have facilitated this research effort. The Pacific species may also have benefited by being remote from the first centres of industrialization. This isolation has preserved many of these salmon stocks into the era of modern research, but continued survival will depend on sound environmental legislation based on adequate biological understanding of their requirements. For the less-studied species, often those of lesser economic value or less accessible to researchers, the present state of salmon research can serve as a guide for research and environmental legislation.

One of the salient points to emerge from this survey of controlling mechanisms is the variety of different stimuli which can be involved in fish migration. These stimuli also interact in complex ways. Compass mechanisms, for example, are arranged in hierarchical order, one mechanism being used preferentially when its stimuli are present, then the other mechanism taking over in the absence of the preferred stimulus. This sort of interaction is more complex and difficult to detect than a simple additive interaction, since in the presence of the preferred stimulus, manipulation of the subordinate stimuli or their receptors may not have any effect.

The variety and subtlety of the controlling mechanism means that it is difficult to foresee the consequences of interference with natural ecosystems. Species which are important links in the ecosystem, but which are not economically important enough to warrant special study or protection, may be particularly vulnerable to interference. For example, the passive electroreceptive fishes, a group which includes long-distance migrants such as the sturgeon, may be affected by "electrical pollution" from power grids.

The evidence for genetic adaptation of fish stocks to their specific migration routes means that contamination of these genotypes with fish from other locations may seriously interfere with migratory ability. It also means that the extinction of a stock is a serious loss even when many other stocks of the species still survive.

Better understanding of controlling mechanisms can lead to new benefits from fish stocks as well as to better understanding of how to protect them; the use of delayed release of salmon to enhance local fisheries is a case in point.

Perhaps the strongest conclusion that one could draw from this work is that basic research should not be underestimated. Only through examining many different hypotheses and a variety of species can the diversity of natural migratory patterns be understood.

References

Adler K, Taylor DH (1973) Extraocular perception of polarized light by orienting salamanders. J Comp Physiol 87:203–212

Adrianov GN, Brown HR, Ilyinsky OB (1974) Responses of central neurons to electrical and magnetic stimuli of the ampullae of Lorenzini in the Black Sea skate. J Comp Physiol 93:289–299

Akoev GN, Ilyinsky OB, Zadan PM (1976) Responses of electroreceptor (ampullae of Lorenzini) of skates to electric and magnetic fields. J Comp Physiol 106:127–136

Alabaster JS (1970) River flow and upstream movement and catch of migratory salmonids. J Fish Biol 2:1–13

Alabaster JS, Downing AL (1966) A field and laboratory investigation of the effect of heated effluents on fish. Fish Invest Ser I, vol VI, No 4. Her Majesty's Stationery Office, London, pp iii + 42 1 plate + 21 text-figures

Alderdice DF, Brett JR, Idler DR, Fagerlund U (1954) Further observations on olfactory perception in migrating adult coho and spring salmon–properties in the repellent in mammalian skin. Fish Res Bd Can Progr Rept Pac Coast St 98:10–12

Ali MA (1959) The ocular structure, retinomotor, and photo-behavioural responses of juvenile Pacific salmon. Can J Zool 37:965–995

Ali MA, Anctil M (1976) Retinas of fishes: an atlas. Springer, Berlin Heidelberg New York, p 284

Ali MA, Hoar WS (1959) Retinal responses of pink salmon associated with its downstream migration. Nature (London) 184:106–107

Allen GH (1959) Behaviour of chinook and silver salmon. Ecology 40:108–113

Allendorf FW, Utter FM (1979) Population genetics. In: Hoar WS, Randall DJ, Brett JR (eds) Fish physiology, vol VIII. Academic Press, London New York, pp 407–454

Allendorf FW, Ryman N, Stennck A, Stahl G (1976) Genetic variation in Scandinavian brown trout (*Salmo trutta*) evidence of distinct sympatric populations. Hereditas 83:73–82

Arnold GP (1974) Rheotropism in fishes. Biol Rev 49:515–576

Aronson LR (1951) Orientation and jumping behaviour in the gobiid fish *Bathygobius soporator*. Am Mus Nov No 1486:1–21

Aronson LR (1971) Further studies on orientation and jumping behaviour in the gobiid fish, *Bathygobius soporator*. Ann NY Acad Sci 188:378–407

Atema J, Jacobson S, Todd J, Boylan D (1973) The importance of chemical signals in stimulating behavior of marine organisms: effects of altered environmental chemistry on animal communication. In: Glass GE (ed) Bioassay techniques and environmental chemistry. Mich Ann Arbor Sci Publ, pp 177–197

Baggerman B (1957) An experimental study on the timing of breeding and migration in the three-spined stickleback (*Gasterosteus aculeatus*). Arch Neerl Zool 12:105–317

Baggerman B (1960a) Salinity preferences, thyroid activity, and the seaward migration of four species of Pacific salmon (*Oncorhynchus*). J Fish Res Board Can 17:295–322

Baggerman B (1960b) Factors in the diadromous migrations of fish. Symp Zool Soc (London) 1:33–60

Baggerman B (1962) Some endocrine aspects of fish migration. Gen Comp Endocrinol Suppl 1:188–205

Baggerman B (1963) The effect of TSH and antithyroid substances on salinity preference and thyroid activity in juvenile Pacific Salmon. Can J Zool 41:307–309

Baker RR (1978) The evolutionary ecology of animal migration. Holmes & Meier, New York, 1012 pp

Baker RR (1980) Goal orientation by blindfolded humans after long-distance displacements: possible involvement of a magnetic sense. Science 210:555–557

Bams RA (1969) Adaptations of sockeye salmon associated with incubation in stream gravels. In: Northcote TG (ed) Symposium on salmon and trout in streams, HR MacMillan lectures in fisheries. Inst Fish, UBC, Vancouver, pp 71–87
Bams RA (1972) A quantitative evaluation of survival to the adult stage and other characteristics of pink salmon (*Oncorhynchus gorbuscha*) produced by a revised hatchery method which simulates optimal natural conditions. J Fish Res Board Can 29:1151–1167
Bams RA (1976) Survival and propensity for homing as affected by presence or absence of locally adapted paternal genes in two transplanted populations of pink salmon (*Oncorhynchus gorbuscha*). J Fish Res Board Can 33:2716–2725
Banks JW (1969) A review of the literature on the upstream migration of adult salmonids. J Fish Biol 1:35–136
Bannister LH (1965) The fine structure of the olfactory surface of teleostean fishes. Q J Microsc Sci 106:333–342
Bardach JE (1956) The sensitivity of the goldfish (*Carassius auratus* L.) to point heat stimulation. Am Nat 90:309–317
Bardach JE, Atema J (1971) The sense of taste in fish. In: Beidler LM (ed) Taste. Handbook of sensory physiology IV. Springer, Berlin Heidelberg New York, pp 293–336
Bardach JE, Case J (1965) Sensory capabilities of the modified fins of squirrel hake (*Urophycis chuss*) and searobins (*Prionotus carolinus* and *P. evolans*). Copeia 1965:194–206
Bardach JE, Villars T (1974) The chemical senses of fishes. In: Grant PT, Mackie AM (eds) Chemoreception in marine organisms. Academic Press, London New York, pp 49–104
Bardach JE, Todd JH, Crickmer R (1967) Orientation by taste in fish of the genus *Ictalurus*. Science 155:1276–1278
Barlow GW (1958) Daily movements of the desert pupfish *Cyprinodon macularius*, in shore pools of the Salton Sea, California. Ecology 39:580–587
Barlow JS (1964) Inertial navigation as a basis for animal navigation. J Theor Biol 6:76–117
Barnett C (1977) Aspects of chemical communication with special reference to fish. Biosci Commun 3:331–392
Bastian J (1976) The range of electrolocation: a comparison of electroreceptor responses and the responses of cerebellar neurons in a gymnotid fish. J Comp Physiol A 108:193–210
Bayley PB (1973) Studies on the migratory characin, *Prochilodus platensis* Holmburg 1889, (Pisces: Characoidei) in the River Pitcomayo, South America. J Fish Biol 5:25–40
Beckett JS (1970) Swordfish, shark, and tuna tagging. Fish Res Bd Can Tech Rept 193
Beiningen KT, Ebel WJ (1970) Effect of John Day Dam on dissolved nitrogen concentrations and salmon in the Columbia River. Trans Am Fish Soc 99:664–671
Beitinger TL, Fitzpatrick LC (1979) Physiological and ecological correlates of preferred temperature in fish. Am Zool 19:319–329
Bell CC, Myers JP, Russell CJ (1974) Electric organ discharge patterns during dominance-related behavioural displays in *Gnathonemus petersii* (Mormyridae). J Comp Physiol 92:201–228
Belyy ND (1972) Downstream migration of the Pike-Perch *Lucioperca lucioperca* (L.) and its food in the early development stages in the lower reaches of the Dnieper. J Ichthyol 12:465–482
Bennett MVL (1971 a) Electric organs. In: Hoar WS, Randall DJ (eds) Fish physiology, vol V, Academic Press, London New York, pp 347–491
Bennett MVL (1971 b) Electroreception. In: Hoar WS, Randall DJ (eds) Fish physiology, vol V. Academic Press, London New York, pp 493–574
Bennett MVL, Clusin WT (1978) Physiology of the ampulla of Lorenzini, the electroreceptors of elasmobranchs. In: Hodgson ES, Mathewson RF (eds) Sensory biology of shark, skates, and rays. Off Nav Res, Arlington, Va, pp 483–505
Berge JA (1979) The perception of weak electric A.C. currents by the European eel, *Anguilla anguilla*. Comp Biochem Physiol 62A:915–920
Bergeijk van WA (1964) Directional and non-directional hearing in fish. In: Tavolga WN (ed) Marine bioacoustics. Pergamon Press, Oxford New York, pp 281–299
Bergeijk van WA (1967) Introductory comments on lateral live function. In: Cahn P (ed) Lateral line detectors. Indiana Univ Press, Bloomington, pp 73–81
Berra TM, Gunning GE (1970) Repopulation of experimentally decimated sections of streams by longear sunfish, *Lepomis megalotis megalotis* (Rafinesque). Trans Am Fish Soc 99:776–781

Bertmar G, Toft R (1969) Sensory mechanisms of homing in salmonid fish I. Introductory experiments on the olfactory sense in grilse of Baltic salmon (*Salmo salar*). Behaviour 35:235–241
Bjornn TC (1971) Trout and salmon movements in two Idaho streams as related to temperature, food, stream flow, cover, and population density. Trans Am Fish Soc 100:423–438
Blackburn M (1970) Conditions related to upwelling which determine distribution of tropical tunas off Western Baja, California. Fish Bull 68:147–176
Black-Cleworth P (1970) The role of electric discharges in the non-reproductive social behavior of *Gymnotus carapo*. Anim Behav Monogr 3:1–77
Blackett RF (1979) Establishment of sockeye (*Oncorhynchus nerka*) and Chinook (*Oncorhynchus tshawytscha*) salmon runs at Frazier Lake, Kodiak Island, Alaska. J Fish Res Bd Can 36:1265–1277
Blakemore R (1975) Magnetotactic bacteria. Science 190:377–379
Blaxter JHS (1970) Response of fish to light. In: Kinne O (ed) Marine ecology, vol I. Wiley Interscience, New York, p 681
Blaxter JHS, Jones MP (1967) The development of the retina and retinomotor responses in the herring. J Mar Biol Assoc UK 47:677–697
Blaxter JHS, Tytler P (1972) Pressure discrimination in teleost fish. Symp Soc Exp Biol 26:417–443
Blaxter JHS, Tytler P (1978) Physiology and function of the swimbladder. In: Lowenstein O (ed) Adv Comp Physiol Biochem 7:311–367
Blaxter JHS, Denton EJ, Gray JAB (1979) The herring swimbladder as a gas reservoir for the acousticolateralis system. J Mar Biol Assoc UK 57:1–10
Bleckmann H (1980) Reaction time and stimulus frequency in prey localization in the surface-feeding fish *Aplocheilus lineatus*. J Comp Physiol A 140:163–172
Bodznick D (1975) The relationship of the olfactory EEG evoked by naturally occurring stream waters to the homing behavior of sockeye salmon (*Oncorhynchus nerka* Walbaum). Comp Biochem Physiol 52A:487–495
Bodznick D (1978a) Water source preference and lakeward migration of sockeye salmon fry (*Oncorhynchus nerka*). J Comp Physiol A 127:139–146
Bodznick D (1978b) Characterization of olfactory bulb units of sockeye salmon with behaviorally relevant stimuli. J Comp Physiol A 127:147–156
Bodznick D (1978c) Calcium ion: an odorant for natural water discriminations and the migratory behavior of sockeye salmon. J Comp Physiol A 127:157–166
Bodznick D, Northcutt RG (1981) Electroreception in lampreys: evidence that the earliest vertebrates were electroreceptive. Science 212:465–467
Boetius J (1967) Experimental indication of lunar activity in European silver eels, *Anguilla anguilla* (L.). Meddr Danm Fisk Havunders 6:1–6
Bohun S, Winn HE (1966) Locomotor activity of the American eel (*Anguilla rostrata*). Chesapeake Sci 7:137–147
Bowlen B (1975) Factors influencing genetic control in lakeward migrations of cutthroat trout fry. Trans Am Fish Soc 104:474–482
Braemer W (1957) Verhaltensphysiologische Untersuchungen am optischen Apparat bei Fischen. Z Vergl Physiol 39:374–398
Braemer W (1959) Versuche zu der in Richtungsgehen der Fische enthaltenen Zeitschätzung. Verh Dtsch Zool Ges Anz 23, Suppl 276–288
Brannon EL (1967) Genetic control of migrating behaviour of newly emerged sockeye salmon fry. Int Pac Salmon Fish Comm Progr Rep 16:1–31
Brannon EL (1972) Mechanisms controlling migration of sockeye salmon fry. Bull Int Pac Salmon Fish Comm 21:1–86
Brannon EL, Quinn IP, Lucchetti GL (1981) Compass orientation of sockeye salmon fry from a complex river system. Can J Zool 59:1548–1553
Branover GG, Vasil'yev AS, Gleiser SI, Tsinober AB (1971) A study of the behavior of eels in natural and artificial magnetic fields and an analysis of its reception mechanism. J Ichthyol 11:608–614
Bray RN, Hixon MA (1978) Night-shocker: predatory behavior of the Pacific electric ray (*Torpedo californica*). Science 200:333–334
Breder CM (1967) On the survival value of fish schools. Zoologica 52:25–40
Breder CM, Nigrelli RF (1935) The influence of temperature and other factors on the winter aggregations of the sunfish, *Lepomis auritus*, with critical remarks on the social behavior of fishes. Ecology 16:33–47

Breder CM, Rasquin P (1947) Comparative studies in the light sensitivity of blind characins from a series of Mexican caves. Bull Am Mus Nat Hist 89:325–351
Breder CM, Rasquin P (1950) A preliminary report on the role of the pineal organ in the control of pigment cells and light reactions in recent teleost fishes. Science 111:10–12
Bregnballe F (1961) Plaice and flounders as consumers of the microscopic bottom fauna. Meddr Danm Fish Havunders 3:133–182
Brett JR (1956) Some principles in the thermal requirements of fishes. Q Rev Biol 31:75–87
Brett JR (1957) Salmon research and hydro-electric power development. Fish Res Bd Can Bull 114
Brett JR (1970) Temperature. Animals, fishes. In: Kinne O (ed) Marine ecology, vol I, Pt 1. John Wiley & Sons, New York, pp 515–560
Brett JR (1971) Energetic responses of salmon to temperature. A study of some thermal relations in the physiology and freshwater ecology of sockeye salmon (*Oncorhynchus nerka*). Am Zool 11:99–113
Brett JR, Alderdice DF (1958) Research on guiding young salmon at two British Columbia field stations. Fish Res Bd Can Bull 117, p 75
Brett JR, Ali MA (1958) Some observations on the structure and photomechanical responses of the Pacific salmon retina. J Fish Res Board Can 15:815–829
Brett JR, Groot C (1963) Some aspects of olfactory and visual responses in Pacific Salmon. J Fish Res Bd Can 20:287–303
Brett JR, MacKinnon D (1952) Some observations on olfactory perception in migrating adult coho and spring salmon. Fish Res Bd Can Progr Rep Pac Coast Stn 90:21–23
Brett JR, MacKinnon D (1954) Some aspects of olfactory perception in migrating adult coho and spring salmon. J Fish Res Bd Can 11:310–318
Brown BE, Inman I, Jearld A (1970) Schooling and shelter-seeking tendencies in fingerling channel catfish. Trans Am Fish Soc 99:540–545
Brown FA, Bennet MF, Webb HM (1960) A magnetic response of an organism. Biol Bull 119:65–74
Brown HR, Ilyinsky OB (1978) The ampullae of Lorenzini in the magnetic field. J Comp Physiol 126:333–341
Brown HR, Andrianov GN, Ilyinsky OB (1974) Magnetic field perception by electroreceptors in Black Sea skate. Nature (London) 249:178–179
Brown HR, Ilyinsky OB, Muravejko VM, Corshkov ES, Fonarev GA (1979) Evidence that geomagnetic variations can be detected by Lorenzinian ampullae. Nature (London) 277:648–649
Brown SB, Hara TJ (1982) Biochemical aspects of amino acid receptors in olfaction and taste. In: Hara T (ed) Chemoreception in fishes. Elsevier, Amsterdam, pp 159–180
Buckland J (1880) Natural history of British fishes. Unwin, London
Bull HO (1930) Studies on conditioned responses in fishes II. Mar Biol Assoc UK 16:615–637
Bull HO (1936) Studies on conditioned responses in fishes VII. Temperature perception in teleosts. J Mar Biol Assoc UK 21:1–27
Burgner RL (1962 a) Studies of red salmon smolts from the Wood River Lakes, Alaska. In: Koo SY (ed) Studies of Alaska Red Salmon. Univ Washington Press, Washington, pp 251–314
Burgner RL (1962 b) Sampling red salmon fry by lake trap in the Wood River Lakes, Alaska. In: Koo SY (ed) Studies of Alaska Red Salmon. Univ Washington Press, Washington, pp 317–348
Burner CJ, Moore HL (1962) Attempts to guide small fish with underwater sound. US Fish Wildl Serv, Spec Sci Rep Fish 403:1–30
Bustard DR, Narver DW (1975 a) Aspects of the winter ecology of juvenile coho salmon (*Oncorhynchus kisutch*) and steelhead trout (*Salmo gairdneri*). J Fish Res Bd Can 32:667–680
Bustard DR, Narver DW (1975 b) Preferences of juvenile coho salmon (*Oncorhynchus kisutch*) and cutthroat trout (*Salmo clarki*) relative to simulated alteration of winter habitat. J Fish Res Bd Can 32:681–687
Cahn P (1967) Lateral line detectors. Indiana Univ Press, Bloomington, 496 pp
Caprio J (1974) Extreme sensitivity and specificity of catfish gustatory receptors to amino acids and derivatives. In: Denton DA, Coghlan JP (eds) Olfaction and taste, vol V. Academic Press, London New York, pp 157–161
Caprio J (1982) High sensitivity and specificity of olfactory and gustatory receptors of catfish to amino acids. In: Hara TJ (ed) Chemoreception in fishes. Elsevier, Amsterdam, pp 109–134
Carscadden JE, Leggett WC (1975) Meristic differences in spawning populations of American shad, *Alosa sapidissima:* evidence for homing to tributaries in the St. John River, New Brunswick. J Fish Res Bd Can 32:653–660

Carthy JD (1958) An introduction to the behaviour of invertebrates. Allen & Unwin, London, 380 pp
Chapman CJ, Sand O (1974) Field studies of hearing in two species of flatfish, *Pleuronectes platessa* (L.) and *Limanda limanda* (L.) (Family Pleuronectidae). Comp Biochem Physiol 47A:371–385
Chapman DW (1962) Aggressive behaviour in juvenile coho salmon as a cause of emigration. J Fish Res Bd Can 19:1047–1080
Chapman DW, Bjornn TC (1969) Distribution of salmonids in streams, with special reference to food and feeding. In: Northcote TG (ed) Symposium on salmon and trout in streams, HR MacMillan lectures in fisheries. Inst Fish, UBC, Vancouver, pp 153–176
Child AR, Burnell AM, Wilkins NP (1976) The existence of two races of Atlantic salmon (*Salmo salar* L.) in the British Isles. J Fish Res Bd Can 8:35–43
Churmasov AV, Stepanov AS (1977) Sun orientation and guideposts of the humpback salmon. Sov J Mar Biol 3:55–63
Claridge PN, Gardner DC (1978) Growth and movements of the twaite shad, *Alosa fallax* (Lacepede) in the Severn Estuary. J Fish Biol 12:203–212
Clark WC, Smith HD (1972) Observation on the migration of sockeye salmon fry (*Oncorhynchus nerka*) in the Lower Babine River. J Fish Res Bd Can 29:151–159
Coburn A, McCart P (1967) A hatchery release tank for pink salmon fry with notes on behaviour of the fry in the tank after release. J Fish Res Bd Can 24:77–85
Cole LC (1954) The population consequences of life history phenomena. Q Rev Biol 29:103–137
Collins GB (1952) Factors influencing the orientation of migrating anadromous fishes. Fish Bull 52:375–396
Collins GB (1976) Effects of dams on Pacific salmon and steelhead trout. Mar Fish Rev 38:39–46
Cooper JC (1982) Comment on electroencephalographic responses to morpholine and their relationship to homing. Can J Fish Aquat Sci 39:1544–1546
Cooper JC, Hasler AD (1974) Electroencephalographic evidence for retention of olfactory cues in homing coho salmon. Science 183:336–338
Cooper JC, Hasler AD (1976) Electrophysiological studies of morpholine-imprinted coho salmon (*Oncorhynchus kisutch*) and rainbow trout (*Salmo gairdneri*). J Fish Rev Bd Can 33:688–694
Cooper JC, Hirsch PJ (1982) The role of chemoreception in salmonid homing. In: Hara TJ (ed) Chemoreception in fishes. Elsevier, Amsterdam, pp 343–362
Cooper JC, Scholz AJ (1976) Homing of artificially imprinted steelhead (rainbow) trout, *Salmo gairdneri*. J Fish Res Bd Can 33:826–829
Cooper JC, Scholz AJ, Horrall RM, Hasler AD, Madison DM (1976) Experimental confirmation of the olfactory hypothesis with homing, artificially imprinted coho salmon (*Oncorhynchus kisutch*). J Fish Res Bd Can 33:703–710
Coutant CC (1977) Compilation of temperature-preference data. J Fish Res Bd Can 34:739–745
Cox P (1916) Are migrating eels deterred by a range of lights? Contr Can Biol 1916:115–118
Creutzberg F (1959) Descrimination between ebb and flood tides in migrating elvers (*Anguilla vulgaris* Turt.) by means of olfactory perception. Nature (London) 184:1961–1962
Creutzberg F (1961) On the orientation of migrating elvers (*Anguilla vulgaris* Turt.) in a tidal area. Neth J Sea Res 1:257–338
Creutzberg F (1963) The role of tidal streams in the navigation of migrating elvers (*Anguilla vulgaris* Turt.). Erg Biol 26:118–127
Creutzberg F, Eltink ATGW, Noort GJ van (1978) The migration of plaice larvae *Pleuronectes platessa* into the western Wadden Sea. In: McLusky DS, Berry AJ (eds) Physiology and behaviour of marine organisms. Pergamon Press, Oxford New York, pp 243–251
Davitz MA, McKaye KR (1978) Discrimination between horizontally and vertically polarized light by the cichlid fish *Pseudotropheus macrophthalmus*. Copeia 1978:333–334
Deelder CL (1952) On the migration of the elver (*Anguilla vulgaris* Turt.) at sea. J Cons Perm Int Explor Mer 18:187–218
Deelder CL (1954) Factors affecting the migration of the silver eel in Dutch inland waters. J Cons Perm Int Explor Mer 20:177–185
Deelder CL (1958) On the behaviour of elvers (*Anguilla vulgaris* Turt.) migrating from the sea into freshwater. J Cons Perm Int Explor Mer 24:135–146
Denton EJ, Gray JAB, Blaxter JHS (1979) The mechanics of the clupeid acoustico-lateralis system: Frequency responses. J Mar Biol Assoc UK 59:27–47

Deubler E, Posner GS (1963) Response of postlarval flounders, *Paralichthys lethostigma,* to water of low oxygen concentration. Copeia 1963:312–317
Dijkgraaf S (1940) Untersuchungen über den Temperatursinn der Fische. Z Vergleich Physiol 27:587–605
Dijkgraaf S (1943) Berichtigung und Ergänzung zu meiner Arbeit „Untersuchungen über den Temperatursinn der Fische". Z Vergleich Physiol 30:252
Dijkgraaf S (1960) Hearing in bony fishes. Proc R Soc London Ser B 152:51–54
Dill PA (1971) Perception of polarized light by yearling sockeye salmon (*Oncorhynchus nerka*). J Fish Res Bd Can 28:1319–1322
Ditton RB, Graete AR, Fedler AJ, Schwartz JD (1979) Access to and usage of offshore liberty ship reefs in Texas. Mar Fish Rev 41:25–31
Dizon AE, Horrall RM, Hasler AD (1973) Long term olfactory "memory" in coho salmon, *Onchorhynchus kisutch.* Fish Bull 71:315–317
Dizon AE, Stevens ED, Neill WH, Magnuson JJ (1974) Sensitivity of restrained skipjack tuna (*Katsuwonus pelamis*) to abrupt changes in temperature. Comp Biochem Physiol A 49:291–299
Dodge DP, MacCrimmon HR (1971) Environmental influences on extended spawning of rainbow trout (*Salmo gairdneri*). Trans Am Fish Soc 100:312–318
Dodson JJ, Leggett WC (1974) Role of olfaction and vision in the behaviour of American shad (*Alosa sapidissima*) homing to the Connecticut River from Long Island Sound. J Fish Res Bd Can 31:1607–1619
Dodson JJ, Young JC (1977) Temperature and photoperiod regulation of rheotropic behavior in prespawning common shiners, *Notropis cornutus* (Mitchill). J Fish Res Bd Can 34:341–346
Dodson JJ, Leggett WC, Jones RA (1972) The behaviour of adult American shad (*Alosa sapidissima*) during migration from salt to freshwater as observed by ultrasonic tracking techniques. J Fish Res Bd Can 29:1445–1449
Donaldson R, Allen GH (1957) Return of silver salmon *Oncorhynchus kisutch* (Walbaum) to point of release. Trans Am Fish Soc 87:13–22
Dovel WL, Mihursky JA, McErlean AJ (1969) Life history aspects of the hogchoker, *Trinectes maculatus,* in the Patuxent River estuary, Maryland. Chesapeake Sci 10:104–119
Døving KB, Nordeng H, Oakley B (1974) Single-unit discrimination of fish odour released by char (*Salmo alpinus* L.). Comp Biochem Physiol 47A:1051–1063
Døving KB, Selset R, Thommesen G (1980) Olfactory sensitivity to bile acids in salmonid fishes. Acta Physiol Scand 108:123–131
Dubrov AP (1978) The geomagnetic field and life. Plenum, New York
Ducharme LJA (1972) An application of louver deflectors for guiding Atlantic salmon (*Salmo salar*) smolts from power turbines. J Fish Res Bd Can 29:1397–1404
Duntley SQ (1962) Underwater visibility. In: Hill MN (ed) The sea, vol I. Wiley Interscience, New York, pp 452–455
Eales JG (1963) A comparative study of thyroid function in migrant juvenile salmon. Can J Zool 41:811–824
Eales JG (1968) The eel fisheries of Eastern Canada. Fish Res Bd Can Bull 166
Ebel WJ (1980) Transportation of chinook salmon, *Oncorhynchus tshawytscha,* and steelhead, *Salmo gairdneri,* smolts in the Columbia River and effects on adult returns. Fish Bull 78:491–505
Ebel WJ, Dawley EM, Mark B (1971) Thermal tolerance of juvenile salmon in relation to nitrogen supersaturation. Fish Bull 69:833–843
Edel RK (1975a) The induction of maturation of female American eels through hormone injections. Helgol Wiss Meeresunters 27:131–138
Edel RK (1975b) The effect of shelter availability on the activity of male silver eels. Helgol Wiss Meeresunters 27:167–174
Ellis DV (1962) Preliminary studies on the visible migration of adult salmon. J Fish Res Bd Can 19:137–148
Ellis DV (1966) Swimming speeds of sockeye and coho salmon on spawning migration. J Fish Res Bd Can 23:181–187
Emlen ST (1975) Migration: orientation and navigation. In: Farner DS, King JR (eds) Avian biology, vol V. Academic Press, London New York, pp 129–219
Enger PS, Kristensen L, Sand O (1976) The perception of weak electric d.c. currents by the European eel (*Anguilla anguilla*). Comp Biochem Physiol 54A:101–103

Eriksson L-O (1978a) A laboratory study of diel and annual activity rhythms and vertical distribution in the perch, *Perca fluviatilis*, at the arctic circle. Environ Biol Fish 3:301–307
Eriksson L-O (1978b) Nocturnalism versus diurnalism – dualism within fish individuals. In: Thorpe J (ed) Rhythmic activity of fishes. Academic Press, London New York, pp 69–89
Erman DC, Leidy GR (1975) Downstream movement of rainbow trout fry in a tributary of Sagehen Creek, under permanent and intermittent flow. Trans Am Fish Soc 104:467–473
Everest FH, Chapman DW (1972) Habitat selection and spatial interaction by juvenile chinook salmon and steelhead trout in two Idaho streams. J Fish Res Bd Can 29:91–100
Falcon J, Meissl H (1981) The photosensory function of the pineal organ of the pike (*Esox lucius* L.) correlation between structure and function. J Comp Physiol 144A:127–137
Fava JA, Tsai C (1975) Chlorinated sewage effluents and avoidance reaction of stream fish. Water Resources Research Center. Md Univ Tech Rep No 35, 59 pp
Fava JA, Tsai C (1976) Immediate behavioural reactions of blacknose dace, *Rhinichthys atratulus*, to domestic sewage and its toxic constituents. Trans Am Fish Soc 105:430–441
Favorite F (1969a) Fishery oceanography – II Salinity front at entrance to Washington's Straight of Juan de Fuca. Comm Fish Rev 31(8–9):36–40
Favorite F (1969b) Fishery oceanography – IV Ocean salinity and distribution of Pacific Salmon. Comm Fish Rev 31(10):29–32
Fay RR (1978) Sound detection and sensory coding by the auditory systems of fishes. In: Mostofsky DI (ed) The behaviour of fish and other aquatic animals. Academic Press, London New York, pp 197–236
Fay RR, Popper AN (1980) Structure and function in teleost auditory systems. In: Popper AN, Fay RR (eds) Comparative studies of hearing in vertebrates. Springer, Berlin Heidelberg New York, pp 1–42
Fenwick JC (1970) The pineal organ; photoperiod and reproductive cycles in the goldfish. J Endocrinol 46:101–111
Fessard A (ed) (1974) Handbook of sensory physiology, vol III, Pt 3. Springer, Berlin Heidelberg New York
Fields PE (1954) The effect of electric lights upon the upstream passage of sockeye salmon through the University of Washington fish ladder. Univ Washington, School Fish Tech Rep No 10, 11 pp
Fields PE, Finger GL, Adkins RJ (1955) The effect of electric lights upon the upstream passage of three species of adult salmon through the University of Washington fish ladder. Univ Washington, School Fish, Seattle Tech Rep No 12, 13 pp
Fields PE, Sainsbury JP, Kenoyer DD (1964) Effect of changing light intensity and spillway flow pattern upon adult salmon passage at the Dalles Dam. Res Fish Seattle 1963 Contr No 166:37–38
Fields RD, Lange GD (1980) Electroreception in the ratfish (*Hydrolagus collei*). Science 207:547–548
Fisknes B, Døving KB (1982) Olfactory sensitivity to group-specific substances in Atlantic salmon (*Salmo salar* L.). J Chem Ecol 8:1083–1092
Flock A (1971) The lateral line mechanoreceptors. In: Hoar WS, Randall DJ (eds) Fish physiology, vol V. Academic Press, London New York, pp 241–263
Foerster RE (1937a) The return from the sea of sockeye salmon (*Oncorhynchus nerka*) with special reference to percentage survival, sex proportions and progress. J Biol Bd Can 3:26–42
Foerster RE (1937b) The relation of temperature to the seaward migration of young sockeye salmon (*Oncorhynchus nerka*). J Biol Bd Can 3:421–438
Foerster RE (1968) The sockeye salmon, *Oncorhynchus nerka*. Fish Res Bd Can Bull 162
Forrester CR (1961) A note on a practical use of a salmon repellent. Can Fish Cult 30:61–62
Forselius S (1957) Studies of anabantid fishes. Zool Bid Uppsala 32:93–593
Forward RB, Horch KW, Waterman TH (1972) Visual orientation at the water surface by the teleost *Zenarchopterus*. Biol Bull Mar Biol Lab Woods Hole 143(1):112–126
Fraenkel GS, Gunn DL (1940) The orientation of animals. 1st edn. Oxford University Press, Oxford
Fraenkel GS, Gunn DL (1961) The orientation of animals. 2nd edn. Dover Publ, New York
Frank KT, Leggett WC (1981) Wind regulation of emergence times and early larval survival in capelin (*Mallotus villosus*). Can J Fish Aquat Sci 38:215–223
Frank KT, Leggett WC (1982) Coastal water mass replacement: its effect on zooplankton dynamics and the predator–prey complex associated with larval capelin (*Mallotus villosus*). Can J Fish Aquat Sci 39:991–1003

French RR, McAlister WB (1970) Winter distribution of salmon in relation to currents and water masses in the Northeastern Pacific Ocean and migrations of sockeye salmon. Trans Am Fish Soc 99:649–663

Frey DF, Miller RJ (1972) The establishment of dominance relationships in the blue gourami, *Trichogaster trichopterus* (Pallas). Behaviour 42:8–62

Fried SM, McCleave JD, LaBar GW (1978) Seaward migration of hatchery-reared Atlantic salmon, *Salmo salar,* smolts in the Penobscot River Estuary, Maine: riverine movements. J Fish Res Bd Can 35:76–87

Frisch von K (1911) Beiträge zur Physiologie der Pigmentzellen in der Fischhaut. Pflügers Arch Ges Physiol 138:319–387

Frisch von K (1923) Ein Zwergwels, der kommt, wenn man ihm pfeift. Biol Zentralbl 43:439–436

Frisch von K (1938) The sense of hearing in fish. Nature (London) 141:8–11

Frisch von K (1949) Die Polarisation des Himmelslichtes als orientierender Faktor bei den Tänzen der Bienen. Experientia 5:142–148

Frost WE (1950) The eel fisheries of the River Bann. J Cons Perm Int Explor Mer 16:358–383

Fry FEJ (1971) The effect of environmental factors on the physiology of fish. In: Hoar WS, Randall DJ (eds) Fish physiology, vol VI. Academic Press, London New York, pp 1–98

Fujihara MP, Hungate FP (1971) *Chondrococcus columnaris* disease of fishes: Influence of Columbia River fish ladders. J Fish Res Bd Can 28:533–536

Fujii T (1975) On the relation between the homing migration of the western Alaska sockeye salmon, *Oncorhynchus nerka* (Walbaum) and Oceanic conditions in the eastern Bering Sea. Mem Fac Fish Hokkaido Univ 22:99–192

Gabriel WL, Leggett WC, Carscadden JE, Glebe BD (1976) Origin and characteristics of "fall-run" American shad (*Alosa sapidissima*) from the St. John River, New Brunswick. J Fish Res Bd Can 33:1764–1770

Gallaway BJ, Strawn K (1974) Seasonal abundance and distribution of marine fishes at a hot-water discharge in Galveston Bay, Texas. Contr Sci, Univ Texas 18:71–137

Gee JH, Machniak K (1972) Ecological notes on a lake-dwelling population of longnose dace (*Rhinichthys cataractae*). J Fish Res Bd Can 29:330–332

Gerking SD (1953) Evidence for the concepts of home range and territory in stream fishes. Ecology 34:347–365

Gerking SD (1959) The restricted movements of fish populations. Biol Rev 34:221–242

Gibson RN (1965) Rhythmic activity in littoral fish. Nature (London) 207:544–545

Gibson RN (1967) Experiments on the tidal rhythm of *Blennius pholis*. J Mar Biol Assoc UK 47:97–111

Gibson RN (1971) Factors affecting the rhythmic activity of *Blennius pholis* L. (Teleostei). Anim Behav 19:336–343

Gibson RN (1973) Tidal and circadian activity rhythms in juvenile plaice, *Pleuronectes platessa*. Mar Biol 22:379–386

Gibson RN (1976) Comparative studies on the rhythms of juvenile flatfish. In: DeCoursey PJ (ed) Biological rhythms in the marine environment. Univ S Car Press

Gibson RN (1978) Lunar and tidal rhythms in fish. In: Thorpe E (ed) Rhythmic activity of fishes. Academic Press, London New York, pp 201–213

Gillet C, Billard R (1977) Stimulation of gonadotropin secretion in goldfish by elevation of rearing temperature. Ann Biol Anim Biochim Biophys 17:673–678

Ginetz RM, Larkin PA (1976) Factors affecting rainbow trout (*Salmo gairdneri*) predation in migrant fry of sockeye salmon (*Oncorhynchus nerka*). J Fish Res Bd Can 33:19–24

Glaser D (1966) Untersuchungen über die absoluten Geschmacksschwellen von Fischen. Z Vergl Physiol 52:1–25

Gleizer SI, Khodorkovsky VA (1971) An experimental determination of geomagnetic reception in the European eel. Rep Acad Sci USSR 201:964–967 (in Russian)

Godin J-GJ (1980a) Temporal aspects of juvenile pink salmon (*Oncorhynchus gorbuscha* Walbaum) emergence from a simulated gravel redd. Can J Zool 58:735–744

Godin J-GJ (1980b) Ontogenetic changes in the daily rhythms of swimming activity and of vertical distribution in juvenile pink salmon (*Oncorhynchus gorbuscha* Walbaum). Can J Zool 58:745–753

Godin J-GJ (1981) Circadian rhythm of swimming activity in juvenile pink salmon (*Oncorhynchus gorbuscha*). Mar Biol 64:341–349

Godin J-GJ (1982) Migrations of salmonid fishes during early life history phases: daily and annual timing. In: Brannon EL, Salo EO (eds) Proceedings: salmon and trout migratory behavior symposium. Univ Washington Press, Seattle, pp 22–50
Goff GP, Green JM (1978) Field studies of the sensory basis of homing and orientation to the home site in *Ulvaria subbifurcata* (Pisces: Stichaeidae). Can J Zool 56:2220–2224
Gooding RM, Magnuson JJ (1967) Ecological significance of a drifting object to pelagic fishes. Pac Sci 21:486–497
Goodyear CP (1970) Terrestrial and aquatic orientation in the starhead topminnow, *Fundulus notti*. Science 168:603–605
Goodyear CP (1973) Learned orientation in the predator avoidance behavior of the mosquitofish *Gambusia affinis*. Behaviour 45:191–224
Goodyear CP, Bennett DH (1979) Sun-compass orientation of immature bluegill. Trans Am Fish Soc 108:555–559
Goodyear CP, Ferguson DE (1969) Sun-compass orientation in the mosquitofish, *Gambusia affinis*. Anim Behav 17:636–640
Gould JL (1982) The map sense of pigeons. Nature (London) 296:205–211
Gould JL, Able KP (1981) Human homing: an elusive phenomenon. Science 212:1061–1063
Gould JL, Kirschvink JL, Deffeyes KS (1978) Bees have magnetic remanence. Science 201:1026–1028
Grau EG, Dickhoff WW, Nishioka RS, Bern HA, Folmar LC (1981) Lunar phasing of the thyroxine surge prepatory to seaward migration of salmonid fish. Science 211:607–609
Gray J (1937) Pseudorheotropism in fishes. J Exp Biol 14:95–103
Gray JAB, Denton EJ (1979) The mechanics of the clupeid acoustico-lateralis system: low frequency measurements. J Mar Biol Assoc UK 59:11–26
Gray RH, Haynes JM (1977) Depth distribution of adult chinook salmon (*Oncorhynchus tshawytscha*) in relation to season and gas-supersaturated water. Trans Am Fish Soc 106:617–620
Gray RH, Genoway RG, Barraclough SA (1977) Behaviour of juvenile chinook salmon (*Oncorhynchus tshawytscha*) in relation to simulated thermal effluent. Trans Am Fish Soc 106:367–370
Green JM (1971) Field and laboratory activity patterns of the tidepool cottid *Oligocottus maculosus* Girard. Can J Zool 49:255–264
Green JM, Fisher R (1977) A field study of homing and orientation to the home site in *Ulvaria subbifurcata* (Pisces: Stichaedae). Can J Zool 55:1551–1556
Greer Walker M, Mitson RB, Storeton-West T (1971) Trials with a transponding acoustic fish tag tracked with electronic sector scanning sonar. Nature (London) 229:196–198
Greer Walker M, Harden Jones FR, Arnold GP (1978) The movements of plaice *Pleuronectes platessa* tracked in the open sea. J Cons Perm Int Explor Mer 3:58–86
Gribanov VI (1948) Kizuch [*Oncorhynchus kisutch* (Walb)]:(biologichestii ocherk). (The coho salmon *Oncorhynchus kisutch* (Walbaum) – A biological sketch). Izv Tikhookean. Nauchn-issled. Inst Rybn Khoz Okeanogr 28:43–101. Fish Res Bd Can (Transl Ser 370)
Griffen DR (1964) Bird migration. Natural History Press, Garden City, NY, p 180
Groot C (1965) On the orientation of young sockeye salmon (*Oncorhynchus nerka*) during their seasonal migration out of lakes. Behav Suppl 14, p 198
Groot C (1972) Migration of yearling sockeye salmon (*Oncorhynchus nerka*) as determined by time-lapse photography of sonar observations. J Fish Res Bd Can 29:1431–1444
Groot C, Wiley WL (1965) Time lapse photography of an ASDIC echo-sounder PPI-scope as a technique for recording fish movements during migration. J Fish Res Bd Can 22:1025–1034
Groot C, Simpson K, Todd J, Murray PD, Buxton GA (1975) Movements of sockeye salmon (*Onchorhynchus nerka*) in the Skeena River estuary as revealed by ultrasonic tracking. J Fish Res Bd Can 32:233–242
Groot de SJ (1964) Diurnal activity and feeding habits of plaice. Rapp Proc Verb Cons Int Explor Mer 155:48
Gross MG, Barnes CA, Riel CK (1965) Radioactivity of the Columbia River effluent. Science 149:1078–1090
Groves AB, Collins GB, Trefethen PS (1968) Roles of olfaction and vision in choice of spawning site by homing adult chinook salmon (*Oncorhynchus tshawytscha*). J Fish Res Bd Can 25:867–876
Gunning GE (1959) The sensory basis for homing in the longear sunfish. Invest Indiana Lakes Streams 5:103–130

Gunning GE (1963) The concepts of home range and homing in stream fishes. Ergebn Biol 26:202–215

Gunning GE (1965) A behavioural analysis of the movement of tagged longear sunfish. Prog Fish Cult 27:211–215

Hain JHW (1975) The behaviour of migrating eels *Anguilla rostrata,* in response to current, salinity, and lunar period. Helgol Wiss Meeresunters 27:211–233

Hale EB (1956) Social facilities and forebrain function in maze performance of green sunfish, *Lepomis cyanellus*. Physiol Zool 29:93–127

Hallock RJ, Elwell RF, Fry DH (1970) Migrations of adult king salmon *Oncorhynchus tshawytscha* in the San Joaquin Delta as demonstrated by the use of sonic tags. Fish Bull Cal 151:1–92

Hammel HT, Stromme SB, Myhre K (1969) Forebrain temperature activates behavioral thermoregulatory response in Arctic sculpins. Science 165:83–85

Hanson CH, White JR, Li HW (1977) Entrapment and impingement of fishes by power plant cooling water intakes: an overview. Mar Fish Rev 39(10):7–17

Hara TJ (1967) Electrophysiological studies of the olfactory system of the goldfish, *Carrassius auratus* L. III. Effects of sex hormones on the electrical activity of the olfactory bulb. Comp Biochem Physiol 22:209–226

Hara TJ (1982) Chemoreception in fishes. Elsevier, Amsterdam, p 433

Hara TJ, Brown SB (1979) Olfactory bulbar electrical responses of rainbow trout (*Salmo gairdneri*) exposed to morpholine during smoltification. J Fish Res Bd Can 36:1186–1190

Hara TJ, Brown SB (1982) Reply. [to Cooper (1982)] Can J Fish Aquat Sci 39:1546–1548

Hara TJ, MacDonald S (1975) Morpholine as olfactory stimulus in fish. Science 187:81–82

Hara TJ, Ueda K, Gorbman A (1965) Electroencephalographic studies of homing salmon. Science 149:884–885

Harden Jones FR (1968) Fish migration. Arnold, London

Harden Jones FR (1977) Performance and behaviour on migration. In: Steele JH (ed) Fisheries mathematics. Academic Press, London New York, pp 145–170

Harden Jones FR, Arnold GP, Greer Walker M, Scholes P (1979) Selective tidal stream transport and the migration of plaice (*Pleuronectes platessa* L.) in the Southern North Sea. J Cons Perm Int Explor Mer 38:331–337

Hartman GF (1965) The role of behavior in the ecology and interaction of underyearling coho salmon (*Oncorhynchus kisutch*) and steelhead trout (*Salmo gairdneri*). J Fish Res Bd Can 22:1035–1081

Hartman WL, Strickland CW, Hoopes DT (1962) Survival and behavior of sockeye salmon fry migrating into Brooks Lake, Alaska. Trans Am Fish Soc 91:133–139

Hartman WL, Heard WR, Drucker B (1967) Migratory behavior of sockeye salmon fry and smolts. J Fish Res Bd Can 24:2069–2099

Hasler AD (1966) Underwater guideposts. Univ Wisc Press, Madison, 155 pp

Hasler AD (1971) Orientation and fish migration. In: Hoar WS, Randall DJ (eds) Fish physiology, vol VI. Academic Press, London New York, pp 429–510

Hasler AD, Scholz AT (1978) Olfactory imprinting in coho salmon (*Oncorhynchus kisutch*). In: Schmidt-Koenig K, Keeton WT (eds) Animal migration, navigation, and homing. Springer, Berlin Heidelberg New York, pp 356–369

Hasler AD, Scholz AT (1983) Olfactory imprinting and homing in salmon. Springer, Berlin Heidelberg New York

Hasler AD, Wisby W (1950) Use of fish for olfactory assay of pollutant phenols in water. Trans Am Fish Soc 79:64–70

Hasler AD, Wisby WJ (1951) Discrimination of stream odors by fishes and its relation to parent stream behavior. Am Nat 85:22–238

Hasler AD, Wisby W (1959) Repelling fish by treatment with potassium phenylacetate. US patent 2,880, 113, March 31, 1959

Hasler AD, Horrall RM, Wisby WJ, Braemer W (1958) Sun orientation and homing in fishes. Limnol Oceanogr 3:353–361

Hasler AD, Gardella EJ, Horrall RM, Henderson FF (1969) Open-water orientation of white bass, *Roccus chrysops*, as determined by ultrasonic tracking methods. J Fish Res Bd Can 26:2173–2192

Hasler AD, Scholz AT, Horrall RM (1978) Olfactory imprinting and homing in salmon. Am Sci 66:347–355

Haven DS (1957) Distribution, growth and availability of juvenile croaker, *Micropogon undulatus,* in Virginia. Ecology 38:88–97

Hawkins AD, Horner K (1981) Directional characteristics of primary auditory neurons from the cod ear. In: Tavolga WN, Popper AN, Fay RR (eds) Hearing and sound communication in fishes. Springer, Berlin Heidelberg New York, pp 311–328

Hayden P, Lindberg RG (1969) Circadian rhythm in mammalian body temperature entrained by cyclic pressure changes. Science 164:1288–1289

Hayes FR (1953) Artificial freshets and other factors controlling the ascent and population of Atlantic salmon in the La Have River, Nova Scotia. Fish Res Bd Can Bull 99, 47 pp

Haynes JM, Gray RH (1980) Influence of Little Goose Dam on upstream movements of adult chinook salmon, *Oncorhynchus tshawytscha.* Fish Bull 78:185–190

Healey MC (1967) Orientation of pink salmon (*Oncorhynchus gorbuscha*) during early marine migration from Bella Coola River system. J Fish Res Bd Can 24:2321–2338

Heard WR (1964) Phototactic behaviour of emerging sockeye salmon. Anim Behav 12:382–388

Heard WR, Crane RA (1976) Raising coho salmon from fry to smolts in estuarine pens, and returns of adults from two smolt releases. Prog Fish Cult 38:171–174

Heath WG (1963) Thermoperiodism in sea-run cutthroat trout (*Salmo clarki clarki*). Science 142:486–488

Heiligenberg W (1977) Principles of electrolocation and jamming avoidance in electric fish. Springer, Berlin Heidelberg New York, 85 pp

Hemmings CC (1966) Factors influencing the visibility of objects underwater. In: Bainbridge R, Evans GC, Rackham O (eds) Light as an ecological factor. Blackwell, Oxford, pp 359–374

Hidaka I (1970) The effect of carbon dioxide on carp palatal chemoreceptors. Bull Jpn Soc Sci Fish 36:1034–1039

Hinde RA (1965) Interaction of internal and external factors in integration of canary reproduction. In: Beach FA (ed) Sex and behavior. John Wiley, New York, pp 381–415

Hirsch PJ (1977) Conditioning of heart rate of coho salmon (*Oncorhynchus kisutch*) to odors. PhD Diss, Univ Wisc, Madison

Hiyama Y, Taniuchi T, Suyama K, Isoka K, Sato R, Kajihara T, Maiwa R (1967) A preliminary experiment on the return of tagged chum salmon to the Otsuchi River, Japan. Jpn Soc Sci Fish Bull 33:18–19

Hoar WS (1951) The behaviour of chum, pink, and coho salmon in relation to their seaward migration. J Fish Res Bd Can 8:241–263

Hoar WS (1953) Control and timing of fish migration. Biol Rev 28:437–452

Hoar WS (1955) Phototactic and pigmentary responses of sockeye salmon smolts following injury to the pineal organ. J Fish Res Bd Can 12:178–185

Hoar WS (1975) General and comparative physiology. Prentice-Hall Inc, Englewood Cliffs, NJ, 848 pp

Hoar WS (1976) Smolt transformation: evolution, behaviour, and physiology. J Fish Res Bd Can 33:1234–1252

Hoar WS, Bell GM (1950) The thyroid gland in relation to the seaward migration of Pacific salmon. Can J Res (D) 28:126–136

Hoar WS, MacKinnon D, Redlich A (1952) Effects of some hormones on the behaviour of salmon fry. Can J Zool 30:273–286

Hoar WS, Keenleyside MHA, Goodall RG (1955) The effect of thyroxine and gonadal steroids on the activity of salmon and goldfish. Can J Zool 33:428–439

Hoar WS, Keenleyside MHA, Goodall RG (1957) Reactions of juvenile Pacific salmon to light. J Fish Res Bd Can 14:815–830

Hochachka PW (1980) Living without oxygen: closed and open systems in hypoxia tolerance. Harvard Univ Press, Cambridge

Hodgson ES, Mathewson RF (1971) Chemosensory orientation in sharks. In: Adler H (ed) Orientation: sensory basis. Ann New York Acad Sci 188:175–181

Hoglund LB, Astrand M (1973) Preferences among juvenile char (*Salvelinus alpinus* L.) to intraspecific odours and water currents studied with the fluvarium technique. Rep Inst Freshw Res Drottningholm 53:21–30

Hoglund LB, Hardig J (1969) Reactions of young salmonids to sudden changes of pH, carbon dioxide tension and oxygen content. Rep Inst Freshw Res Drottningholm 49:76–119

Hokanson KEF, Kleiner CF, Thorslund TW (1977) Effects of constant temperatures and diel temperature fluctuations on specific growth and mortality rates and yield of juvenile rainbow trout, *Salmo gairdneri*. J Fish Res Bd Can 34:639–648
Holland GA (1957) Migration and growth of the dogfish shark, *Squalus acanthias* (L.) of the eastern North Pacific. Wash Dept Fish, Fish Res Pap 2:1–17
Holst von E (1950) Die Arbeitsweise des Statolithenapparates bei Fischen. Z Vergl Physiol 32:60–120
Hontela A, Peter RE (1978) Daily cycles in serum gonadotropin levels in the goldfish: effects of photoperiod, temperature, and sexual condition. Can J Zool 56:2430–2442
Hopkins CD (1973) Lightning as background noise for communication among electric fish. Nature (London) 242:268–270
Hopkins CD, Heiligenberg WF (1978) Evolutionary designs for electric signals and electroreceptors in gymnotid fishes of Surinam. Behav Ecol Sociobiol 3:113–134
Houde ED, Forney JL (1970) Effects of water currents on distribution of walleye larvae in Oneida Lake, NY. J Fish Res Bd Can 27:445–456
Houston AH (1957) Responses of juvenile chum, pink, and coho salmon to sharp sea-water gradients. Can J Zool 35:371–383
Hunter JR (1969) Communication of velocity changes in jack mackerel (*Trachurus symmetricus*) schools. Anim Behav 17:507–514
Huntsman AC (1948) Freshets and fish. Trans Am Fish Soc 75:257–266
Ichikawa M, Ueda K (1977) Fine structure of the olfactory epithelium in goldfish, *Carassius auratus*. A study of retrograde degeneration. Cell Tissue Res 183:445–455
Idler DR, Fagerlund U, Mayoh H (1956) Olfactory perception in migration salmon. I. L-serine, a salmon repellant in mammalian skin. J Gen Physiol 39:889–892
Idler DR, McBride JR, Jonas REE, Tomlinson N (1961) Olfactory perception in migrating salmon. II. Studies on a laboratory bioassay for homestream water and mammalian repellent. Can J Biochem Physiol 39:1575–1584
Iersel van JJA (1953) An analysis of the parental behaviour of the male three-spined stickleback (*Gasterosteus aculeatus* L.). Behav Suppl III: 159 pp
Iriki M, Seiko M, Nagai M, Tsuchiya K (1976) Effects of thermal stimulation to the spinal cord on the heart rate in cyprinid fishes. Comp Biochem Physiol A 53:61–63
Ivanoff A, Waterman TH (1957) Elliptical polarization of submarine illumination. J Mar Res 16:255–282
Jahn LA (1966) Open-water movements of the cutthroat trout (*Salmo clarki*) in Yellowstone Lake after displacement from spawning streams. J Fish Res Bd Can 23:1475–1485
Jahn LA (1969) Movements and homing of cutthroat trout (*Salmo clarki*) from open-water areas of Yellowstone Lake. J Fish Res Bd Can 26:1243–1261
Jenkins TM (1971) Role of social behaviour in dispersal of introduced rainbow trout (*Salmo gairdneri*). J Fish Res Bd Can 28:1019–1027
Jens G (1953) Über den lunaren Rhythmus der Blankaalwanderung. Arch J Fischereiwiss 4:94–110
Jensen AL, Duncan RN (1971) Homing of transplanted coho salmon. Prog Fish Cult 33:216–218
Johnsen PB, Hasler AD (1980) The use of chemical cues in the upstream migration of coho salmon *Oncorhynchus kisutch* Walbaum. J Fish Biol 17:67–73
Johnson DE, Fields PE (1959) The orientation of sexually mature male salmon to simulated star patterns. Coll Fish Univ Washington, Tech Rep 48, 10 pp
Johnson WE, Groot C (1963) Observations on the migration of young sockeye salmon (*Oncorhynchus nerka*) through a large, complex lake system. J Fish Res Bd Can 20:919–938
Johnston CE, Eales JG (1968) Influence of temperature and photoperiod on guanine and hypoxanthine levels in skin and scales of Atlantic salmon (*Salmo salar*) during parr-smolt transformation. J Fish Res Bd Can 25:1901–1909
Johnston CE, Eales JG (1970) Influence of body size on silvering of Atlantic salmon (*Salmo salar*) at parr-smolt transformation. J Fish Res Bd Can 27:983–987
Jones FRH (1956) The behaviour of minnows in relation to light intensity. J Exp Biol 37:271–281
Jones FRH (1969) Observations on the behaviour of fish made with the bifocal sector scanner. Proc 1967 FAO Conf Fish Behav
Jones JRE (1964) Fish and river pollution. Butterworths, London, 203 pp
Jorgensen JM, Flock A, Wersall J (1972) The Lorenzinian ampullae of *Polyodon spathula*. Z Zellforsch 130:362–377

Kalmijn AJ (1971) The electric sense of sharks and rays. J Exp Biol 55:371–383
Kalmijn AJ (1974) The detection of electric fields from inanimate and animate sources other than electric organs. In: Fessard A (ed) Handbook of sensory physiology, vol II, Pt 3: Electroreceptors and other specialized receptors. Springer, Berlin Heidelberg New York, pp 147–200
Kalmijn AJ (1978 a) Experimental evidence of geomagnetic orientation in elasmobranch fishes. In: Schmidt-Koenig K, Keeton WT (eds) Animal migration, navigation, and homing. Springer, Berlin Heidelberg New York, pp 347–353
Kalmijn AJ (1978 b) Electric and magnetic sensory world of sharks, skates, and rays. In: Hodgson ES, Mathewson RF (eds) Sensory biology of sharks, skates, and rays. Off Nav Res, Arlington, VA, pp 507–528
Kalmijn AJ (1979) Electromagnetic guidance systems in fishes. In: Tenforde TS (ed) Magnetic field effects on biological systems. Plenum, New York, pp 15–17
Kalmijn AJ (1982) Electrical and magnetic field detection in elasmobranch fishes. Science 218:916–918
Karlson P, Luscher M (1959) "Pheromones": a new term for a class of biologically active substances. Nature (London) 183:55–56
Karmanova IG, Belich AI, Lazarev SG (1981) An electrophysiological study of wakefulness and sleep-like states in fish and amphibians. In: Laming PR (ed) Brain mechanisms of behaviour in lower vertebrates. Cambridge Univ Press, Cambridge, pp 181–202
Katsuki Y (1973) The ionic receptive mechanism in the acoustico-lateralis system. In: Moller AR (ed) Basic mechanisms in hearing. Academic Press, London New York, pp 307–334
Katsuki Y, Yanagisawa K (1982) Chemoreception in the lateral-line organ. In: Hara T (ed) Chemoreception in fishes. Elsevier, Amsterdam, pp 227–242
Katsuki Y, Yanagisawa K, Tester K, Kendall JI (1969) Shark pit organ: responses to chemicals. Science 163:405–407
Katsuki Y, Hashimoto T, Kendall J (1971) The chemoreception in the lateral-line organs of teleosts. Jpn J Physiol 21:99–118
Kawamura G, Shibata A, Yonemori T (1981) Response of teleosts to the plane of polarized light as determined by the heart beat rate. Bull Jpn Soc Sci Fish 47:727–729
Keenleyside MHA, Hoar WS (1955) Effects of temperature on the responses of young salmon to water current. Behaviour 7:77–87
Keeton WT (1979) Avian orientation and navigation. Annu Rev Physiol 41:353–366
Kelso BW, Northcote TG, Wehrhahn CF (1981) Genetic and environmental aspects of the response to water current by rainbow trout (*Salmo gairdneri*) originating from inlet and outlet streams of two lakes. Can J Zool 59:2177–2185
Kerns OE (1961) Abundance and age of Kvichak River red salmon smolts. Fish Bull 61:301–320
Kholodov YA (1966) Vliyaniye electromagnitnykh i magnitnykh poley na tsentral'nuyu nervnuyu sistema. Moscow, Nauka Press
Khoo HW (1974) Sensory basis of homing in the intertidal fish *Oligocottus maculosus* Girard. Can J Zool 52:1023–1029
Kirschvink JL, Gould JL (1981) Biogenic magnetite as a basis for magnetic field detection in animals. Biosystems 13:181–201
Kissil GW (1974) Spawning of the anadromous alewife, *Alosa pseudoharengus,* in Bride Lake, Connecticut. Trans Am Fish Soc 103:312–317
Kleerekoper H (1969) Olfaction in fishes. Indiana Univ Press, Bloomington, p 222
Kleerekoper H (1972) Orientation through chemoreception in fishes. In: Galler SK, Schmidt-Koenig K, Jacobs GJ, Belleville RE (eds) Animal orientation and navigation. NASA SP-262:459–468
Kleerekoper H, Timms AM, Westlake GF, Davy FB, Marlar T, Anderson VM (1969) Inertial guidance system in the orientation of the goldfish (*Carassius auratus*). Nature (London) 223:501–502
Kleerekoper H, Timms AM, Westlake GF, Davy FB, Marlar T, Anderson VM (1970) An analysis of locomotor behaviour of goldfish (*Carassius auratus*). Anim Behav 18:317–330
Kleerekoper H, Matis JH, Timms AM, Gensler P (1973) Locomotor response of the goldfish to polarized light and its e-vector. J Comp Physiol 86:27–36
Klima EF, Wickham DA (1971) Attraction of coastal pelagic fishes with artificial structures. Trans Am Fish Soc 100:86–99
Kreithen ML, Keeton WT (1974) Detection of changes in atmospheric pressure by the homing pigeon, *Columba livia*. J Comp Physiol 89:73–82

Kristensen I (1963) Hypersaline bays as an environment of young fish. Proc Gulf Carib Fish Inst, 16th Annu Session 139–142

Krogius PV (1954) The relation of the upstream migration run of sockeye and seaward migrations of the young to daily trends in temperature, pH, and content of dissolved gases. Izv Tikhookean, nauchno-issled, Inst ryb Khoz Okeanogr 45:201–202 (Fish Res Bd Can Transl 197, 1959)

LaBar GW (1971) Movement and homing of cutthroat trout (*Salmo clarki*) in Clear and Bridge Creeks, Yellowstone National Park. Trans Am Fish Soc 100:41–49

LaBar GW, McCleave JD, Fried SM (1978) Seaward migration of hatchery-reared Atlantic salmon (*Salmo salar*) smolts in the Penobscot River estuary, Maine: open-water movement. J Cons Perm Int Explor Mer 38:257–269

Lam TJ, Hoar WS (1967) Seasonal effects of prolactin on freshwater osmoregulation of the marine form (*trachurus*) of the stickleback *Gasterosteus aculeatus*. Can J Zool 45:509–516

Larkin PA, Walton A (1969) Fish school size and migration. J Fish Res Bd Can 26:1372–1374

Leask MJM (1977) Primitive models of magnetoreception. In: Schmidt-Koenig K, Keeton WI (eds) Animal migration navigation and homing. Springer, Berlin Heidelberg New York, pp 318–322

Leggett WC (1973) The migrations of shad. Sci Amer 228:92–98

Leggett WC (1976) The American shad (*Alosa sapidissima*) with special reference to its migration and population dynamics in the Connecticut River. In: Merriman D, Thorpe LM (eds) The Connecticut River ecological study: The impact of a nuclear power plant. Am Fish Soc Monogr No 1, pp 169–225

Leggett WC (1977) The ecology of fish migrations. Annu Rev Ecol Syst 8:285–308

Leggett WC, Carscadden JE (1978) Latitudinal variation in reproductive characteristics of american shad (*Alosa sapidissima*): evidence for population specific life history strategies in fish. J Fish Res Bd Can 35:1469–1478

Leggett WC, Trump CL (1977) Energetics of migration in American shad. In: Schmidt-Koenig K, Keeton WT (eds) Animal migration, navigation, and homing. Springer, Berlin Heidelberg New York, pp 370–377

Leggett WC, Whitney RR (1972) Water temperature and the migration of American shad. Fish Bull 70:659–670

Lehrman DS (1965) Interaction between internal and external environments in the regulation of the reproductive behaviour of the ring dove. In: Beach FA (ed) Sex and behavior. John Wiley, New York, pp 355–380

Liebelt JE (1969) A serological study of cutthroat trout (*Salmo clarki lewisi*) from tributaries and the outlet of Yellowstone Lake. Proc Mont Acad Sci 29:31–39

Liermann K (1933) Über den Bau des Geruchsorgans der Teleostiers. Z Anat Entwicklungsgesch 100:1–39

Liley NR (1982) Chemical communication in fish. Can J Fish Aquat Sci 39:22–35

Lindauer M, Martin H (1972) Magnetic effects on dancing bees. In: Galler SR, Schmidt-Koenig K, Jacobs GJ, Belleville RE (eds) Animal orientation and navigation. Symp, Washington, NASA, pp 559–567

Lindsey CC, Northcote TG (1963) Life history of redside shiners, *Richardsonias balteatus*, with particular reference to movements in and out of Sixteen Mile Lake streams. J Fish Res Bd Can 20:1001–1030

Lissmann HW (1958) On the function and evolution of electric organs in fish. J Exp Biol 35:156–191

Lissmann HW, Machim KE (1963) Electric receptors in non-electric fish (*Clarias*). Nature (London) 199:88–89

Locket NA (1977) Adaptations to the deep sea environment. In: Crescitelli F (ed) Handbook of sensory physiology, vol VII/5. Springer, Berlin Heidelberg New York, pp 67–192

Long CW (1968) Diel movement and vertical distribution of juvenile anadromous fish in turbine intakes. Fish Bull 66:599–609

Lorz HW, Northcote TG (1965) Factors affecting stream location and timing and intensity of entry of spawning kokanee (*Oncorhynchus nerka*) into an inlet of Nicola Lake, BC. J Fish Res Bd Can 23:665–687

Lowe RH (1952) The influence of light and other factors on the seaward migration of the silver eel *Anguilla anguilla* L. J Anim Ecol 21:275–309

Lowenstein O (1971) The labyrinth. In: Hoar WS, Randall DJ (eds) Fish physiology, vol V. Academic Press, London New York, pp 207–240

Loyacano HA, Chapell JA, Gauthreaux A (1977) Sun-compass orientation in juvenile large-mouth bass *Micropterus salmoides*. Trans Am Fish Soc 106:77–79

Lund WA Jr, Maltezos GC (1970) Movements and migration of the bluefish, *Pomatomus saltatrix*, tagged in waters of New York and Southern New England. Trans Am Fish Soc 99:719–725

Lyon EP (1904) On rheotropism I. rheotropism in fishes. Am J Physiol 12:149–161

Lythgoe JN (1966) Visual pigments and underwater vision. In: Bainbridge R, Evans GC, Rackham O (eds) Light as an ecological factor. Blackwell, Oxford, pp 375–391

Lythgoe JN (1979a) Vision in fishes: ecological adaptations. In: Ali MA (ed) Environmental physiology of fishes. Plenum, New York, pp 431–445

Lythgoe JN (1979b) The ecology of vision. Clarendon, Oxford Univ Press, New York

Lythgoe JN, Hemmings CC (1967) Polarized light and underwater vision. Nature (London) 213:893

MacClean JA, Gee JH (1971) Effects of temperature on movements of prespawning brook sticklebacks, *Culaea inconstans*, in the Roseau River, Manitoba. J Fish Res Bd Can 28:919–923

Machniak K, Gee JH (1975) Adjustment of buoyancy by tadpole madtom; *Noturus gyrinus* and black bullhead, *Ictalurus melas*, in response to a change in water velocity. J Fish Res Bd Can 32:303–307

MacKinnon D, Brett JR (1955) Some observations on the movement of Pacific salmon fry through a small inpounded water basin. J Fish Res Bd Can 12:362–368

Madison DM, Horrall RM, Stasko AB, Hasler AD (1972) Migratory movements of adult sockeye salmon (*Oncorhynchus nerka*) in coastal British Columbia as revealed by ultrasonic tracking methods. J Fish Res Bd Can 29:1025–1033

Madison DM, Scholz AT, Cooper JC, Horrall RM (1973) Olfactory hypotheses and salmon migration: a synopsis of recent findings. Tech Rep Fish Res Bd Can 414:1–35

Magnuson JJ, Crowder LB, Medvick PA (1979) Temperature as an ecological resource. Am Zool 19:331–343

Mahnken C, Joyner T (1973) Salmon for New England fisheries. III. Developing a coastal fishery for Pacific salmon. Mar Fish Rev 35:9–13

Manzer JL (1964) Preliminary observations on the vertical distribution of Pacific salmon (Genus *Oncorhynchus*) in the Gulf of Alaska. J Fish Res Bd Can 21:891–903

Marcström A (1959) Reaction thresholds of roaches (*Leuciscus rutilus* L.) to some aromatic substances. Arkiv Zool (Ser 2) 12:335–338

Marcy BC, Beck AD, Vlanowicz RE (1978) Effects and impacts of physical stress on entrained organisms. In: Schubel JR, Marcy BC (eds) Power plant entrainment a biological assessment. Academic Press, London New York, pp 135–188

Marquette WM, Long CW (1971) Laboratory studies of screens for diverting juvenile salmon and trout from turbine intakes. Trans Am Fish Soc 100:439–447

Marshall NB (1966) The life of fishes. World Publ Co, New York

Mason JC (1969) Hypoxial stress prior to emergence and competition among coho salmon fry. J Fish Res Bd Can 26:63–91

Mason JC (1974) Movements of fish populations in Lymn Creek, Vancouver Island: a summary from weir operations during 1971 and 1972, including comments on species life histories. Tech Rep Environ Can Fish Mar Serv No 483, 35 pp

Mason JC (1975) Seaward movement of juvenile fishes including lunar periodicity in the movement of coho salmon (*Oncorhynchus kisutch*) fry. J Fish Res Bd Can 32:2542–2547

Mason JC (1976) Response of underyearling coho salmon to supplemental feeding in a natural stream. J Wildl Manage 40:775–788

Mason JC, Chapman DW (1965) Significance of early emergence, environmental rearing capacity and behavioural ecology of juvenile coho salmon in stream channels. J Fish Res Bd Can 22:173–190

Matty AJ (1978) Pineal and some pituitary hormone rhythms in fish. In: Thorpe J (ed) Rhythmic activity in fishes. Academic Press, London New York, pp 21–30

McBride JR, Idler DR, Jones EE, Tomlinson N (1962) Olfactory perception in juvenile salmon. I. Observations on response of juvenile sockeye to extracts of food. J Fish Res Bd Can 19:327–334

McCart P (1967) Behaviour and ecology of sockeye salmon fry in the Babine River. J Fish Res Bd Can 24:375–428

McCauley RW (1977) Laboratory methods for determining temperature preference. J Fish Res Bd Can 34:749–752

McCleave JD (1967) Homing and orientation of cutthroat trout (*Salmo clarki*) in Yellowstone Lake, with special reference to olfaction and vision. J Fish Res Bd Can 24:2011–2044
McCleave JD (1978) Rhythmic aspects of estuarine migration of hatchery-reared Atlantic salmon (*Salmo salar*) smolts. J Fish Biol 12:559–570
McCleave JD, Horrall RM (1970) Ultrasonic tracking of homing cutthroat trout (*Salmo clarki*) in Yellowstone Lake. J Fish Res Bd Can 27:715–730
McCleave JD, Kleckner RC (1982) Selective tidal stream transport in the estuarine migration of glass eels of the American eel (*Anguilla rostrata*). J Cons Int Explor Mer 40:262–271
McCleave JD, LaBar GW (1972) Further ultrasonic tracking and tagging studies of homing cutthroat trout (*Salmo clarki*) in Yellowstone Lake. Trans Am Fish Soc 101:44–54
McCleave JD, Power JH (1978) Influence of weak electric and magnetic fields on turning behaviour in elvers of the American eel *Anguilla rostrata*. Mar Biol 46:29–34
McCleave JD, LaBar GW, Kirchus FW (1977) Within season homing movements of displaced mature Sunapee trout (*Salvelinus alpinus*) in Floods Pond, Maine. Trans Am Fish Soc 106:156–162
McCracken FD (1963) Seasonal movements of the winter flounder, *Pseudopleuronectes americanus* (Walbaum) on the Atlantic coast. J Fish Res Bd Can 20:551–586
McCutcheon FH (1966) Pressure sensitivity, reflexes, and buoyancy responses in teleosts. Anim Behav 14:204–217
McDonald J (1960) The behaviour of Pacific salmon fry during their downstream migration to freshwater and saltwater nursery areas. J Fish Res Bd Can 17:665–685
McInerney JE (1964) Salinity preference: an orientation mechanism in salmon migration. J Fish Res Bd Can 21:995–1018
Meehan WR (1966) Growth and survival of sockeye salmon introduced into Ruth Lake after removal of resident fish populations. US Fish Wildl Serv Spec Rep Fish No 532, 18 pp
Meehan WR, Siniff DB (1962) A study of the downstream migrations of anadromous fishes in the Taku River, Alaska. Trans Am Fish Soc 91:399–411
Meyer PF, Kuhl H (1953) Welche Milieu-Faktoren spielen beim Glasaalaufstieg eine Rolle? Arch Fischereiwiss 4:87–94
Mighell J (1975) Some observations of imprinting of juvenile salmon in fresh and saltwater. 1975 Symposium on Salmon Homing (summary notes). Nat Mar Fish Serv Seattle Washington
Miles SG (1968) Rheotaxis of elvers of the American eel (*Anguilla rostrata*) in the laboratory to water from different streams in N.S. J Fish Res Bd Can 25:1591–1602
Miller RJ, Brannon EL (1982) The origin and development of life history patterns in Pacific salmonids. In: Brannon EL, Salo EO (eds) Salmon and trout migratory behavior symposium. School Fish, Univ Washington, Seattle, pp 296–309
Moller P (1976) Electric signals and schooling behavior in a weakly electric fish, *Marusenius cyprinoides* L. (Mormyriformes). Science 193:697–699
Moore HL, Newman HW (1956) Effects of sound waves on young salmon. US Fish Wildf Serv, Spec Sci Rep Fish 172:1–19
Morgan RP, Carpenter EJ (1978) Biocides. In: Schabel JR, Marcy BC (eds) Power plant entrainment: a biological assessment. Academic Press, London New York, pp 95–134
Morris RW, Kittleman LR (1967) Piezoelectric property of otoliths. Science 158:368–370
Moss SA (1970) The response of young American shad to rapid temperature changes. Trans Am Fish Soc 99:381–384
Moulton JM (1960) Swimming sounds and the schooling of fishes. Biol Bull 119:210–223
Müller K (1978a) Locomotor activity of fish and environmental oscillations. In: Thorpe JE (ed) Rhythmic activity of fishes. Academic Press, London New York, pp 1–19
Müller K (1978b) Locomotor activity in whitefish shoals. In: Thorpe JE (ed) Rhythmic activity of fishes. Academic Press, London New York, pp 225–233
Mundie JH (1969) Ecological implications of the diet of juvenile coho in streams. In: Northcote TG (ed) Symposium on salmon and trout in streams. HR MacMillan lectures in fisheries. Inst Fish, UBC, Vancouver, pp 135–152
Munroe WR, Balmain KH (1956) Observations on the spawning runs of brown trout in the South Queich, Loch Leven. Sci Invest Freshwat Fish Scot 13:1–17
Munz FW (1971) Vision: visual pigments. In: Hoar WS, Randall DJ (eds) Fish physiology, vol V. Academic Press, London New York, pp 1–32

Munz FW, McFarland WN (1977) Evolutionary adaptations of fishes to the photic environment. In: Crescitelli F (ed) Handbook of sensory physiology, vol VII/5. Springer, Berlin Heidelberg New york, pp 193–274

Murray RW (1971) Temperature receptors. In: Hoar WS, Randall DJ (eds) Fish physiology, vol V. Academic Press, London New York, pp 121–133

Musik JA, Mercer LP (1977) Seasonal distribution of Black Sea bass *Centropristis striata* in the mid-Atlantic bight with comments on the ecology and fisheries of the species. Trans Am Fish Soc 106:12–25

Neave F (1949) Game fish populations of the Cowichan River. Bull Fish Res Bd Can 84:1–32

Neave F (1955) Notes on the seaward migration of pink and chum salmon fry. J Fish Res Bd Can 12:369–374

Neill WH (1979) Mechanisms of fish distribution in heterothermal environments. Am Zool 19:305–317

Neill WH, Magnuson JJ, Chipman GD (1972) Behavioural thermoregulation by fishes: a new experimental approach. Science 176:1443–1445

Neurath W (1949) Über die Leistung des Geruchssinnes bei Elritzen. Z Vergl Physiol 31:609–626

Neves RJ, Depres L (1979) The oceanic migration of American shad, *Alosa sapidissima*, along the Atlantic coast. Fish Bull 77:199–212

Newcombe C, Hartman G (1973) Some chemical signals in the spawning behaviour of rainbow trout (*Salmo gairdneri*). J Fish Res Bd Can 30:995–997

Nisbet ICT (1955) Atmospheric turbulence in bird flight. Br Birds 48:557–559

Nordeng H (1971) Is the local orientation of anadromous fishes determined by pheromones? Nature (London) 233:411–413

Nordeng H (1977) A pheromone hypothesis for homeward migration in anadromous salmonids. Oikos 28:155–159

Norris KS (1963) The functions of temperature in the ecology of the percoid fish (*Girella nigricans* Ayres). Ecol Monogr 33:23–62

Northcote TG (1958) Effect of photoperiodism on response of juvenile trout to water currents. Nature (London) 181:1283–1284

Northcote TG (1962) Migratory behaviour of juvenile rainbow trout, *Salmo gairdneri*, in outlet and inlet streams of Loon Lake, British Columbia. J Fish Res Bd Can 19:201–270

Northcote TG (1969a) Lakeward migration of young rainbow trout (*Salmo gairdneri*) in the upper Lardeau River, BC. J Fish Res Bd Can 26:33–45

Northcote TG (1969b) Patterns and mechanisms in the lakeward migratory behaviour of juvenile trout. In: Northcote TG (ed) Symposium on salmon and trout in streams, HR MacMillan lectures in fisheries. Inst Fish, UBC, Vancouver, pp 183–203

Northcote TG (1978) Chapter 13: migratory strategies and production in freshwater fishes. In: Gerking SD (ed) Ecology of freshwater fish production. Wiley, New York, pp 326–359

Northcote TG (1981) Juvenile current response, growth, and maturity of above and below waterfall stocks of rainbow trout, *Salmo gairdneri*. J Fish Res Bd Can 38:741–751

Northcote TG, Kelso BW (1981) Differential response to water current by two homozygous LDH phenotypes of young rainbow trout (*Salmo gairdneri*). Can J Fish Aquat Sci 38:348–352

Novotny AJ (1980) Delayed release of salmon. In: Thorpe JE (ed) Salmon ranching. Academic Press, London New York, pp 325–369

Nyman L (1972) Some effects of temperature on eel (*Anguilla*) behaviour. Rep Inst Freshw Res Drottningholm 52:90–102

Obara S (1976) Mechanism of electroreception in ampullae of Lorenzini of the marine catfish, *Plotosus*. In: Reuben JP, Purpura DP, Bennett MLV, Kendel ER (eds) Electrobiology of nerve synapse and muscle. Raven Press, New York, pp 129–147

O'Connor JM (1972) Tidal activity rhythm in the hogchoker *Trinectes maculatus* (Bloch and Schneider). J Exp Mar Biol Ecol 9:173–177

O'Leary DP, Vilches-Troya J, Dunn RF, Campos-Munoz A (1981) Magnets in guitarfish vestibular receptors. Experienta (Basel) 37:86–88

Olla BL, Studholme AL (1978) Comparative aspects of the activity rhythms of tautog, *Tautoga onitis*, bluefish *Pomatomus saltatrix*, and Atlantic mackerel, *Scomber scombrus*, as related to their life habits. In: Thorpe JE (ed) Rhythmic activity of fishes. Academic Press, London New York, pp 131–151

Olla BL, Studholme AL, Bejda AT, Samet C (1980) The role of temperature in triggering migratory behavior of the adult tautog *Tautoga onitis*. Mar Biol 59:23–30
Orians GH (1962) Natural selection and ecological theory. Am Nat 96:257–264
Osborne MFM (1961) The hydrodynamical performance of migratory salmon. J Exp Biol 38:365–390
Oshima K, Gorbman A (1966a) Olfactory responses in the forebrain of goldfish and their modification by thyroxine treatment. Gen Comp Endocrinol 7:398–409
Oshima K, Gorbman A (1966b) Influence of thyroxine and steroid hormones on spontaneous and evoked unitary activity in the olfactory bulb of goldfish. Gen Comp Endocrinol 7:482–491
Oshima K, Gorbman A (1968) Modification by sex hormones of the spontaneous and evoked bulbar electrical activity in goldfish. J Endocrinol 40:409–420
Oshima K, Gorbman A (1969) Effect of estradiol on NaCl-evoked olfactory bulbar potentials in goldfish: Dose response relationships. Gen Comp Endocrinol 13:92–97
Oshima K, Hahn WE, Gorbman A (1969a) Olfactory discrimination of natural waters by salmon. J Fish Res Bd Can 26:2111–2121
Oshima K, Hahn WE, Gorbman A (1969b) Electroencephalographic olfactory responses in adult salmon to waters traversed in the homing migration. J Fish Res Bd Can 26:2123–2133
Otto RG (1971) Effects of salinity on the survival and growth of pre-smolt coho salmon (*Oncorhynchus kisutch*). J Fish Res Bd Can 28:343–349
Otto RG, McInerney JE (1970) The development of salinity preference in pre-smolt coho salmon, *Oncorhynchus kisutch*. J Fish Res Bd Can 27:783–800
Palmer JD (1973) Tidal rhythms: the clock control of the rhythmic physiology of marine organisms. Biol Rev 48:377–418
Pankhurst NW (1982) Relation of visual changes to the onset of sexual maturation in the European eel *Anguilla anguilla* (L.). J Fish Biol 21:127–140
Papi F (1960) Orientation at night: the moon. Cold Spring Harbor Symp Quant Biol 25:475–480
Parker GH (1903) The sense of hearing in fishes. Am Nat 37:185–204
Parker GH (1912) The relation of smell, taste, and the common chemical sense in vertebrates. Proc Natl Acad Sci USA 15:219–234
Parker GH, Heusen van AP (1917) The response of the catfish, *Amiurus nebulosus* to metallic and nonmetallic rods. Am J Physiol 44:405–420
Parker RA, Hasler AD (1959) Movements of some displaced centrarchids. Copeia 1959:11–18
Parker RO, Stone RB, Buchanan CC, Steimle FW Jr (1974) How to build marine artificial reefs. US Dept of Comm NOAA/NMFS Fish Facts 10:47 pp
Parker RO, Stone RB, Buchanan CC (1979) Artificial reefs off Murrells Inlet, South Carolina. Mar Fish Rev 41(9):12–24
Pesaro M, Balsamo M, Gandolfi G, Tongiorgi P (1981) Olfactory discrimination in juvenile eels. Monit Zool Ital 15:323
Peter RE, Crim LW (1979) Reproductive endocrinology of fishes: gonadal cycles and gonadotropin in teleosts. Annu Rev Physiol 41:323–335
Peter RE, Hontela A (1978) Annual gonadal cycles in teleosts: Environmental factors and gonadotropin levels in the blood. In: Farner DS, Assenmacher I (eds) Environmental endocrinology. Springer, Berlin Heidelberg New York, pp 20–25
Peterman RM, Gatto M (1978) Estimation of functional responses of predators on juvenile salmon. J Fish Res Bd Can 35:797–808
Peters M (1971) Sensory mechanisms of homing in salmonids: a comment. Behaviour 39:18–19
Peters RC, Bretschneider F (1972) Electric phenomena in the habitat of the catfish *Ictulurus nebulosus* L. J Comp Physiol 81:345–362
Peters RC, Wijland van F (1974) Electro-orientation in the passive electric catfish *Ictalurus nebulosus* Les. J Comp Physiol 92:273–280
Peterson DA (1972) Barometric pressure and its effect on spawning activities of rainbow trout. Progr Fish Cult 34:110–112
Pettersson O (1926) Currents and fish migration in the transition area. J Cons Perm Int Explor Mer 1:322–326
Pfeiffer W (1960) Über die Schreckreaktion bei Fischen und die Herkunft des Schreckstoffes. Z Vergl Physiol 43:578–614
Pfeiffer W (1964) The morphology of the olfactory organ of *Hoplopagrus guentheri* Gill 1862. Can J Zool 42:235–237

Pfeiffer W (1968) Die Fahrenholzschen Organe der Dipnoi und Brachiopterygii. Z Zellforsch 90:127–147
Pfeiffer W (1982) Chemical Signals in communication. In: Hara TJ (ed) Chemoreception in fishes. Elsevier, Amsterdam, pp 307–326
Pfleiger WL (1975) Fishes of Missouri. Missouri Dept Conserv, 343 pp
Phillips JB (1977) Use of the earth's magnetic field by orienting cave salamanders (*Eurycea lucifuga*). J Comp Physiol 121:273–288
Pitcher TJ, Magurran AE, Winfield IJ (1982) Fish in larger schools find food faster. Behav Ecol Sociobiol 10:149–151
Poddubnyi AG (1969) Sonic tags and floats as a means of studying fish response to natural environmental changes and to fishing gear. FAO Fish Rep 62:793–801
Popper AN, Coombs S (1980) Auditory mechanisms in teleost fishes. Am Sci 68:429–440
Powers EB, Clark RT (1943) Further evidence on chemical factors affecting the migratory movements of fishes, especially the salmon. Ecology 24:109–113
Pressley PH (1980) Lunar periodicity in the spawning of yellowtail damselfish, *Microspathodon chrysurus*. Environ Biol Fish 5:155–159
Prest VK (1969) Retreat of Wisconsin and recent ice in North America. Geol Surv Can Map 1257A
Presti D, Pettigrew JD (1980) Ferromagnetic coupling to muscle receptors as a basis for geomagnetic field sensitivity in animals. Nature (London) 285:99–101
Pumphrey RJ (1950) Hearing. Symp Soc Exp Biol 4:3–18
Qasim SZ, Rice AL, Knight-Jones EW (1963) Sensitivity to pressure changes in teleosts lacking swimbladders. J Mar Biol Assoc India 5:289–293
Quinn TP (1980) Evidence for celestial and magnetic-compass orientation in lake migrating sockeye salmon fry. J Comp Physiol 137:243–248
Quinn TP (1982a) Intraspecific differences in sockeye salmon fry compass orientation mechanisms. In: Brannon EL, Salo EO (eds) Salmon and trout migratory behavior symposium. School Fish, Univ Washington, Seattle, pp 79–85
Quinn TP (1982b) A model for salmon navigation on the high seas. In: Brannon EL, Salo EO (eds) Salmon and trout migratory behavior symposium. School Fish, Univ Washington, Seattle, pp 229–237
Quinn TP, Brannon EL (1982) The use of celestial and magnetic cues by orienting sockeye salmon smolts. J Comp Physiol 147A:547–552
Quinn TP, Groot C (1983) Orientation of chum salmon (*Oncorhynchus keta*) after internal and external magnetic field alteration. Can J Fish Aquat Sci 40:1598–1606
Quinn TP, Merrill RT, Brannon EL (1981) Magnetic field detection in sockeye salmon *Oncorhynchus nerka*. J Exp Zool 217:137–142
Quinn TP, Brannon EL, Whitman JRP (1983) Pheromones and the water source preferences of adult coho salmon *Oncorhynchus kisutch* Walbaum. J Fish Biol 22:677–684
Qutob Z (1960) Pressure perception in Ostariophysii. Experientia 16:426
Qutob Z (1962) The swimbladder of fishes as a pressure receptor. Arch Neerl Zool 15:1–67
Raleigh RF (1967) Genetic control in the lakeward migrations of sockeye salmon (*Oncorhynchus nerka*) fry. J Fish Res Bd Can 24:2613–2622
Raleigh RF (1971) Innate control of migrations of salmon and trout fry from natal gravels to rearing areas. Ecology 52:291–297
Raleigh RF, Chapman DW (1971) Genetic control in lakeward migrations of cutthroat fry. Trans Am Fish Soc 100:33–40
Ramster JW, Medler KJ (1978) Notes on the tidal streams off Orfordness in relation to fish tracking. J Cons Perm Int Explor Mer 38:87–91
Randolph KN, Clemens HP (1976) Some factors influencing the feeding behavior of channel catfish in culture ponds. Trans Am Fish Soc 105:718–724
Raney EC, Menzel BW (1969) Heated effluents and effects on aquatic life, with emphasis on fishes. A bibliography. Cornell Univ Water Res Mar Sci Center. Phil Electr Comp Ichythyol Assoc Bull 2:470 pp
Rawson DS (1957) The life history and ecology of the yellow walleye, *Stizostedion vitreum*, in Lac La Ronge, Saskatchewan. Trans Am Fish Soc 86:15–37
Raymond HL (1968) Migration rates of yearling chinook salmon in relation to flows and impoundments in the Columbia and Snake Rivers. Trans Am Fish Soc 97:356–359

Raymond HL (1969) Effect of John Day reservoir on the migration rate of juvenile chinook salmon in the Columbia River. Trans Am Fish Soc 98:513–514

Regnart HC (1931) The lower limits of perception of electrical currents by fish. J Mar Biol Assoc UK 17:415–420

Reisman HM (1968) Effects of social stimuli on the secondary sex characters of male three-spined sticklebacks, *Gasterosteus aculeatus*. Copeia 1968:816–826

Reynolds WW, Casterlin ME (1976a) Thermal preferenda and behavioral thermoregulation in three centrarchid fishes. In: Thermal ecology, vol II. Tech Inf Serv Springfield VA, pp 185–190

Reynolds WW, Casterlin ME (1976b) Locomotor activity rhythms in the bluegill sunfish, *Lepomis macrochirus*. Am Midl Nat 96:221–225

Reynolds WW, Casterlin ME (1978a) Behavioral thermoregulation in the rock bass *Ambloplites rupestris*. Comp Biochem Physiol 60A:263–264

Reynolds WW, Casterlin ME (1978b) Ontogenetic change in preferred temperature and diel activity of the yellow bullhead, *Ictalurus natalis*. Comp Biochem Physiol 59A:409–411

Reynolds WW, Casterlin ME (1979) Effect of temperature on locomotor activity in the goldfish (*Carassius auratus*) and the bluegill (*Lepomis macrochirus*) presence of an "activity well" in the region of the final preferendum. Hydrobiologia 65:3–5

Reynolds WW, Casterlin ME, Millington ST (1978) Circadian rhythm of preferred temperature in the bowfin *Amia calva*, a primitive holostean fish. Comp Biochem Physiol 60A:107–109

Rice AL (1964) Observations on the effects of changes of hydrostatic pressure on the behaviour of some marine animals. J Mar Biol Assoc UK 44:163–175

Rich WH, Holmes HB (1929) Experiments on marking young chinook salmon on the Columbia River, 1916 to 1922. Bull US Bur Fish 44:215–264

Richards FA (1975) The Cariaco Basin. Oceanogr Mar Biol Rev 13:11–67

Richkus WA (1974) Factors influencing the seasonal and daily patterns of alewife migration in a Rhode Island river. J Fish Res Bd Can 31:1485–1497

Richkus WA (1975a) Migratory behaviour and growth of juvenile anadromous alewives, *Alosa pseudoharengus*, in a Rhode Island drainage. Trans Am Fish Soc 104:483–493

Richkus WA (1975b) The response of juvenile alewives to water currents in an experimental chamber. Trans Am Fish Soc 104:494–498

Ricker WE (1972) Hereditary and environmental factors affecting certain salmonid populations. In: Simon RC, Larkin PA (eds) The stock concept in Pacific salmon. HR MacMillan Lectures in Fisheries, Seattle WA, April 1970. Univ BC, Vancouver, BC, pp 19–160

Riddell BE, Leggett WC (1981) Evidence of an adaptive basis for geographic variation in body morphology and time of downstream migration in juvenile Atlantic salmon (*Salmo salar*). Can J Fish Aquat Sci 38:308–320

Riddell BE, Leggett WC, Saunders RL (1981) Evidence of adaptive polygenic variation between two populations of Atlantic salmon (*Salmo solar*) native to tributaries of the SW Miramichi River, NB. Can J Fish Aquat Sci 38:321–333

Riley JD (1973) Movements of O-group plaice *Pleuronectes platessa* L. as shown by latex tagging. J Fish Biol 5:323–343

Rivas LR (1953) The pineal apparatus of tunas and related scombird fishes as a possible light receptor controlling phototactic movements. Bull Mar Sci Gulf Carib 3:168–180

Rommel SA, McCleave JD (1973a) Sensitivity of American eels (*Anguilla rostrata*) and Atlantic salmon (*Salmo salar*) to weak electric and magnetic fields. J Fish Res Bd Can 30:657–663

Rommel SA, McCleave JD (1973b) Prediction of oceanic electric fields in relation to fish migration. J Cons Int Explor Mar 35:27–31

Roth A (1968) Electroreception in the catfish *Ictalurus nebulosus* Les. Z Vergl Physiol 61:196–202

Roth A (1969) Elektrische Sinnesorgane beim Zwergwels *Ictalurus nebulosus* (*Amiurus nebulosus*). Z Vergl Physiol 65:368–388

Roth A (1973) Electroreceptors in Brachiopterygii and Dipnoi. Naturwissenschaften 60:106

Rounsefell GA, Kelez GB (1938) The salmon and salmon fisheries of Swiftshore Bank, Puget Sound, and the Fraser River. Fish Bull 49:693–823

Rowan WM (1926) Photoperiodism, reproductive periodicity, and annual migration of birds and certain fishes. Proc Boston Soc Nat Hist 38:147–189

Royce WF, Smith LS, Hartt AC (1968) Models of oceanic migrations of Pacific salmon and comments on guidance mechanisms. Fish Bull 66:441–462

Ruggles CP, Ryan P (1964) An investigation of louvers as a method of guiding juvenile Pacific salmon. Can Fish Cult 33:1–68
Ryman N (ed) (1981) Fish gene pools. Ecol Bull (Stockholm) 34:111 pp
Ryman N, Allendorf FW, Stahl G (1979) Reproductive isolation with little genetic divergence in sympatric populations of brown trout (*Salmo trutta*). Genetics 92:247–262
Sale PF (1975) Patterns of use of space in a guild of territorial reef fishes. Mar Biol 29:89–97
Sale PF (1977) Maintenance of high diversity in coral reef communities. Am Nat 111:337–359
Sand O (1974) Directional sensitivity of microphonic potentials from the perch ear. J Exp Biol 60:881–889
Sand O (1976) Microphonic potentials as a tool for auditory research in fish. In: Schuijf A, Hawkins AD (eds) Sound reception in fish. Elsevier, Amsterdam, pp 27–48
Sand O, Enger PS (1973) Evidence for the auditory function of the swimbladder in the cod. J Exp Biol 59:405–414
Saunders RL, Henderson EB (1970) Influence of photoperiod on smolt development and growth of Atlantic salmon (*Salmo salar*). J Fish Res Bd Can 27:1295–1311
Saunders RL, Sprague JB (1967) Effects of copper-zinc mining pollution on a spawning migration of Atlantic salmon (*Salmo salar*). Water Res 1:419–432
Schaffer WM (1979) The theory of life history evolution and its application to atlantic salmon. Symp Zool Soc (London) 44:307–326
Schaffer WM, Elson PF (1975) The adaptive significance of variations in life history among local populations of Atlantic salmon in North America. Ecology 56:577–590
Scharrer E, Smith SW, Palay SL (1947) Chemical sense and taste in the fishes, *Prionotus* and *Trichogaster*. J Comp Neurol 86:183–198
Scheich H, Bullock TH (1974) The role of electroreceptors in the animals life. II. The detection of electric fields from electric organs. In: Fessard A (ed) Handbook of sensory physiology, vol III/3. Springer, Berlin Heidelberg New York, pp 201–256
Schmidt-Koenig K, Keeton WT (1977) Animal migration, navigation, and homing. Springer, Berlin Heidelberg New York
Scholz AT (1980) Hormonal regulation of smolt transformation and olfactory imprinting in coho salmon. PhD Thes, Univ Madison, Wisc, 363 pp
Scholz AT, Madison DM, Stasko AB, Horrall RM, Hasler AD (1972) Orientation of salmon in response to currents in or near the home stream. Am Zool 12(4):654
Scholz AT, Cooper JC, Madison DM, Horrall RM, Hasler AD, Dizon AE, Poff RJ (1973) Olfactory imprinting in coho salmon: Behavioural and electrophysiological evidence. Proc 16th Conf Great Lakes Res 1973:143–153
Scholz AT, Horrall RM, Cooper JC, Hasler AD, Madison DM, Poff RJ, Daly RI (1975) Artificial imprinting of salmon and trout in Lake Michigan, Wisconsin. Dep Nat Res Fish Manage Rep 80:45 pp
Scholz AT, Horrall RM, Cooper JC, Hasler AD (1976) Imprinting to chemical cues: the basis for home-stream selection in salmon. Science 192:1247–1249
Scholz AT, Cooper JC, Horrall RM, Hasler AD (1978a) Homing of morpholine-imprinted brown trout, *Salmo trutta*. Fish Bull 76:293–295
Scholz AT, Gosse CK, Cooper JC, Horrall RM, Hasler AD, Daly RI, Poff RJ (1978b) Homing of rainbow trout transplanted in Lake Michigan: A comparison of three procedures used for imprinting and stocking. Trans Am Fish Soc 107:439–443
Schreck CB (1982) Parr-smolt transformation and behavior. In: Brannon EL, Salo EO (eds) Proceedings: salmon and trout migratory behavior symposium. Univ Washington, Seattle, pp 164–172
Schuijf A (1975) Directional hearing of cod (*Gadus morhua*) under approximate free field conditions. J Comp Physiol 98:307–332
Schuijf A (1981) Models of acoustic localization. In: Tavolga WN, Popper AN, Fay RR (eds) Hearing and sound communication in fishes. Springer, Berlin Heidelberg New York, pp 267–310
Schuijf A, Buwalda RJA (1980) Underwater localization – a major problem in fish acoustics. In: Popper AN, Fay RR (eds) Comparative studies of hearing in vertebrates. Springer, Berlin Heidelberg New York, pp 43–77
Schuijf A, Hawkins AD (1976) Sound reception in fish. Elsevier, New York
Schulz D (1975) Salinity preferences of elvers and young yellow eels (*Anguilla anguilla*). Helgol Wiss Meeresunters 27(2):199–210

Schutz F (1956) Vergleichende Untersuchungen über die Schreckreaktion bei Fischen und deren Verbreitung. Z Vergl Physiol 38:84–135

Schwartz E (1965) Bau und Funktion der Seitenlinie des Streifenhechtlings (*Aplocheilus lineatus* Cuv. u. Val). Z Vergl Physiol 50:55–87

Schwartz E, Hasler AD (1966) Perception of surface waves by the blackstripe topminnow *Fundulus notatus*. J Fish Res Bd Can 23:1331–1352

Schwassmann HO (1962) Experiments on sun orientation in some freshwater fish. PhD Thes, Univ Wisconsin, 153 pp

Schwassmann HO (1971) Biological rhythms. In: Hoar WS, Randall DJ (eds) Fish physiology, vol VI. Academic Press, London New York, pp 371–428

Schwassmann HO, Braemer W (1961) The effect of experimentally changed photoperiod on the sun-orientation rhythm of fish. Physiol Zool 34:273–326

Schwassmann HO, Hasler AD (1964) The role of the sun's altitude in the sun orientation of fish. Physiol Zool 37:163–178

Scott WB, Crossman EJ (1973) Freshwater fishes of Canada. Fish Res Bd Can Bull 184

Seckel GR (1972) Hawaiian caught skipjack tuna and their physical environment. Fish Bull 70:763–787

Selset R, Døving KB (1980) Behaviour of mature anadromous char (*Salmo alpinus* L.) towards odorants produced by smolts of their own population. Acta Physiol Scand 108:113–122

Semple JR, McLeod CL (1975) Experiments related to directing Atlantic salmon smolts, *Salmo salar*, around hydroelectric turbines. In: Saila SB, Heath DC (eds) Fisheries and energy production: a symposium. Lexington, Mass, pp 141–165

Shaklee JB, Tamaru CS (1977) Biochemical and morphological evidence of sibling species of bonefish *"Albula vulpes"*. Am Zool 17:973

Sharma RK (1973) Fish protection at water diversions and intakes: A bibliography of published and unpublished references. Environ Statem Proj Argonne Natl Lab, Argonne Ill

Shaw E (1969) The duration of schooling among fish separated and those not separated by barriers. Am Mus Nov 2373:1–13

Shelford VE, Allee WC (1913) The reactions of fishes to gradients of dissolved atmospheric gases. J Exp Zool 14:207–266

Shrode JB, Gerking SD (1977) Effects of constant and fluctuating temperatures on reproductive performance of a desert pupfish, *Cyprinodon n. nevadensis*. Physiol Zool 50:1–10

Simpson KS (1979) Orientation differences between populations of juvenile sockeye salmon. Fish Mar Serv Tech Rep 717:p 114

Smith MH, Chesser RK (1981) Rationale for conserving genetic variation of fish gene pools. Ecol Bull 34:13–20

Smith RJF (1982) The adaptive significance of the alarm substance – fright reaction system. In: Hara TJ (ed) Chemoreception in fishes. Elsevier, Amsterdam, pp 327–342

Smith SB (1969) Reproductive isolation in summer and winter races of steelhead trout. In: Northcote TG (ed) Symposium on salmon and trout in streams. HR MacMillan lectures in fisheries. Inst Fish, UBC, Vancouver, pp 21–38

Solomon PJ (1973) Evidence for pheromone-influenced homing by migrating Atlantic salmon, *Salmo salar* (L.). Nature (London) 244:231–232

Sorensen I (1951) An investigation of some factors affecting the upstream migration of the eel. Rep Inst Freshw Res Swed 32:126–132

Spieler RE, Noeske TA, DeVlaming V, Meier AH (1977) Effects of thermocycles on body weight gain and gonadal growth in the goldfish, *Carassius auratus*. Trans Am Fish Soc 106:440–444

Sprague JB (1968) Avoidance reactions of salmonid fish to representative pollutants. Water Res 2:23–24

Sprague JB, Drury DE (1969) Avoidance reactions of salmonid fishes to representative pollutants, vol I. Proc 4th Inst Conf Wat Pollut Res, Prague 1969, pp 169–179

Sprague JB, Elson PF, Saunders RL (1965) Sublethal copper-zinc pollution in a salmon river: A field and laboratory study. Int J Air Water Pollut 9:531–543

Srivastava CBL, Seal M, Das PK, Goposh A (1978) Anatomical identification of the presumed electroreceptors of two air-breathing catfishes, *Clarias batrachus* and *Heteropneustes fossilis*. Experientia 34:1345–1346

Stabell OB, Selset R, Sletten K (1982) A comparative study on population specific odorants from Atlantic salmon. J Chem Ecol 8:201–217

Stacey NE, Cook AF, Peter RE (1979) Spontaneous and gonadotropin induced ovulation in the goldfish, *Carassius auratus L.*: effects of external factors. J Fish Biol 15:349–361

Stasko AB, Horrall RM, Hasler AD, Stasko D (1973) Coastal movements of mature Fraser River pink salmon (*Oncorhynchus gorbuscha*) as revealed by ultrasonic tracking. J Fish Res Bd Can 30:1309–1316

Steimle FW, Stone RB (1973) Bibliography on artificial reefs. Coast Plains Center Mar Dev Serv, Wilmington, NC, p 129

Stepanov AS, Churmasov AV, Cherkashin SA (1979) Migration direction finding by chum [sic] salmon according to the sun. Soviet J Mar Biol 5:92–99

Stewart K (1962) Observations on the morphology and optical properties of the adipose eyelid of fishes. J Fish Res Bd Can 19:1161–1162

Stober QJ (1969) Underwater noise spectra, fish sounds, and response to low frequencies of cutthroat trout (*Salmo clarki*) with reference to orientation and homing in Yellowstone Lake. Trans Am Fish Soc 98:652–663

Stober QJ, Dinnel PA, Harlburt EF (1980) Acute toxicity and behavioral responses of coho salmon (*Oncorhynchus kisutch*) and shiner perch (*Cymatogaster aggregata*) to chlorine in heated seawater. Water Res 14:347–354

Stone RB, Pratt HL, Parker RD, Davis GE (1979) A comparison of fish populations on an artifical and natural reef in the Florida Keys. Mar Fish Rev 41(9):1–11

Stott B, Buckley BR (1979) Avoidance experiments with homing shoals of minnows, *Phoxinus phoxinus* in a laboratory stream channel. J Fish Biol 14:135–146

Stuart TA (1957) The migrations and homing behaviour of brown trout. Freshw Salmon Fish Res 18:27 pp

Stuart TA (1962) The leaping behaviour of salmon and trout at falls and obstructions. Freshw Salmon Fish Res 28:1–44

Sullivan CM (1954) Temperature reception and responses in fish. J Fish Res Bd Can 11:153–170

Sutterlin AM (1975) Chemical attraction of some marine fish in their natural habitats. J Fish Res Bd Can 32:729–738

Sutterlin AM, Gray R (1973) Chemical basis for homing in Atlantic salmon (*Salmo salar*). J Fish Res Bd Can 30:983–989

Sutterlin AM, Waddy S (1975) Possible role of the posterior lateral line in the obstacle entrainment by brook trout (*Salvelinus fontinalis*). J Fish Res Bd Can 32:2441–2446

Sutterlin AM, Saunders RL, Henderson EB, Harmon PR (1982) The homing of Atlantic salmon (*Salmo salar*) to a marine site. Can Tech Rep Fish Aquat Sci 1058:6 pp

Symons PEK (1971) Behavioural adjustment of population density to available food by juvenile atlantic salmon. J Anim Ecol 40:569–587

Szamier RB (1974) Fine structure of electroreceptors in marine catfish. Anat Rec 178:473

Szamier RB, Bennett MVL (1980) Ampullary electroreceptors in the freshwater ray, *Potamotrygon*. J Comp Physiol 138:225–230

Tabata M (1982) Persistence of pineal photosensory function in blind cave fish, *Astyanax mexicanus*. Comp Biochem Physiol 73A:125–128

Tarrant RM (1966) Threshold of perception of eugenol in juvenile salmon. Trans Am Fish Soc 95:112–115

Tavolga WN (1956) Visual chemical and sound stimuli: as cues in the sex discriminatory behaviour of the gobiid fish, *Bathygobius soporator*. Zoologica 41:49–64

Tavolga WN (1971) Acoustic orientation in the sea catfish *Galeichthys felis*. In: Adler H (ed) Orientation: sensory basis. Ann NY Acad Sci 188:80–97

Tavolga WN (1976) Acoustic obstacle detection in the sea catfish (*Arius felis*). In: Schuijf A, Hawkins AD (eds) Sound reception in fish. Elsevier, Amsterdam, pp 185–204

Tavolga WN (1977) Mechanisms for directional hearing in the sea catfish (*Arius felis*). J Exp Biol 67:97–115

Tavolga WN, Popper AN, Fay RR (1981) Hearing and sound communication in fishes. Springer, Berlin Heidelberg New York

Teeter JH, Szamier RB, Bennett MVL (1980) Ampullary receptors in the sturgeon *Scaphirhynchus platorynchus* (Rafinesque). J Comp Physiol 138:213–223

Teichmann H (1959) Über die Leistung des Geruchssinnes beim Aal (*Anguilla anguilla* L.). Z Vergl Physiol 42:206–254

Templeman W (1966) Marine resources of Newfoundland. Fish Res Bd Can Bull 154

Templeman W (1976) Transatlantic migrations of spiny dogfish (*Squalus acanthias*). J Fish Res Bd Can 33:2605–2609

Tesch FW (1974) Influence of geomagnetism and salinity on the directional choice of eels. Helgol Wiss Meeresunters 26, 1974:382–395

Tesch FW (1977) The eel. Chapman, London

Thomas AE (1973) Spawning migration and intragravel movement of the torrent Sculpin *Cottus rhotheus*. Trans Am Fish Soc 102:620–622

Thomas AE (1975) Migration of chinook salmon fry from simulated incubation channels in relation to water temperature, flow, and turbidity. Prog Fish Cult 37:219–223

Thorpe JE (1978) Rhythmic activity of fishes. Academic Press, London New York, p 312

Thorpe JE (1982) Migration in salmonids, with special reference to juvenile movements in freshwater. In: Brannon EL, Salo EO (eds) Salmon and trout migratory behavior symposium. School Fish, Univ Washington, Seattle, pp 86–97

Thunberg BE (1971) Olfaction in parent stream selection by the alewife (*Alosa pseudoharengus*). Anim Behav 19:217–225

Tinbergen N (1942) An objectivistic study of the innate behaviour of animals. Bibl Biotheor 1:39–98

Todd PR (1981) Timing and periodicity of migrating New Zealand and freshwater eels (*Anguilla* sp.). NZ J Mar Freshw Res 15:225–235

Treviranus (1822) Biologie oder Philosophie der lebenden Natur für Naturforscher und Ärzte, vol VI. Power, Cattingen (from Kleerekoper 1969)

Tsvetkov VI (1969) On the threshold sensibility of some freshwater fishes to rapid changes in pressure (in Russian). Vop Ikhtiol 9:928–935

Tsvetkov VI (1972) The sensitivity of the loach (*Misgurnus fossilis* (L.) to pressure changes. J Ichthyol 12(5):871–874

Tucker DW (1959) A new solution to the Atlantic eel problem. Nature (London) 183:495–401

Tyler AV (1971) Surges of winter flounder, *Pseudopleuronectes americanus*, into the intertidal zone. J Fish Res Bd Can 28:1727–1732

Tytler P, Blaxter JHS (1977) The effect of swimbladder deflation on pressure sensitivity in the saithe *Pollachius virens* (L.). J Mar Biol Assoc UK 57:1057–1064

Ueda K, Hara TJ, Gorbman A (1967) Electroencephalographic studies on olfactory discrimination in adult spawning salmon. Comp Biochem Physiol 21:133–143

Ueda K, Hara TJ, Satu M, Kaji S (1971) Electrophysiological studies of olfactory discrimination of natural waters by hime salmon, a land-locked Pacific salmon, *Oncorhynchus nerka*. J Fac Sci Tokyo Univ Sec IV 12(2):167–182

Unger I (1966) Artificial reefs – a review. Am Litt Soc Spec Publ 4:1–74

Vanderwalker JG (1967) Response of salmonids to low-frequency sound. In: Tavolga WN (ed) Marine biacoustics, vol II. Pergamon Press, Oxford New York, pp 45–58

Varanelli CC, McCleave JD (1974) Locomotor activity of atlantic salmon parr (*Salmo salar* L.) in various light conditions and in weak magnetic fields. Anim Behav 22:178–186

Vasil'yev AS, Gleiser SI (1973) Changes in the activity of the freshwater eel (*Anguilla anguilla* L.) in magnetic fields. J Ichthyol 13:322–333

Veen de JF (1963) On the phenomenon of soles swimming near the surface of the sea. Rapp Cons Explor Mer 155:51

Veen de JF (1967) On the phenomenon of soles (*Solea solea* L.) swimming at the surface. J Cons Perm Int Explor Mer 31:207–236

Veen de JF (1970) On the orientation of the plaice (*Pleuronectes platessa* L.) I evidence for orienting factors derived from the ICES transplantation experiments in the years 1904–1909. J Cons Perm Int Explor Mer 33:192–227

Veen van T, Haitevg HG, Müller K (1976) Light-dependent motor activity and photonegative behaviour in the eel (*Anguilla anguilla* L.). J Comp Physiol A 111:209–219

Verheijen FJ, Groot de SJ (1967) Diurnal activity pattern of plaice and flounder (Pleuronectidae) in aquaria. Neth J Sea Res 3:383–390

Vernon EH (1957) Morphometric comparison of three races of kokanee (*Oncorhynchus nerka*) within a large British Columbia lake. J Fish Res Bd Can 14:573–598

Viancour TA (1979) Electroreceptors of a weakly electric fish. II. Individually tuned receptors. J Comp Physiol A 133:327–338

Viehman W (1979) The magnetic compass of blackcaps (*Sylvia atricapilla*). Behaviour 68:24–30

Vince MA (1969) Embryonic communication, respiration, and the synchronization of hatching. In: Hinde RA (ed) Bird vocalizations. Cambridge Univ Press, London, pp 233–260
Vinnikov YA (1965) Structural and cytochemical organization of receptor cells of the sense organs in the light of their functional evolution. Zh Evol Biokhim Fiziol 1:67 [Transl: Fed Proc Am Soc Exp Biol Suppl 25(2):134–142]
Vladykov VD (1955) Fishes of Quebec. Eels. Dep Fish Que Alb 6:1–12
Vlaming de VL (1972) The effects of diurnal thermoperiod treatments on reproductive function in the estuarine gobiid fish *Gillichthys mirabilis* Cooper. J Exp Mar Biol Ecol 9:155–163
Vreeland RR, Wahle RJ, Arp AH (1975) Homing behaviour and contribution to Columbia River fisheries of marked coho salmon released at two locations. Fish Bull 73:717–725
Wachtel AW, Szamier RB (1969) Special cutaneous receptor organs of fish. IV. Ampullary organs of the non-electric catfish, *Kryptopterus*. J Morphol 128:291–308
Wagner HH (1970) The parr-smolt metamorphosis in steelhead trout as affected by photoperiod and temperature. PhD Thes, Oregon State Univ, Corvallis
Wagner HH (1974) Photoperiod and temperature regulation of smolting in steelhead trout (*Salmo gairdneri*). Can J Zool 52:219–234
Walcott C, Gould JL, Kirshvink JL (1979) Pigeons have magnets. Science 205:1027–1029
Wales W (1975) Extraretinal control of vertical migration in fish larvae. Nature (London) 253:42–43
Walker BW (1949) Periodicity of spawning by the grunion, an atherine fish. Thes, Univ Los Angeles
Walker BW (1952) A guide to the grunion. Cal Fish Game 38:409–420
Walker MM, Dizon AE, Kirschvink JL (1982) Geomagnetic field detection by yellowfin tuna. In: Oceans 82 conference record: Industry, government, education – partners in progress – Washington D.C. Sept 20–22 1982. IEEE, NY, USA, IEEE 82Ch 1827-5 Oceans 82, pp 755–758
Walker TJ (1967) History, histological methods, and details of the structure of the lateral line of the walleye surfperch. In: Cahn P (ed) Lateral line detectors. Indiana Univ Press, Bloomington, pp 13–25
Walls GL (1942) The vertebrate eye and its adaptive radiation. Cranbrook Inst Sci Bull 19. Bloomfield Hills, Mich, p 785
Ward HB (1932) The origin of the landlocked habit in salmon. Proc Natl Acad Sci USA 18:569–580
Waterman TH (1954) Polarization patterns in submarine illumination. Science 120:927–932
Waterman TH (1958) Polarized light and plankton navigation. In: Buzzati-Traverso AA (ed) Perspectives in marine biology. Univ Cal Press, Berkeley, pp 429–450
Waterman TH (1959) Animal navigation in the sea. Gunma J Med Sci 8:243–262
Waterman TH (1972) Visual direction finding by fishes. In: Galler SR, Schmidt-Koenig K, Jacobs GJ, Belleville RE (eds) Animal orientation and navigation, a symposium. NASA, Wash., pp 437–456
Waterman TH (1975) Natural polarized light and e-vector discrimination by vertebrates. In: Evans GC, Bainbridge R, Rockham O (eds) Light as an ecological factor II. Blackwell, Oxford, pp 305–335
Waterman TH (1981) Chapter 3, Polarization sensitivity. In: Land MG, Laughlin SB, Nassel DR, Strausfeld NJ, Waterman TH (eds) Comparative physiology and evolution of vision in invertebrates, B: Invertebrate visual centers and behavior I, chap 3. Springer, Berlin Heidelberg New York, pp 282–469
Waterman TH, Aoki K (1974) E-vector sensitivity patterns in goldfish optic tectum. J Comp Physiol 95A:13–28
Waterman TH, Forward RB (1970) Field evidence for polarized light sensitivity in the fish *Zenarchopterus*. Nature (London) 228:85–87
Waterman TH, Forward RB (1972) Field demonstration of polarotaxis in the fish *Zenarchopterus*. J Exp Zool 180:33–54
Waterman TH, Hashimoto H (1974) E-vector discrimination by goldfish optic tectum. J Comp Physiol 95A:1–12
Waterman TH, Westell WE (1956) Quantitative effect of sun's position on submarine light polarization. J Mar Res 15:149–169
Weel van PB (1952) Reaction of tuna and other fish to stimuli. 1951, Pt 2. Observations on the chemoreception of tuna. US Dep Int Fish Wildlife Serv Rep Fish 91:8–35
Weihs D (1978) Tidal stream transport as an efficient model for migration. J Cons Perm Int Explor Mer 38:92–99
Wenz GM (1964) Curious noises and the sonic environment in the ocean. In: Tavolga WN (ed) Marine bioacoustics. Pergamon Press, Oxford New York, pp 101–119
Werner RG (1979) Homing mechanism of spawning white suckers in Wolf Lake, NY. NY Fish Game J 26(1):48–58

Westby GWM (1974) Assessment of the signal value of certain discharge patterns in the electric fish, *Gymnotus carapo*, by means of playback. J Comp Physiol 92:327–341
White HC (1934) A spawning migration of Salmon in E. Apple River. Annu Rep Biol Bd Can 1933:4
White HC (1936) The homing of salmon in Apple River N.S. J Biol Bd Can 2:391–400
Whitear M (1965) Presumed sensory cells in fish epidermis. Nature (London) 208:703–704
Whitear M (1971a) Free nerve endings in fish epidermis. J Zool 163:231–236
Whitear M (1971b) Cell specialization and sensory function in fish epidermis. J Zool 163:237–264
Whitman RP, Quinn TP, Brannon EL (1982) Influence of suspended volcanic ash on homing behaviour of adult chinook salmon. Trans Am Fish Soc 111:63–69
Wickett WP (1959) Note on the behaviour of pink salmon fry. Prog Rep Pac Coast Stn Fish Res Bd Can 113:8–9
Wickham DA, Russell GM (1974) An evaluation of mid-water artificial structures for attracting coastal pelagic fishes. Fish Bull 72:181–191
Williams F (1972) Consideration of three proposed models of the migration of young skipjack tuna (*Katsuwonus pelamis*) into the eastern Pacific Ocean. Fish Bull 70:741–762
Williams GC (1957) Homing behaviour of California rocky shore fishes. Univ Cal, Berkeley. Publ Zool 59:249–284
Williams GC (1975) Sex and evolution. Princeton Univ Press, Princeton, NJ
Wilson JAF, Westerman RA (1967) The fine structure of the olfactory mucosa and nerve in the teleost *Carassius carassius* L. Z Zellforsch Mikrosk Anat 83:196–206
Wiltschko W, Wiltschko R (1972) Magnetic compass of European robins. Science 176:62–64
Winn HE (1955) Formation of a mucous envelope at night by parrotfishes. Zool NY 40:145–147
Winn HE, Bardach JE (1959) Differential food selection by moray eels and a possible role of the mucous envelope of parrot fishes in reduction of predation. Ecology 40(2):296–298
Winn HE, Bardach JE (1960) Some aspects of the comparative biology of parrot fishes at Bermuda. Zool NY 45(1):29–34
Winn HE, Marshall JA, Hazlett B (1964a) Behaviour, diel activities, and stimuli that elicit sound production and reactions to sounds in the longspine squirrelfish. Copeia 1964:413–425
Winn HE, Salmon M, Roberts N (1964b) Sun-compass orientation by parrot fishes. Z Tierpsychol 21:798–812
Wisby WJ (1952) Olfactory responses of fish as related to parent stream behaviour. PhD Thes, Univ Wisc, Madison
Wisby WJ, Hasler AD (1954) The effect of olfactory occlusion on migrating silver salmon (*Oncorhynchus kisutch*). J Fish Res Bd Can 11:472–478
Wolke RE, Bouck GR, Stroud RK (1975) Gas-bubble disease: a review in relation to modern energy production. In: Saila SB (ed) Fisheries and energy production: a symposium. DC Heath, Lexington, pp 239–265
Woodhead AD (1975) Endocrine physiology of fish migration. Oceanogr Mar Biol Annu Rev 13:287–382
Woodhead PMJ (1966) The behaviour of fish in relation to light in the sea. Oceanogr Mar Biol Annu Rev 4:337–403
Yamamoto M (1982) Comparative morphology of the peripheral olfactory organ in teleosts. In: Hara TJ (ed) Chemoreception in fishes. Elsevier, Amsterdam, pp 39–59
Young RE, Roper CFE, Walters JF (1979) Eyes and extraocular photoreceptors in midwater cephalopods and fishes: their roles in detecting downwelling light for counterillumination. Mar Biol 51:371–380
Zaugg WS (1981) Advanced photoperiod and water temperature effects on gill Na^+-K^+ adenosine triphosphatase activity and migration of juvenile steelhead (*Salmo gairdneri*). J Fish Res Bd Can 38:758–764
Zaugg WS, Adams Bl, McLain LR (1972) Steelhead migration: potential temperature effects as indicated by gill adenosine triphosphatase activities. Science 176:415–416
Zimmerman F (1980) Effect of tagging on rainbow trout. Underwat Telem Newslett 10(1):6–8
Zimmerman MA, McCleave JD (1975) Orientation of American eels (*Anguilla rostrata*) in weak magnetic and electric fields. Helgol Wiss Meeresunters 27:175–189
Zolotov V, Frantsevich L (1973) Orientation of bees by the polarized light of a limited area of the sky. J Comp Physiol 85:25–36

Systematic Index

Subject Index

Zoophysiology (formerly **Zoophysiology and Ecology**)

Coordinating Editor: **D.S. Farner**
Editors: **B. Heinrich, K. Johansen, H. Langer, G. Neuweiler, D.J. Randall**

Volume 1: **P.J. Bentley**

Endocrines and Osmoregulation

A Comparative Account of the Regulation of Water and Salt in Vertebrates

1971. 29 figures. XVI, 300 pages. ISBN 3-540-05273-9

"The author... has, with competence and insight, succeeded in the difficult task of covering two fields ... Bentley presents a thoroughly competent synthesis, and the result is a wellintegrated and balanced book. The book follows the zoological point of view, not only in outline, but also in the integration of physiological function with the natural life of the animal. The coherent viewpoint makes the text **readable and interesting,** and a large number of clear tables makes the materials **easily accessible.** The adequate coverage can serve as an introduction to the research literature in both fields treated in this book. If future volumes are of equal quality and value, **the series will be a significant contribution."**

Quarterly Review of Biology

Volume 2: **L. Irving**

Arctic Life of Birds and Mammals

Including Man

1972. 59 figures. XI, 192 pages. ISBN 3-540-05801-X

"The author's intense and unabated interest in arctic biology over the last three decades is reflected in the content and perspective of this volume. His unusually keen insight into the life of arctic birds and mammals (including man), which has led to this competent synthesis, is based on a familiarity conceivable only in a person who has experienced arctic life..."

Quarterly Review of Biology

Volume 3: **A.E. Needham**

The Significance of Zoochromes

1974. 54 figures. XX, 429 pages. ISBN 3-540-06331-5

"Dr. Needham's book is doubly welcome, for it not only considers animal pigments from the points of view of structure and function, but it also presents information and concepts which have never been assembled in one volume before...
The format of the book is very pleasing, with particulary high qualitiy typeface and paper. It was a good idea to preface each chapter with a brief synopsis of its subject matter and to include a conclusion section at the end. There are many tables which collect data not easily found elsewhere: the book is **a valuable and unique contribution** to the literature on pigments."

Nature

Volume 4/5: **A.C. Neville**

Biology of the Arthopod Cuticle

1975. 233 figures. XVI, 448 pages. ISBN 3-540-07081-8

"...The layout is clear and orderly throughout... As the text has a clear and economical style, as there are numerous figures and electron micrographs and an extensive but selective bibliography, this is an essential work of reference. But the treatment throughout is of a critical review... the book is admirably produced and printed..."

Quarterly J. Exp. Physiology

Volume 6: **K. Schmidt-Koenig**

Migration and Homing in Animals

1975. 64 figures, 2 tables. XII, 99 pages.
ISBN 3-540-07433-3

"The author... has provided a valuable service in collecting together examples of homing and migration in a diversity of animal groups. The plan is an excellent one: each chapter, devoted to a single taxonomic group, is subdivided into examples of field performance in orientation and its experimental analysis..."

The IBIS

Volume 7: **E. Curio:**

The Etiology of Predation

1976. 70 figures, 16 tables. X, 250 pages.
ISBN 3-540-07720-0

"... It is good because is stimulating, exhaustive and logical. No important aspect of the subject is missed. The author illustrates all his main points with a multiplicity of examples drawn from recent research. The reference list of nearly 700 items is evidence of the thoroughness of the treatment and the marshalling of examples used in explanation. Curio is an enthusiast and conveys the excitement to be found in much of the research on this subject; he also draws pointed attention to the gaps in our knowledge. For all these reasons **the book is a must for ethologists, ecologists, experimental psychologists and university libraries.** As a first treatment of seminal quality the book could well become a reference classic and inspire numerous research projects.
...The illustrations are clear and relevant..."

The Quart. Review Biology

Springer-Verlag Berlin Heidelberg New York Tokyo

Volume 8: **W. Leuthold**

African Ungulates

A Comparative Review of Their Ethology and Behavioral Ecology

1977. 55 figures, 7 tables. XIII, 307 pages.
ISBN 3-540-07951-3

·...Dr. Leuthold displays a masterly command of his subject.... The work is basically a review of published knowledge with an original approach, enlivened by the author's interpretations and based on his intimate first-hand knowledge of the subject. The first chapter, on the application of ethological knowledge to wildlife management, covers an important area... a wealth of references is given so that the chapter provides useful guide to the literature. the illustrations are good and well chosen to demonstrate points made verbally and not first to embellish the text. The book will provide **excellent background reading for undergraduates and research students as well as for anyone seriously interested in African wildlife.** On the whole, **it can be thoroughly recommended.**"

J. Applied Ecology

Volume 9: **E. B. Edney**

Water Balance in Land Arthropods

1977. 109 figures, 36 tables. XII, 282 pages.
ISBN 3-540-08084-8

... Dr. Erdney has provided a wealth of organized information on prior work and ideas for needed research, **all of which make the book a bargain.** The volume should prove useful, not only to those who work in arthopod water relations (it is a must for them), but also those of us interested in invertebrate and general ecology, entomology, comparative physiology, and biophysics." *AWRA Water Res. Bull.*

Volume 10: **H.-U. Thiele**

Carabid Beetles in Their Environments

A Study on Habitat Selection by Adaptations in Physiology and Behaviour

Translated from the German by J. Wieser
1977. 152 figures, 58 tables. XVII, 369 pages.
ISBN 3-540-08306-5

"...Because the book is comparative both in method and interpretation, it is a contribution to systematics as well as to ecology... **a fine synthesis of current knowledge** of homeostatic aspects of ecological relationships of carabids, and it is a fitting tribute to the man to whom it is dedicated: Carl H. Lindroth, who was instrumental in formulating the approaches and techniques that are commonly used in ecological research on these fine beetles. The materials is **well organized** and the text is **easily readable, thanks to the clarity of thought and expression** of the author and to the skill of an able translator." *Science*

Volume 11: **M. H. A. Keenleyside**

Diversity and Adaptation in Fish Behaviour

1979. 67 figures, 15 tables. XIII, 208 pages.
ISBN 3-540-09587-X

"... it is important as the first serious attempt by a senior researcher to produce an overview of the discipline. Previous works have all been symposium volumes or collections of papers haphazardly assembled, and Keenleyside has produced a volume that is of substantially greater value than these. In clearly perceiving that the unique and valuable features of fish behavior are its diversity of form and circumstance, he has charted a course that future authors would be wise to follow.
The book is well produced, well written, and easy to read. The illustrations are clear and straightforward." *Science*

Volume 12: **E. Skadhauge**

Osmoregulation in Birds

1981. 42 figures. X, 203 pages. ISBN 3-540-10546-8

Contents: Introduction. – Intake of Water and Sodium Chloride. – Uptake Through the Gut. – Evaporation. – Function of the Kidney. – Function of the Cloaca. – Function of the Salt Gland. – Interaction Among the Excretory Organs. – A Brief Survey of Hormones and Osmoregulation. – Problems of Life in the Desert, of Migration, and of Egg-Laying. – References. – Systematic and Species Index. – Subject Index.

Volume 13: **S. Nilsson**

Autonomic Nerve Function in the Vertebrates

1983. 83 figures. XIV, 253 pages. ISBN 3-540-12124-2

Contents: Introduction. – Anatomy of the Vertebrate Autonomic Nervous Systems. – Neurotransmission. – Receptors for Transmitter Substances. – Chemical Tools. – Chromaffin Tissue. – The Circulatory System. – Spleen. – The Alimentary Canal. – Swimbladder and Lung. – Urinary Bladder. – Iris. – Chromatophores. – Concluding Remarks. – References. – Subject Index.

Volume 14: **A. D. Hasler, A. T. Scholz**

Olfactory Imprinting and Homing in Salmon

Investigations into the Mechanism of the Imprinting Process

In collaboration with R. W. Goy
1983. 25 figures. XIX, 134 pages. ISBN 3-540-12519-1

Contents: Olfactory Imprinting and Homing in Salmon: Notes on the Life History of Coho Salmon. Imprinting to Olfactory Cues: The Basis for Home-Stream Selection by Salmon. – Hormonal Regulation of Smolt Transformation and Olfactory Imprinting in Salmon: Factors Influencing Smolt Transformation: Effects of Seasonal Fluctuations in Hormone Levels on Transitions in Morphology, Physiology, and Behavior. Fluctuations in Hormone Levels During the Spawning Migration: Effects on Olfactory Sensitivity to Imprinted Odors. Thyroid Activation of Olfactory Imprinting in Coho Salmon. Endogenous and Environmental Control of Smolt Transformation. – Postscript. – References. – Subject Index.

Springer-Verlag
Berlin
Heidelberg
New York
Tokyo